Perovskite Materials for Energy and Environmental Applications

Scrivener Publishing
100 Cummings Center, Suite 541J
Beverly, MA 01915-6106

Publishers at Scrivener
Martin Scrivener (martin@scrivenerpublishing.com)
Phillip Carmical (pcarmical@scrivenerpublishing.com)

Perovskite Materials for Energy and Environmental Applications

Edited by
Khursheed Ahmad
Department of Chemistry, Indian Institute of Technology, Indore, India
and
Waseem Raza
Department of Chemistry, Indian Institute of Technology, New Delhi, India

Scrivener
Publishing

WILEY

This edition first published 2022 by John Wiley & Sons, Inc., 111 River Street, Hoboken, NJ 07030, USA and Scrivener Publishing LLC, 100 Cummings Center, Suite 541J, Beverly, MA 01915, USA

For more information about Scrivener publications please visit www.scrivenerpublishing.com.

Wiley Global Headquarters
111 River Street, Hoboken, NJ 07030, USA

For details of our global editorial offices, customer services, and more information about Wiley products visit us at www.wiley.com.

Library of Congress Cataloging-in-Publication Data

ISBN 978-1-119-76027-6

Cover image: Pixabay.Com
Cover design by Russell Richardson

Set in size of 11pt and Minion Pro by Manila Typesetting Company, Makati, Philippines

Printed in the USA

10 9 8 7 6 5 4 3 2 1

Contents

Preface

In recent years, global energy demand has increased dramatically. This continuously increasing demand brought on by rapid population growth and industrialization represents one of the biggest challenges facing mankind. Currently, this demand is mainly being fulfilled by fossil fuels, which are limited and will be depleted in the near future due to the very high consumption rate. Moreover, the burning of fossil fuels produces a large amount of harmful gases, such as CO_2, SO_x, and NO_x, which are being released into the atmosphere and polluting the environment. The higher concentration of harmful gases in the environment increases the risk of global warming and acid rain. In addition to this, direct dumping of different organic and inorganic waste into water systems also pollutes water reservoirs and causes waterborne diseases, leading to environmental problems. Therefore, since the ultimate global challenge is to achieve environmental sustainability for our society, it is very important to develop alternative long-lasting, clean and renewable energy sources to replace fossil fuels.

Among all the alternative energy sources, solar energy is considered one of the most promising alternatives to fossil fuels due to its easy availability, cleanness, and accessibility. Moreover, the sun is the fundamental and ultimate source of renewable energy for our planet, which can provide endless carbon emission-free energy for mankind. Around 3.85 yottajoules of solar energy reaches the Earth every year, which is approximately 10^4 times higher than the world energy consumption annually. However, the availability of solar light depends on the time of day, weather, and seasonal conditions. Therefore, it is critical to convert sunlight to usable or storable energy that can be used as required. In this context, different approaches, such as solar heating, solar thermal electricity, photocatalysis, and photovoltaics, have been involved in harvesting solar energy. Among all these approaches, photovoltaics (PV) is regarded as the most efficient and promising approach, which can directly convert solar radiation into electric power using the photovoltaic effect. Therefore, PV is a successful case of effective utilization of solar energy to generate electricity using semiconducting materials, which can fulfill global energy demand and ease environmental problems. Moreover, PV is the crucial machinery that can provide a carbon-free (decarbonized economy) power supply to achieve a sustainable society. A PV system is made up of solar arrays (solar panels), a solar inverter, and other electrical accessories, and solar cells are building blocks of the PV system. Therefore, solar cells are the electrical device in the PV system which can directly convert solar radiation to electric power. The first PV solar cell was built by Edmond Becquerel in 1839 at age 19. The first solid-state PV solar cell with 1% efficiency was created by Charles Fritts in 1883 with the help of a selenium semiconductor with a thin layer of gold. However, the first practical PV solar cell was constructed by Daryl M. Chapin, Calvin S. Fuller, and Gerald L. Pearson in 1954 based on silicon solar cells with 4.5 to 6% efficiency.

Different types of PV solar cells have been developed, including first-generation (1G) solar cells based on single-crystalline silicon; second-generation (2G) solar cells based on cadmium telluride (CdTe), copper indium gallium selenide (CIGS) and amorphous silicon; and third-generation (3G) solar cells based on organic solar cells (OSCs), dye-sensitized solar cells (DSSCs), multijunction cells, quantum-dot-sensitized solar cells (QDSCs) and perovskite solar cells (PSCs). The commercial 1G solar cells based on crystalline silicon that are currently dominating the PV market exhibit remarkable stability with a maximum power conversion efficiency (PCE) of about 25% in the lab, but the theoretical limit of the PCE is about 31%. The PCE of 2G solar cells is lower than 1G but they are significantly less expensive than 1G solar cells due to their lower processing cost and materials. Since 1G solar cells made up of nontoxic crystalline silicon have about a 25-year performance warranty and show fairly high PCE with negligible degradation, they are leading the PV solar cell market. Although 1G solar cells exhibit high PCE, they are not free from some fundamental drawbacks such as high manufacturing and installation costs. Although there is a relatively large global demand for silicon solar cells, the device formation at high temperature is very expensive due to the high melting point of silicon, limiting their future development. Therefore, it is crucial to explore new PV materials with high PCE and low production costs to replace or complete the state-of-the-art PV technology for the future development of solar cells. In order to fulfill the gap, scientists have been on a quest to develop 3G solar cells with the aim of high PCE with lower cost. Among all 3G solar cells, perovskite solar cells (PSCs) have recently been attracting much attention and have also emerged as a hot research area of competing materials for silicon PV due to their comparable PCE, easy fabrication, long charge-carrier lifetime, low binding energy, low defect density, and low cost.

Perovskite is a mineral consisting largely of calcium titanate ($CaTiO_3$), which was discovered by the German mineralogist Gustav Rose in the Ural Mountains of Russia in 1839 and named after the Russian mineralogist Lev Aleksevich Perovski. Since then, the term perovskite is generally applied to the class of materials having the same type of crystal structure and stoichiometry as $CaTiO_3$. A perovskite is an organic/inorganic hybrid and possesses a general formula of ABO_3, in which A and B stand for cations of rare-earth or alkaline-earth metals and transition metals of 3d, 4d, and 5d configurations, and O stands for oxygen, which bonds to both cations. The ABO_3 structure of perovskite is ideally cubic, whereas A atom is bigger than B and forms the corner of cubic cells, coordinated by a 12-fold cuboctahedral, B atoms are in the center and surrounded by an octahedral of anions with 6-fold coordination and the oxygen atom at the centers of the faces of the unit cells. The oxide perovskites show good activity in various fields such as dielectrics, piezoelectrics, photocatalysis, ferroelectrics and pyroelectrics, etc., but are not good for PV applications. Some compounds, like $LiNbO_3$, $BiFeO_3$, and $PbTiO_3$, exhibit some PV activity due to the ferroelectric polarization effect, which makes them suitable for PV applications. However, halide perovskite, a new class of perovskite, differs from oxide perovskite in that it has a halide anion instead of oxide anions with general formula ABX_3, presenting the excellent semiconducting properties that are desired for PV solar cells. Moreover, the A is a monovalent metallic cation from group I of the periodic table instead of divalent in the case of oxide perovskite, B is divalent cations that form transition metals instead of tetravalent, and X is the non-metallic anion (halide).

Halide perovskites were discovered in the 1890s, but they were first comprehensively studied by Wells *et al.* in 1893. Mitzi's group investigated the physical properties of two-dimensional (2D) halide perovskite materials having organic group in the 1990s. In 2006, the Miyasaka group used perovskites as the first photovoltaic absorber, which provided only 2.2% PCE. After that, the development of PSCs continued to boom and exhibited fast improvement in efficiency to 3.8% in 2009, 6.5% in 2011, 20.1% in 2015, 22.1% in 2017, 24.2% in 2019, and 25.5% in 2020. This fast improvement in the performance of PVCs might have been due to their unique properties such as small bandgap, high dielectric constant, lower excitation binding energies, high absorption coefficient, high charge-carrier mobility and length, and so on.

This book provides a state-of-the-art summary and discussion about the recent progress in the development and engineering of PSCs materials along with the future directions it might take. Moreover, the recent advances in perovskite materials for energy conversion and environmental applications have been compiled herein.

Khursheed Ahmad
Waseem Raza
April 2022

1

Computational Approach for Synthesis of Perovskite Solar Cells

A.S. Mathur* **and B.P. Singh**

Department of Physics, Institute of Basic Science, Dr. Bhimrao Ambedkar University, Agra (U.P.), India

Abstract

Computational approach has emerged as an effective technique for analyzing the synthesis of novel structures of perovskite solar cells. This approach involves modeling and simulation of solar cells. With respect to solar cells, numerical simulation has proven to be a viable tool over the years to study and comprehend the features and properties of solar cell devices, like the electrical, optical, and mechanical properties of complicated device structures of solar cell. The advantages of simulation are calculating the electrical behavior before the manufacturing process, calculating and visualizing internal electronic values that cannot be measured. It also helps to reduce processing costs and time spent on manufacturing solar cell devices by providing useful information on how the parameters of production can be varied to improvise the behavior of the device. Quantitative modeling of the photovoltaic response of the solar cell is an important subject for improving both the understanding of operation mechanisms and the device performance. In the present chapter, basic steps involved in the modeling of solar cells have been described along with some simulation approaches being used for perovskite solar cells, such as SCAPS-1D, AMPS, ASA, AFORS-HET, and so on. Results of some recent studies have also been presented, elucidating the importance of modeling and simulation approaches in the design and development of perovskite solar cells.

Keywords: Perovskite solar cells, modeling, simulation, photovoltaic parameters

**Corresponding author*: arpitswarupmathur@gmail.com

Khursheed Ahmad and Waseem Raza (eds.) Perovskite Materials for Energy and Environmental Applications, (1–36) © 2022 Scrivener Publishing LLC

1.1 Introduction

Computational simulation is a technique that uses a software program to research and analyze the behavior of a real device or an imaginary system. Simulation is based on a mathematical model describing the process under study. In many instances of functional interest, numerical simulation of semiconductor devices is an effective tool for chip development. It is a significant technique for optimizing new semiconductor devices.

With respect to solar cells, numerical simulation has proven to be a viable tool over the years to study and comprehend the features and properties of solar cell devices, like the electrical, optical, and mechanical properties of complicated device structures of solar cell. The advantages of simulation are calculating the electrical behavior before the manufacturing process, calculating and visualizing internal electronic values that cannot be measured. It also helps to reduce processing costs and time spent on manufacturing solar cell devices by providing useful information on how the parameters of production can be varied to improvise the behavior of the device.

In this chapter, different simulation softwares are discussed. It presents the physical basis for modeling of semiconductor devices. Basic equations required for semiconductor modeling are also given here. Then after, it discusses in detail various recombination phenomenon taking place in semiconductors. After this, some softwares used for simulation of different types of solar cells are detailed one by one.

1.2 Preliminary Steps

(I) Physical Basis for Semiconductor Device Modeling

A. Semiconductor at Thermal Equilibrium

Every operation of a semiconductor device relies on the carriers carrying charge within the semiconductor and creating electrical currents. It is crucial to understand the exact number of these carriers so as to interpret the functioning of the device. First, we assume the thermal equilibrium of the semiconductor. This means that there are no external forces acting on the semiconductor, such as voltage, electrical fields, magnetic fields, or temperature gradients. In this case, all of the semiconductor's properties are independent of time [1].

Carrier Concentration

First, we consider the intrinsic condition applied to the semiconductor without any impurities. The current in a semiconductor is contributed by two types of charge carrier, the electron and the hole. To evaluate the carrier concentration the function of the density of the allowed electron energy states and the occupation function of the allowed energy states must be established. The function of energy density states, *N(E)*, defines the number of allowed electronic states per unit volume and energy. The function describing occupation is the well-established distribution function given by Fermi-Dirac, *f(E)*, which defines the ratio of filled states having an electron to maximum states allowed with the given energy E.

The density of states at an energy E in the conduction band, which is given as $N_C(E)$ is close to E_C, and in the valance band, $N_V(E)$ is close to E_V and are given by [2]:

$$N_C(E) = \frac{1}{2\pi^2}\left(\frac{2m_e}{h^2}\right)^{\frac{3}{2}}(E - E_C)^{\frac{1}{2}} \tag{1.1a}$$

$$N_V(E) = \frac{1}{2\pi^2}\left(\frac{2m_H}{h^2}\right)^{\frac{3}{2}}(E_V - E)^{\frac{1}{2}} \tag{1.1b}$$

with m_e and m_h being electron and hole effective mass respectively. E_C denotes the conduction band bottom, whereas E_V represents the valance band top.

The Fermi-Dirac distribution function is expressed as follows:

$$f(E) = \frac{1}{1 + exp\left(\frac{E - E_F}{k_B T}\right)} \tag{1.2}$$

with k_B representing the Boltzmann constant, the absolute temperature is represented by T and E_F the Fermi energy level.

For energy much greater than k_BT (i.e. $E - E_F \gg k_BT$), the Fermi function can be approximated by the Boltzmann function:

$$f(E) = \frac{1}{1 + exp\left(\frac{E - E_F}{k_B T}\right)} \cong exp\left[-\frac{(E - E_F)}{k_B T}\right] \tag{1.3}$$

The density of free electrons *(n)* (conduction-band levels which have been occupied) is found after multiplying the total number of states $N_C(E)$ with occupation function *f(E)* and then integrating over the conduction band:

$$n = \int_{E_C}^{\infty} N_C(E).f(E).dE \tag{1.4}$$

By replacing equations 1.1(a) and 1.3 into 1.4 the resultant expression for n is found after resolving the equation:

$$n = N_C exp\left(\frac{(E_F - E_C)}{k_B T}\right) \tag{1.5}$$

The parameter N_C is termed the conduction band effective density of states which is:

$$N_C = \left(\frac{2\pi m_e k_B T}{h^2}\right)^{\frac{3}{2}} \tag{1.6}$$

We can derive an expression for the concentration of holes in a similar manner. The probability that a state of energy is captured by a hole is the same as its not being held by an electron, i.e., *1-f(E)*. Thus,

$$p = \int_{-\infty}^{E_V} N_V(E).[1 - f(E)].dE \tag{1.7}$$

Applying the Boltzmann approximation, we can get:

$$1 - f(E) \cong exp\left[-\frac{(E_F - E)}{k_B T}\right] \tag{1.8}$$

Substituting Equations 1.1(b) and 1.8 into Eq. 1.7 yields

$$p = N_V exp\left(\frac{E_V - E_F}{k_B T}\right) \tag{1.9}$$

$$N_V = \left(\frac{2\pi m_h k_B T}{h^2}\right)^{\frac{3}{2}} \tag{1.10}$$

N_V shows the valance band effective density of states. The difference in the values of N_C and N_V is only because of the difference in m_e and m_h. The variation in N_C and N_V from one semiconductor to another is also due to the dissimilarity in the effective masses, also.

Intrinsic carrier concentration

The number of electrons present in the conduction band and the number of holes present in the valance band are equal in an intrinsic (undoped) semiconductor which is in thermal equilibrium; $n = p = n_i$, where n_i is the concentration of carriers of intrinsic semiconductors.

When multiplying the corresponding sides of Equations 1.5 and 1.9 one obtains:

$$np = N_C N_V exp - \frac{(E_C - E_V)}{k_B T} = N_C N_V exp - \frac{E_g}{k_B T} \tag{1.11}$$

where E_g is the forbidden gap:

$$E_g = E_C - E_V \tag{1.12}$$

Equation 1.11 shows that for a given semiconductor and T the product of np is a constant, free of the concentrations of the dopant. It is a significant relation and is generally shown in the following form:

$$np = n_i^2 \tag{1.13}$$

$$n_i = \sqrt{N_C N_V} \exp\left(-\frac{E_g}{2k_B T}\right) \tag{1.14}$$

The equation 1.13 is called law of mass action and it is true for any semiconductor at equilibrium. The intrinsic carrier concentration in a semiconductor is dependant only on the bandgap E_g. It is a material property (at a given temperature).

When in the intrinsic material we represent the Fermi level position E_i one can obtain from equations 1.5 and 1.9:

$$n_i = N_C exp\left(\frac{E_i - E_C}{k_B T}\right) = N_V exp\left(\frac{E_V - E_i}{k_B T}\right) \tag{1.15}$$

From equation 1.15 we can deduce easily the position of E_i, which is given by:

$$E_i = \frac{E_C + E_V}{2} + \frac{k_B T}{2} \ln\left(\frac{N_V}{N_C}\right) \tag{1.16}$$

The Fermi level E_i is nearby to the midgap; a minor alteration is due to the variance in the effective densities of conduction and valance and band.

Also using Equation 1.15 and reference energy as E_i, we have for N-type materials the alternate equations given as:

$$n = n_i exp\left(\frac{E_F - E_i}{k_B T}\right) \text{ or } E_F - E_i = k_B T \ln\frac{n}{n_i} \tag{1.17}$$

And for P-type materials:

$$p = n_i exp\left(\frac{E_i - E_F}{k_B T}\right) \text{ or } E_i - E_F = k_B T \ln\frac{p}{n_i} \tag{1.18}$$

Donors and acceptors

The number of holes and electrons in their bands respectively and therefore the conductivity of the semiconductor, may be regulated by adding different dopants or different types of impurities called acceptors and donors. The doped semiconductor, known as extrinsic material.

A semiconductor has four kinds of charged species: holes, electrons, donor ions which are positive N_d, and acceptor ions which are negative N_a. Generally, there is no net charge in all samples.

The densities of the negative and positive particles should be equal in order to maintain the charge neutrality:

$$n + N_a = p + N_d \tag{1.19}$$

For n and p we can solve equations 1.19 and 1.13:

$$n = \frac{N_d - N_a}{2} + \left[\left(\frac{N_d - N_a}{2}\right)^2 + n_i^2\right]^{\frac{1}{2}} \tag{1.20a}$$

$$p = \frac{N_a - N_d}{2} + \left[\left(\frac{N_a - N_d}{2}\right)^2 + n_i^2\right]^{\frac{1}{2}} \tag{1.20b}$$

The equations 1.20a and 1.20b are seldom used. Alternatively, either of the following cases is mostly valid:

1. $N_d - N_a \gg n_i$ (i.e., N-type material),

$$n = N_d - N_a \tag{1.21}$$

$$p = \frac{n_i^2}{n} \tag{1.22}$$

If additionally, $N_d \gg N_a$, then

$$n = N_d \quad \text{and} \quad p = \frac{n_i^2}{N_d} \tag{1.23}$$

The Fermi energy can then be written as:

$$E_F = E_i + k_B T ln \frac{N_d}{n_i} \tag{1.24}$$

2. $N_a - N_d \gg n_i$ (i.e., P-type material)

$$p = N_a - N_d \tag{1.25}$$

$$n = \frac{n_i^2}{p} \tag{1.26}$$

If furthermore, $N_a \gg N_d$, then

$$p = N_a \quad \text{and} \quad n = \frac{n_i^2}{N_a} \tag{1.27}$$

The Fermi energy can then be written as:

$$E_F = E_i + k_B T ln \frac{N_A}{n_i} \tag{1.28}$$

B. Nonequilibrium carrier concentration

Many semiconductor devices work by producing charge carriers that exceed the thermal equilibrium values, such excess carriers can be generated either by optical excitation or by adding an electrical field.

Through producing excess carriers (electrons and holes), thermal equilibrium in the semiconductor is no longer maintained, and the Fermi energy is no longer strictly specified. For the nonequilibrium semiconductor, we can outline two novel parameters: the quasi-Fermi energy for electrons and holes respectively.

The electron and hole concentrations of for the nonequilibrium are then given by:

$$n = N_C exp\left(\frac{E_{Fn} - E_C}{k_B T}\right) \tag{1.29}$$

$$p = N_V exp\left(\frac{E_V - E_{Fp}}{k_B T}\right) \tag{1.30}$$

E_{Fn} and E_{Fp} are the quasi-Fermi levels for electron and hole.

When electrons and holes are at equilibrium, E_{Fn} and E_{Fp} coincide and this is known as E_F.

(II) Basic equations for semiconductor modeling

The so-called equations of semiconductor devices could explain the activity of most of the devices made up of semiconductor materials, including solar cells.

The elementary equations for the operation of semiconductor-based devices describe the static and dynamic carrier behavior in semiconductors under external factors, like optical excitation or applied field, both causing deviations from conditions of thermal equilibrium.

Poisson's Equation

The electric potential is related to the density of the space-charge by Poisson equation:

$$div(\varepsilon \nabla \psi) = -\rho \tag{1.31}$$

where ψ represents the electrostatic potential, the local permittivity is given by ε, and ρ represents the density of the local space charge.

The total contribution from all charges which may be mobile or fixed, including holes, electrons and ionized impurities is expressed as local space charge density.

$$\rho = q(N_d + N_a + n + p) \tag{1.32}$$

Where the elementary charge is given by q, n gives the electron concentration, p represents the concentration of holes, N_a is the concentration of ionized acceptors and N_d is the concentration of ionized donor.

The electric field is obtained by the gradient of the potential as:

$$\vec{E} = \Delta \psi \tag{1.33}$$

Continuity equations

When the electric fields and density gradients are present, the equation of continuity describes the behavior of excess carriers with time and in space:

$$\frac{\partial n}{\partial t} = \frac{1}{q} div\overrightarrow{J_n} + G_n - R_n \tag{1.34}$$

$$\frac{\partial p}{\partial t} = -\frac{1}{q} div\overrightarrow{J_p} + G_p - R_p \tag{1.35}$$

where electron and hole concentrations are represented by n and p respectively. J_n and J_p are the electron current density and hole current density, G_n and G_p are the electrons and hole generation rates, R_n and R_p are the electron and hole recombination rates.

In steady state $\frac{\partial n}{\partial t} = 0$ and $\frac{\partial p}{\partial t} = 0$.

Current density equations

Derivations focused on the transport theory given by Boltzmann have depicted that the current densities can be approximated in the continuity equations by the drift-diffusion model. In the present case, of the quasi Fermi levels E_{Fn} and E_{Fp} express the current densities:

$$\overrightarrow{J_n} = \mu_n \mathrm{n} \nabla E_{Fn} \tag{1.36}$$

$$\overrightarrow{J_p} = \mu_p \mathrm{p} \nabla E_{Fp} \tag{1.37}$$

Where μ_n and μ_p are the mobilities of the electrons and holes. The quasi-Fermi levels are linked to the carrier concentrations and the potential through the two Boltzmann approximations as follows:

$$n = n_i exp\left(\frac{q\psi - E_{Fn}}{k_B T}\right) \tag{1.38}$$

$$p = n_i exp\left(\frac{-q\psi + E_{Fp}}{k_B T}\right) \tag{1.39}$$

To describe the quasi-Fermi levels these two equations may then be rewritten as:

$$E_{Fn} = q\psi - k_B T ln\left(\frac{n}{n_i}\right) \tag{1.40}$$

$$E_{Fp} = q\psi - k_B T ln\left(\frac{p}{n_i}\right) \tag{1.41}$$

By putting these equations into the expression for current density, the following current relationships can be attained:

$$\overrightarrow{J_n} = qD_n \nabla n - qn\mu_n \nabla \psi - \mu_n n(k_B T \nabla (lnn_i)) \tag{1.42}$$

$$\overrightarrow{J_p} = -qD_p \nabla p - qp\mu_p \nabla \psi - \mu_p p(k_B T \nabla (lnn_i)) \tag{1.43}$$

Where D_n and D_p represent the electron and hole diffusion constants, respectively:

$$D_n = \frac{k_B T}{q}\mu_n \tag{1.44}$$

$$D_p = \frac{k_B T}{q}\mu_p \tag{1.45}$$

The final term in (1.42) and (1.43) reflects the difference in the effective concentration of the intrinsic carriers, taking into account the decreasing influence of the band-gap. Defining active electrical fields:

$$\overrightarrow{E_n} = -\nabla\left(\psi + \frac{k_B T}{q} lnn_i\right) \tag{1.46}$$

$$\overrightarrow{E_p} = -\nabla\left(\psi - \frac{k_B T}{q} lnn_i\right) \tag{1.47}$$

The drift-diffusion equations can then be written in more conventional form as.

$$\vec{J_n} = qn\mu_n \vec{E_n} + qD_n \nabla n \quad (1.48)$$

$$\vec{J_p} = qn\mu_p \vec{E_p} + qD_p \nabla p \quad (1.49)$$

Optical generation of electron-hole pairs

It is necessary to the operation of solar cells to produce electron-hole pairs by absorbing sunlight. Holes and electrons lead to the transition of energy carried by the photons of light into electrical energy.

The number of incident photons S_0 (ν) (per unit area, per unit time and per unit energy) determine the number of generated electron-hole pairs. The photon flux S (x, ν) decreases inside the semiconductor exponentially as:

$$S(x, \nu) = S_0(\nu)\ exp\ (-\alpha x) \quad (1.50)$$

where ν is the frequency. The absorption process in the semiconductor determines the absorption coefficient $\alpha(\nu)$.

At x distance from the surface of the semiconductor the generation rate $G(x, \nu)$ of electron-hole pairs can be obtained as:

$$G(x, \nu) = \alpha(\nu)\ S_0(\nu)(1 - R)\ exp(-\alpha x) \quad (1.51)$$

Where R is the reflectance.

Here an assumption is made that one electron-hole pair is generated by each absorbed photon.

(III) Recombination phenomenon in semiconductors

Recombination is a course of annihilation or destruction of holes and electrons.

There are different mechanisms of recombination significant in the solar cell operation: band-to-band recombination or radiative recombination, Auger recombination, and recombination via defects or traps in the bandgap (often mentioned as Shockley-Read-Hall recombination or SRH). These processes are illustrated if Figure 1.1.

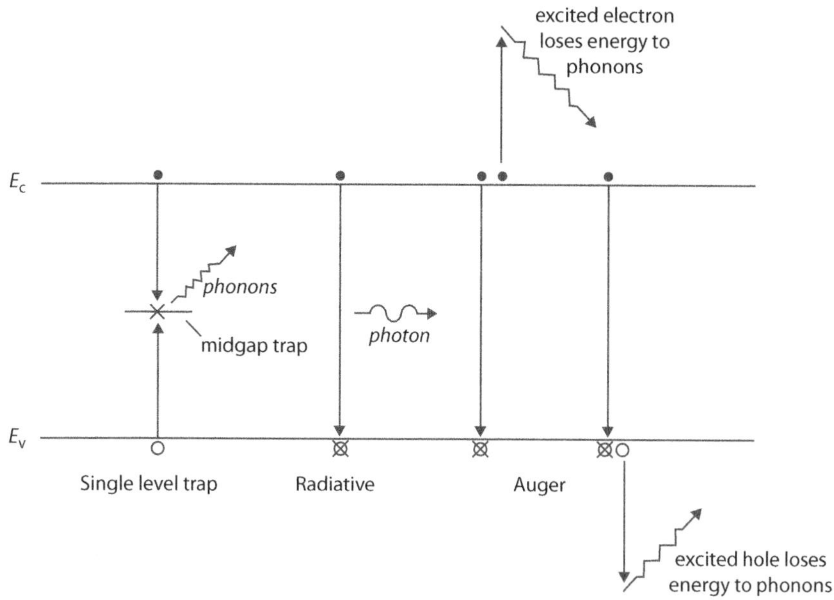

Figure 1.1 The different recombination processes in semiconductors.

A. Radiative recombination

Radiative recombination may be understood as the opposite of optical generation. This recombination takes place if a free electron drops out of the conduction band and recombine with a free hole in the valance band and a photon of energy equivalent to the difference in starting and ending energy states is released. For direct bandgap semiconductors, radiative recombination is very significant, but not as relevant for indirect semiconductors such as silicon, since a photon must also be absorbed or expelled for an electron to complete the transmission.

Due to radiative processes the net recombination rate is expressed as:

$$R_\lambda = B(np - n_i^2) \tag{1.52}$$

Where B is a material constant.

For n-type semiconductors under low injection ($p_0 \le p \ll n_0$) the net rate of recombination can be given as:

$$R_\lambda \approx \frac{p - p_0}{\tau_{\lambda,p}} \tag{1.53}$$

The effective lifetime due to radiative recombination, $\tau_{\lambda,p}$, is given by:

$$\tau_{\lambda,p} = \frac{1}{n_0 B} \tag{1.54}$$

B. Auger recombination

Auger recombination process is relatively similar to the process of radiative recombination with the exception that the transitional energy is transferred to an electron or hole that is excited in the respective conduction or valance band (Figure 1.1). Due to Auger recombination the net recombination is expressed as:

$$R_{Auger} = C_n n^2 p + C_p np^2 \tag{1.55}$$

Where C_n and C_p are the electron and hole Auger coefficients, respectively.

The most widely used Auger coefficient values were calculated by Dziewior and Schmid ($C_n = 2.8 \times 10^{-31}\ cm^6 s^{-1}$ and $C_p = 0.99 \times 10^{-31}\ cm^6 s^{-1}$) for Silicon having a doping concentration more than $5 \times 10^{18}\ cm^{-3}$.

Auger recombination is relevant in semiconductors with high carrier concentration or also under high level injection (e.g. concentrator solar cells).

C. Shockley-Read-Hall recombination (SRH)

The Shockley-Read-Hall recombination process generally occupies the net rate of recombination in low quality materials having high density of defects.

For single-energy level traps the SRH rate of volume recombination, U_{SRH}, is expressed as:

$$U_{SRH} = N_T \frac{np - n_i^2}{\tau_{p0}(n + n_1) + \tau_{n0}(p + p_1)} \tag{1.56}$$

Where τ_{n0} and τ_{p0} are the electron mean carrier lifetime and hole mean carrier lifetime, respectively, and are associated to the charge carrier thermal velocity, v_{th}, the concentration of defect, N_T, and the capture cross-sections of electron and capture cross-sections of hole of the specific defect, σ_n and σ_p as

$$\tau_n = (N_T v_{th} \sigma_n)^{-1} \quad \tau_p = (N_T v_{th} \sigma_p)^{-1}$$

n_1 and p_1 are defined as:

$$n_1 = n_i exp\left(\frac{E_T - E_i}{k_B T}\right), p_1 = N_V exp\left(\frac{E_i - E_T}{k_B T}\right) \tag{1.57}$$

According to definition, n_1 and p_1 are the free-electron and free-hole concentrations when the Fermi level (E_F) lies at the trap energy level (E_T).

Actually, all three recombination processes deliberated here occur simultaneously.

The net recombination can be found by addition of the three rates of recombination as:

$$R_{net} = R_\lambda + R_{Auger} + R_{SRH} \tag{1.58}$$

1.3 Advanced Semiconductor Analysis (ASA)

The ASA program is intended to simulate devices based on amorphous and crystalline semiconductors. The ASA program resolves one dimensional elementary semiconductor equations (Poisson equation and electron continuity equation and hole continuity equation) and utilizes variables such as concentration of free electrons, n, and that of holes, p, and electrostatic potential. Additionally, in order to describe explicit device operation and optoelectronic properties of the material it employs several advanced physical.

The ASA one dimensional (1-D) device simulator is very appropriate for simulating thin film structures of silicon solar cells. The program fulfils the usual requirements for simulating thin-film solar cell devices [3].

For modeling silicon thin-film devices the electronic structure of hydrogenated amorphous silicon (a-Si: H) and hydrogenated microcrystalline silicon (μc-Si: H) need to be considered. The atomic structure of a-Si:H shows spatial disorder leading to the energy band gap with edges of conduction band (CB) and valance band (VB) not well-defined with a continuous density of states (DOS). While studying the transport properties of carriers in a-Si:H, in the DOS distribution, it is necessary to differentiate between the localized states and the extended states. Within the mobility gap, the localized states greatly affect the processes of charge trapping followed by recombination. Thus, as is the case with crystalline semiconductor modeling, the charge trapped inside the localized states cannot be neglected. For the calculation of recombination-generation (R-G) statistics different methods are required for the localized states in the mobility gap of *a*-Si:H as they can be of different nature. The Shockley-Read-Hall (SRH) recombination is not used for amorphous films in device structures in a-Si:H to model recombination as the states introduced by dopants and/or impurities are negligible compared to the recombination via the tail states or dangling bond states. Though, SRH recombination based on the carrier lifetimes can be used for crystalline materials in the ASA program.

From the optical perspective, in order to achieve greater light conversion efficiencies, both the effective use of solar spectrum and light distribution in the solar cells are significant. Light regulation is accomplished in thin film solar cells through the application of techniques for light trapping. The techniques for trapping of light are focused on substrates with textured surface being introduced and using special (back-) reflector layers. The substrates with textured surface provide rough interfaces to the solar cells. The light incident at rough interfaces is scattered and the simulation of solar cells must consider the rough interface scattering mechanisms so as to precisely assess the generation profile for charge carriers within the solar cell. It needs the design of optical models which take into account propagation of both coherent nonscattered (specular) light and incoherent distributed (diffused) light through a system.

To use the solar spectrum efficiently, a multi-junction approach is required in thin-film silicon solar cells. At an interface between two neighbouring junctions, the tunnelling assisted recombination is accountable for the movement of the charge carriers through a solar cell with several junctions. Such an interface is defined as the tunnel-recombination junction (TRJ). Two methods can be used for modeling TRJ in the ASA program. The Delft approach utilizes the improved transport of carriers in the high-field region of TRJ and the trap-assisted tunnelling model. The Pennsylvania approach utilizes the implementation of a strongly defective layer with a sharply compact bandgap at the interface of n/p and the gradation of the n-layer and p-layer mobility gap in the regions next to the defective layer.

The ASA program's key characteristics are outlined [4]:

- Amorphous and/or crystalline multilayer semiconductor device modeling
- Models including both the extended and the localized (tail and defect) state that describe a complete DOS as an energy function.

- Use of the defect-pool model for the distribution of defect-states in a-Si:H
- Basic statistics for generation-recombination of donor and acceptor states and of ambipolar states
- Optical models for measuring the absorption profile in flat and/or rough interfaces
- Constant adjustment (gradation) of virtually all input parameters as a function of position in the device or in the gap energy level
- Tunnel-recombination junction model
- a-Si:H solar cell degradation modeling
- Written in language ANSI C

The ASA program can be installed on computers working under Windows or Linux operating systems. The ASA program is protected against unauthorised use and therefore its operation is allowed only with the simultaneous use of the ASA protection key [5–7].

Here, one calculation using the ASA program is followed on the basis of the flowchart in Figure 1.2 [3, 8–12]:

i. The ASA program has a built-in parser to run through the input file. After one statement is read, the appropriate part of the program is invoked.
ii. The first part of an input file always consists of statements that

    - Define the simulated device, i.e. the layers, including their width and the grid.
    - Set electrical parameters of the material, e.g., the valance tail slope parameters N_{v0}, E_{vo}. These parameters are set for every electrical layer.
    - Set optical parameters, e.g., the refractive index and extinction coefficient of the layers.
    - Set calculation and model settings, e.g., if the Newton method is to be used.

iii. After this initialization, the calculations are started with "solve."

    - The program generates the grid according to the definition earlier in the input file. The arrays of the simulated device are allocated and filled with the (initial) values

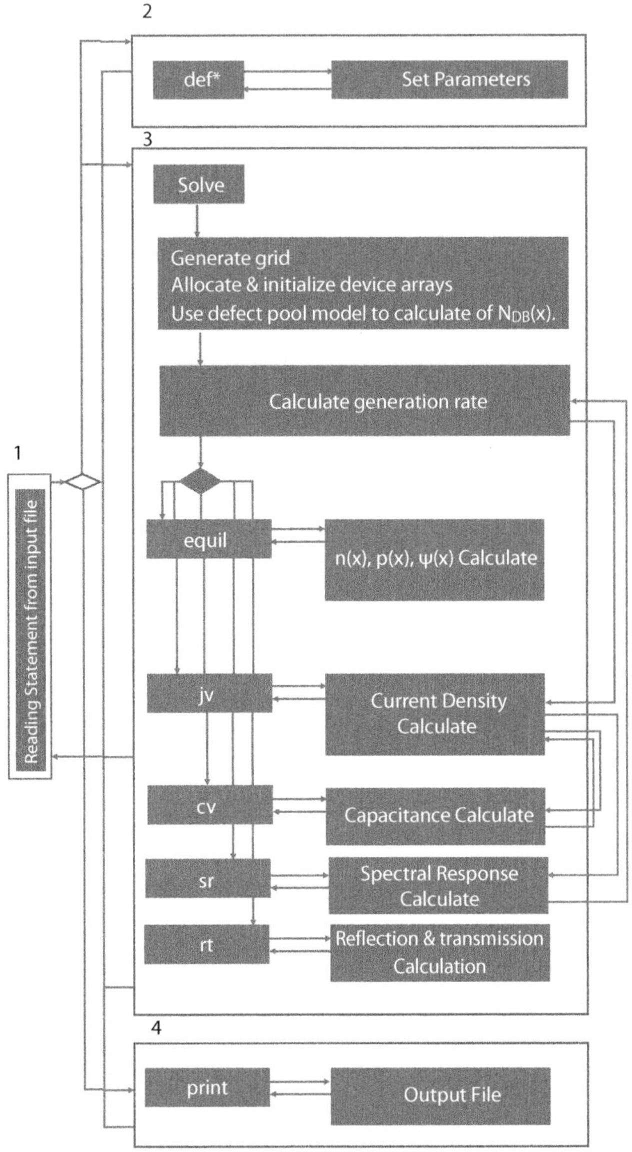

Figure 1.2 Flowchart showing steps for calculation using ASA adapted from [3].

defined by the given (electrical) parameters. When the defect pool model is selected, the defect density $N_{DB}(x)$ is calculated.

- If it is not read from external file, the program calculates the generation rate in the device using the ***GENPRO1***, ***GENPRO2***, or ***GENPRO3*** module.
- One of the five simulation modes can be selected by a "solve" statement:

(a) equil: in this mode, the electron, hole densities and the matching potential are calculated in thermodynamic equilibrium. These calculations are automatically carried out before mode 2.

(b) jv: the current-voltage characteristics of the simulated device are computed, by using the Poisson and continuity equations. A selection can be made between the Newton and Gummel method. The calculations can be carried out with or without carriers generation by illumination.

(c) cv: the capacitance-voltage characteristics of the simulated device are calculated. It is calculated by using mode 2 to find the currents at an applied voltage of $V_{\text{app}} = V - \frac{1}{2}\Delta V$ and $V_{\text{app}} = V + \frac{1}{2}\Delta V$, which gives rise to a difference in charge density $\Delta\rho(x)$. By calculating the difference in surface charge at the front and back contacts, $\Delta\sigma$ is also found. These define the capacitance at voltage V.

(d) sr: the spectral response (i.e. the external quantum efficiency EQE) is calculated by using mode 2 and the generation rate profile calculated by one of the optical models. The EQE is defined by $\Delta I/(q\Delta n_{photon})$, where Δn_{photon} is the differential number of photons between the (optional) bias spectrum S and the bias spectrum plus a monochromatic probe light $S + \Delta n_{photon}$.

(e) rt: the reflection and transmission are calculated. This is carried out by module ***GENPRO1*** (no scattering included).

iv. The desired program output can finally be saved to particular files using the "print" statement.

1.4 Analysis of Microelectronic and Photonic Structures (AMPS)

AMPS is a computer-based program in one-dimensional designed to simulate physics of transport in solid-state semiconductor devices. This applies the first principles of continuity equations and Poisson's equations method to examine transport behavior of electronic and optoelectronic semiconductor device based structures. Such architectures of device can consist of or incorporate amorphous or crystalline, polycrystalline materials. AMPS numerically solve the three controlling equations for semiconductor devices (the Poisson equation, the equations of electron continuity, and the hole continuity) short of having any a-priori statements regarding the processes of regulation of transport in such devices. AMPS can be used with this general and precise numerical treatment to study various device structures, including the following [13]:

- Solar cells having P-n and p-i-n structures and detectors with homojunction and heterojunction;
- Homojunction and heterojunction of microelectronic structures of p-n, p-i-n, n-i-n and p-i-p;
- Designs of solar cells with multijunction;
- Microelectronic multi-junction structures;
- Structures for detectors with graded composition and solar cells;
- Microelectronic structures graded in composition;
- New microelectronic, photovoltaic and optoelectronic devices;
- Optional back layered Schottky barrier devices.

Information, such as J-V characteristics, can be achieved in the dark and under illumination using the solutions given by an AMPS simulation software. These can be measured as a temperature variable. Collection efficiencies can also be obtained for solar cell, as well as detector designs as a function of light bias, voltage, and temperature. Additionally, significant information, like electric field concentrations, free and trapped carrier populations, profiles of recombination, and current densities of individual carriers, can be found from the AMPS program as a function of position. AMPS can be used to evaluate transport in different types of device structures that may include amorphous, crystalline, or polycrystalline layers or combinations of these. AMPS is intended to analyze, model, and optimize microelectronic, photovoltaic, or optoelectronic device architectures.

AMPS incorporate the following:

- a contact treatment allowing thermionic emission, as well as recombination happening at contacts of the device;
- a generalized gap state model for bulk or interface distribution of density of states;
- both recombination processes, i.e., band-to-band recombination process and Shockley-Read-Hall recombination phenomenon;
- a model for recombination that instead of using the frequently-applied single recombination level method calculates Shockley-Read-Hall recombination transport with any inputted general gap state distribution;
- Fermi-Dirac statistics instead of Boltzmann statistics only;
- gap state concentrations calculated with real statistics for temperature instead of frequently used T = 0K method;
- a model for trapped charge, which accounts for charge in any inputted overall gap state distribution;
- a model for gap state, allowing energy variation of capture cross-section;
- distribution of gap states whose properties change with position;
- spatial variation of carrier mobility;
- spatial variation of electron and hole affinities;
- different mobility gaps and optical gaps;
- calculation of characteristics of the device as a function of temperature and also with or without illumination in both forward and reverse bias;
- analysis of device structures made-up utilizing single crystalline, multicrystalline, or amorphous materials or all three.

The transport physics of device can be described in three controlling equations when modeling microelectronic and optoelectronic devices: the equation of Poisson, the equation of continuity for free holes, and the continuity equation for free electrons. So evaluating transport properties turn out to be a challenge to overcome with solving three coupled nonlinear differential equations, each having two boundary conditions associated with it. In AMPS, these three equations together with the suitable boundary conditions are concurrently tackled, so as to achieve a set of three unknown state variables at each device level: the electrostatic potential, the quasi-Fermi level of the hole, and the quasi-Fermi level of the electron. The carrier quantities, currents, fields, and so on, can thus be determined from these three state variables. To ascertain these state variables, the computer uses the method of finite differences and also the Newton-Raphson methodology. Iteratively, the Newton-Raphson Method calculates the roots of a function or roots of a set of functions if these roots are given a suitable initial supposition. Through AMPS, the one-dimensional structure being studied is separated into sections by a network of grid points. Then for each grid point, the three sets of unknowns are solved. We note that, at the user's discretion, AMPS requires the mesh to have adjustable grid spacing. As stated, after obtaining these three state variables as a function of x, band edges, recombination profiles electrical field, carrier populations, trapped charge, current densities, and any other data related to transport may be extracted.

First, AMPS measure the simple band diagram, built-in potential, electric field, trapped carrier populations, and free carrier populations found in a device if there is no bias (voltage or light) of any kind. These solutions obtained at thermodynamic equilibrium permit to "see what the device will look like."

AMPS will then use such solutions obtained at thermodynamic equilibrium as starting presumptions for the iterative scheme that will contribute to the full characterization of a device structure under voltage, illumination, or both voltage and illumination bias. AMPS produce output, such as band diagram (which include quasi-Fermi levels), carrier populations, recombination profiles, currents, current-voltage (I-V) characteristics, and spectral response for device structures with different voltage, illumination or voltage and illumination bias [14–16].

Given the range of voltage bias in the window for specifying the conditions for voltage bias, this range of voltages applies to both the dark I-V and light I-V. If the user wishes to get a biased band diagram, he/she can open the selected biased window to give AMPS the value that must (1) lie in the voltage detailed in the previous window and (2) be constant with the previously selected voltage step.

The only distinction at the user input window between light and dark I-V is in the lighting process, the user has to check "light on." AMPS received AM1.5 by default but the user may specify the photon flux and spectrum. A box called "light-x," is provided as a neutral filter/concentrator. The absorption coefficient must be manually entered by the user. Yet AMPS provide the user with an alternative in the "E_{opt}" box to shift the absorption coefficient linearly. Because absorption coefficient information is not easy to obtain most of the time, this linear shift will help the user check the band gap adjustment. You can see details on the user interface.

If the user wishes to see the current produced at each wavelength, the "spectral response" box must be checked. With and without light bias, AMPS will always produce spectral response. The spectrum and the flux in the window of the spectrum determine the light bias. The user-defined spectrum, therefore, defines the range of the QE graph. Users can also adjust the frequency of the probe laser. At the defined voltage bias, AMPS will produce QE.

1.5 Automat for Simulation of Heterostructures (AFORS-HET)

A lot of different experimental approaches are used to analyze (thin film) heterojunction solar cells, varying from traditional techniques for the characterization of solar cell, such as current-voltage characteristics or quantum efficiency to more sophisticated ones, such as surface photovoltage, photo- and electroluminescence, impedance, capacitance/conductance, decay in photoconductance intensity modulated photocurrent spectroscopy or electrically detected magnetic resonance [17–20].

AFORS-HET allows these measures to be interpreted AFORS-HET simulates solar cells with heterojunction, as well as the observables of the related techniques of measurement. A visual framework enables all simulation information to be displayed, stored, and compared. It is possible to perform arbitrary variations of parameters, multi-dimensional parameter fitting and optimization.

AFORS-HET is used specifically to model heterojunction solar cells based on amorphous/crystalline Si with the TCO/a-Si:H(n, p)/c-Si(p, n)/a-Si:H(p, n)/Al form where ultra-thin (5 nm) a-Si:H sheets of hydrogenated amorphous silicon are mounted on top of a thick (300 μm) crystalline wafer of Silicon. Up to 19.8% of experimental efficiencies have been found.

It is possible to model an arbitrary sequence of semiconducting layers, laying down the properties of corresponding layer and interface, i.e., the defect state distribution (DOS). Utilizing recombination statistics by Shockley-Read-Hall, semiconductor equations are solved in one dimension (1) for thermodynamic equilibrium, (2) for stable conditions under externally applied voltage or current and/or illumination, (3) for small extra sinusoidal externally applied voltage/illumination modulations, (4) for temporary conditions due to general external variations. It is, therefore, possible to determine the internal characteristics of the cell-like band diagrams, local rates of generation and recombination, local currents of the cell, current densities, and phase shifts. In addition, it is possible to simulate a range of characterization methods, i.e., I-V, transient, pulse-length, wavelength, intensity, and voltage-dependent photovoltaic surface (TR-SPV, PD-SPV, WD-SPV, ID-SPV, VD-SPV), transient luminescence (TR-PEL, intTR-PEL), internal and external quantum efficiency (QE), impedance/admittance spectrum, spectral-resolved photo, and electroluminescence (PEL) [21–23].

External users can implement new characterization approaches and new numerical segments (open-source on request). To date, these numerical modules have been established: (a) modules for front contact: metal/semiconductor Schottky- or Schottky-Bardeen- or metal/insulator/semiconductor contact; (b) modules for interface: no interface or drift diffusion or thermionic emission semiconductor/semiconductor heterojunction interface; (c) modules for bulk layer: arbitrary layer (standard) or layer of crystalline silicon; (d) modules for optical layer: Lambert-Beer absorption or coherent/incoherent multiple reflection.

A suitable series of semiconducting layers and interfaces must be specified before measurement. For example, the properties of the semiconductor, viz. the thin film emitter a-Si:H(n) and the silicon wafer c-Si(p), should be specified. Therefore, the defect state distribution (DOS) must be defined for each layer and interfaces if appropriate. In particular, the boundary contacts must be specified: the TCO surface has been demonstrated as an optical layer for the chosen example (requiring the reflectivity and absorption measured). The contact with TCO/a-Si:H is supposed to be a depleted Schottky contact, whereas the contact's calculated barrier height represents an input parameter. The metallic c-Si(p)/Al back contact is supposed to be a flat band with a recombination speed of 10^7 cm/s for simplicity.

Once the external cell parameters (temperature, illumination, and cell voltage or cell current) have been specified, the internal cell results, such as local rates of recombination, densities of carrier, currents, band energies, and so on, can be computed both under steady-state conditions (DC) or by arbitrarily changing external cell parameters using small additional sinusoidal perturbations, and the ratio of amplitude of all these quantities can be observed.

Variation in input parameter and output parameter to monitor can be specified.

In addition, characterization procedures (measurements) can be simulated by changing external variables, as in a specific experimentation and by doing certain postprocessing analysis of the data with the results of internal cells. The following methods of measurement have been used so far: methods of *DC-mode characterization:* I-V, SPV, QE, PEL, ASSPC, C-V, C-f, C-T; methods of *TR-mode characterization:* TR-PEL, TR-SPV, intTR-PEL, TR-PC. External users can insert or implement other characterization methods (open-source on request) [24–26].

The resulting total cell current (I-V) can be calculated by variation in the voltage of the external cell at a specified value of illumination. The number of photons emitted because of band-to-band radiative recombination can be computed using the generalized Planck equation from the quasi-Fermi energy splitting within the cell.

ZnO/a-Si:H(n)/c-Si(p)/Al heterojunction silicon solar cells' solar cell performance is critically dependent on the D_{it} interface state density a-Si:H/c-Si. A $D_{it} = 10^{12}$ cm^{-2} state density at interface lowers the solar cell's open-circuit voltage by more than 100 mV. Open-circuit photoluminescence is a quick and nondestructive characterization method susceptible to D_{it} and needs no contact. The photoluminescence signal is quenched by a growing recombination at interface because of a higher D_{it}.

Using complex numbers, small values of additional sinusoidal disturbances of the external parameters of the cell are treated. The shift in phase and the ratio of the amplitude between the AC cell voltage and the AC cell current can be computed as dependent on the frequency of perturbation if a small additional sinusoidal (AC) external signal voltage is superimposed on the DC voltage of the cell. The device conductance G(ω)/capacitance C(ω) at the specified frequency ω may be determined from the real/imaginary parts of the AC cell current:

$$G(\omega) = \frac{Re\left\{\overrightarrow{j_{ext}}\right\}(\omega)}{\overrightarrow{V_{ext}}} \qquad C(\omega) = \frac{1}{\omega}\frac{Im\left\{\overrightarrow{j_{ext}}\right\}(\omega)}{\overrightarrow{V_{ext}}}$$

When conducted under open-circuit situations, the impedance of a ZnO/a-Si:H(n)/c-Si(p)/Al heterojunction silicon solar cell is quite susceptible to the interface state density D_{it}. For larger D_{it}, 10^{10} $cm^{-2} \leq D_{it} \leq 10^{12}$ cm^{-2}, the resonant frequency (maximum of the phase shift) shifts toward higher frequencies. On the other hand, if operated under dark or short-circuit conditions, the D_{it} sensitivity is low: for instance, the conductance depending on the temperature in the dark changes only for $D_{it} \geq 10^{12}$ cm^{-2} which is related to a change of the band bending at the equilibrium.

Because of a random shift in external parameters, the system's time response may be determined by using a transient. ".ttd" file detailing the variations, for example, it may simulate the increase and decline of the SPV signal or the spectral integrated PL signal because of a short monochromatic laser pulse.

If ZnO/a-Si:H(n)/c-Si(p)/Al silicon heterojunction solar cells are excited by a monochromatic laser pulse at a wavelength of 900 nm, the generation of extra carriers occurs only in the c-Si(p) wafer. Hence, the recombination of the a-Si:H/c-Si interface may be efficiently tested. The SPV signal essentially tracks the shift in band bending at the surface because of excessive generation/recombination of carrier, whereas the PL signal is straight linked to excess generation/recombination of carrier. When laser pulse of 10 ns is used, all signals during the pulse do not enter conditions of steady state. Nevertheless, the two signals' degeneration behavior occur in different time domains. It is inside the range of ms for decay of SPV and within the range of 100 ns for decay of PL. The density of states at a-Si:H/c-Si interface, D_{it} critically relies not only on the early values immediately after the pulse but also on the transient decay.

Until now, the AFORS-HET software has been primarily used to (1) determine total achievable amorphous/crystalline solar cells efficiencies, (2) create criteria for designing for such solar cells, (3) establish calculation approaches for controlling a-Si:H/c-Si recombination interface.

1.6 Solar Cell Capacitance Simulator (SCAPS)

SCAPS is a one-dimensional program for simulation of solar cells established at the University of Ghent's Department of Electronics and Information Systems (ELIS). It analyzes the behavior and characteristics of solar cell structures numerically. Various measurements of the output parameters of solar cells could be performed by SCAPS. It can calculate the open circuit voltage (V_{oc}), the short circuit current (J_{sc}), the output characteristic J-V, the fill factor (FF), the quantum efficiency (QE), the output efficiency of the cell, the generation and recombination profiles, and so on [27–30].

Similar to any other numerical simulation program, SCAPS solves the elementary equations for semiconductors: the Poisson equation, which relate the charge and the electrostatic potential ψ, as well as the electron and hole continuity equations. The maximum length of the cell L is split into N intervals in one direction, and the magnitude of ψ_i and the concentrations of electron and hole n_i and p_i in each interval are the unknown variables of the problem. These may be identified by solution of nonlinear 3N equations numerically, i.e., the elementary equations at each of the i

intervals. Additionally, instead of (ψ_i, n_i, p_i), one can choose ψ_i, E_{Fni}, and E_{Fpi} as independent variables. Here, for electrons and holes, E_{Fn} and E_{Fp} are the quasi-Fermi energy levels. As the continuity equations include a nonlinear term for recombination in n and p, the basic equations become nonlinear [31–35].

The electrical characteristics can be determined according to the defined physical structure and conditions of bias. This can be done by the assumption that the device's function can be approximated into a grid in one dimension composed of a set of grid points also called as nodes. The transport of carriers through the system can be modeled by adding the set of differential equations (Poisson's equation and continuity equations) on this grid (or the discretization of the equation). The grid with finite element can be used to depict the domain of the simulation.

SCAPS is a program for Windows, and some of its chief characteristics are classified here as follows:

- It is possible to include up to seven layers of semiconductor to the solar cell device.
- When required, gradation of nearly all physical parameters can be done in a new window.
- Capacity to approximate band diagram in steady state, profile for recombination, and transport of carrier.
- Variable bias of voltage, different temperature, and various illumination options
- Can be used to measure concentrations and currents, J-V, C-f, C-V, and Q-V, AC characteristics and spectral response.
- Calculation of single and batch values.
- Final, as well as intermediate values, graphs, and other significant data, can be obtained as output and can be exported

SCAPS are arranged in several panels where the parameters can be defined by the user. The key panel is the "action panel" (Figure 1.3), in which one can set an operation point (temperature, voltage, frequency, and illumination) and the action list for calculation to be performed (J-V, C-f, C-V, Q(λ)). The running parameter (V, f, or λ) vary in the definite range in each calculation, whereas the values quantified in the operation point are for all other parameters. The user can also view the results earlier calculated directly, i.e., J-V, C-f, C-V, Q(λ), likewise, band diagrams, electric field, densities of the carrier, partial currents of recombination [36–45].

We click on the "set problem" button for the problem definition, i.e., the geometry, materials, and different characteristics of the solar cell.

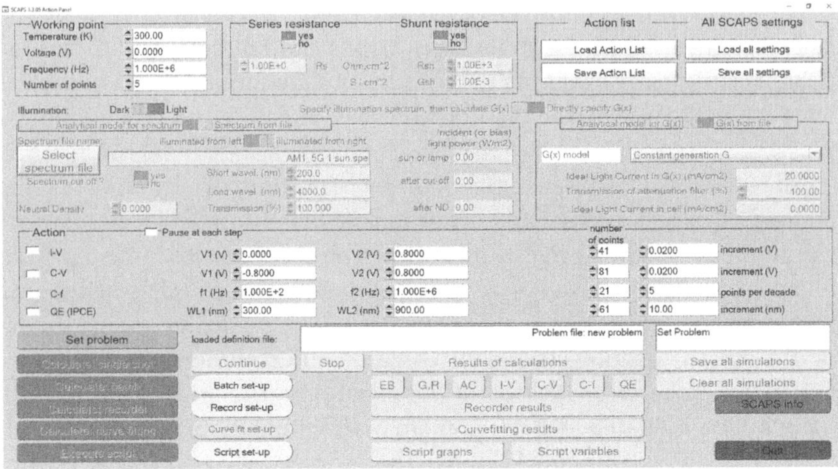

Figure 1.3 Action panel of SCAPS.

As depicted in Figure 1.4, the Solar Cell Definition Panel opens. In this panel, we can define up to 9 layers of structures. The back contact is the first layer; the last one is the front contact. The properties of intermediate semiconductor layers can be specified by the user. Each layer contains the following semiconductor properties, except for front and back contact:

- thickness (μm),
- electron affinity (eV),

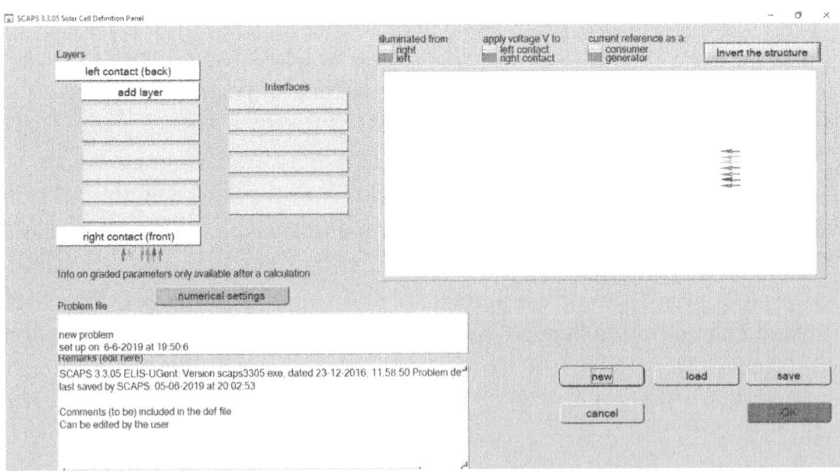

Figure 1.4 Solar cell definition panel of SCAPS.

- band gap (eV),
- dielectric permittivity (relative),
- valance band effective density of states ($1/cm^3$),
- conduction band effective density of states ($1/cm^3$),
- hole thermal velocity (cm/s),
- electron thermal velocity (cm/s),
- acceptor shallow density (N_D),
- donor shallow density (N_D).

The optical absorption of the semiconductor layers can be taken from a user file. Examples of such user files are distributed with the program: *Si.abs*, *CdS.abs*, *GaAs.abs*, and so on. We can add other absorption files for other relevant materials.

We can also change the model for absorption, model for recombination, and apply defects to a surface in addition to these basic parameters. A panel for layer properties snapshot is illustrated in Figure 1.5.

On the panel for cell definition, which opens the panel for the properties of contacts (Figure 1.5), you can configure the contact properties by either selecting the front or back contact button.

The identification of each contact is made as follows:

- electron and hole surface recombination velocities,
- information about the metal work function.

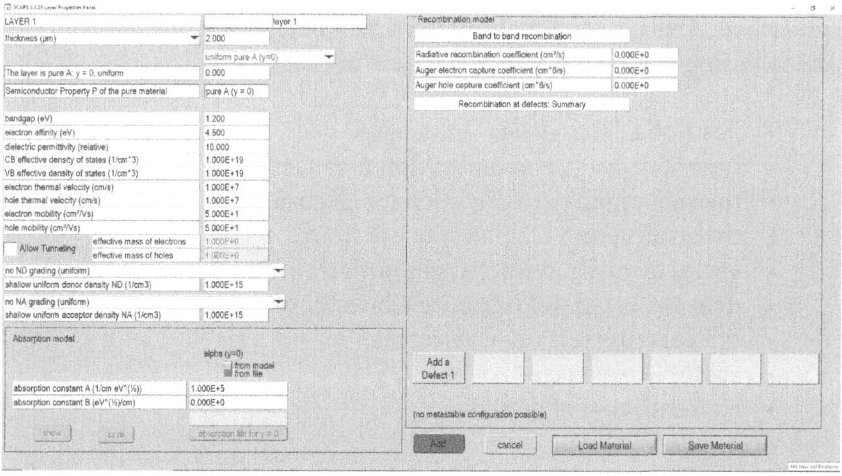

Figure 1.5 Layer properties panel in SCAPS.

After loading the solar cell design with the button for set problem on SCAPS, one can set the working point. The work point specifies the parameters in a measurement that are not varied:

- Temperature, T: appropriate to all measurements. In SCAPS, the only variables that have an explicit dependence on temperature are $N_C(T)$, $N_V(T)$, and thermal voltage k_BT.
- Voltage, V: is discarded in simulation of I-V and C-V. It is the voltage for dc-bias in simulation of C-f and simulation of QE(λ). SCAPS starts at 0V and proceeds through a series of steps at the point of operation, which we can also define.
- Frequency, f: is discarded in simulation of I-V, QE(λ) and C-f. It is the frequency of the simulation of the C-V measurement.
- Illumination, the illumination requirements can be further defined when simulating under illumination. The main options are: dark or light, light side choice (left/right), spectrum option. A single sun(=1000 W/m^3) illumination with the "air mass.5, global" spectrum is the default, but for specialized simulations, we have a wide range of monochromatic light and spectra.

In a diode, the current at n-contact is converted from the p-contact hole current to the electron current. It implies that recombination must occur somewhere in the diode, even in the most ideal device. The user must define recombination at least at one location (in a layer or interface) somewhere. Defects are most important parameters for study of solar cells. The following parameters identify defects in SCAPS:

- position of energy level in the gap,
- type of defect (i.e. acceptor, donor or neutral),
- thermal capture cross-section for electrons,
- thermal capture cross-section for holes,
- energetic distribution (single, uniform ...),
- defect energy level reference (above E_V or below E_C),
- optical cross section of electrons,
- capture cross section of holes,
- concentration of defects.

We can select one or more of the following measurements to be simulated in the action part of the action panel (Figure 1.3): I-V, C-V, C-f and

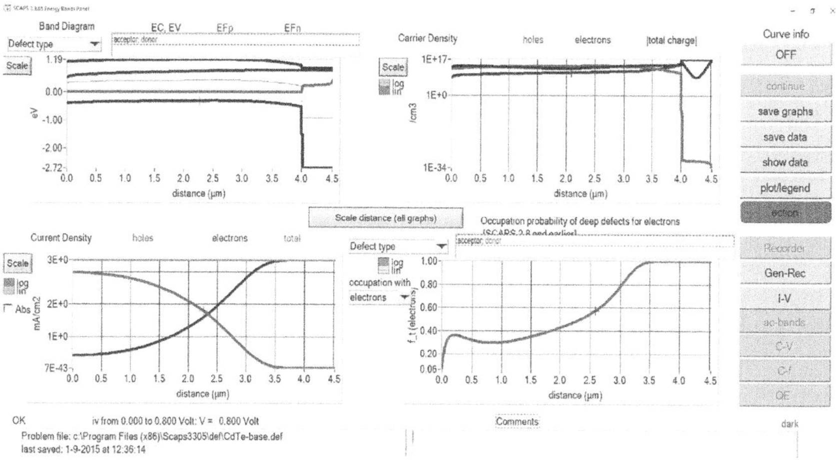

Figure 1.6 Screenshot of energy bands panel window in SCAPS.

QE(λ). The start and end values of the argument and the number of steps can be adjusted if necessary.

We select the "calculate" button in the action panel after implementing all the necessary data (layer properties, solar cell configuration). The panel of the energy bands opens, and the calculations begin. The window for panel for energy bands opens after measurement with the band diagram, density of carriers, density of current, and probability of occupation of deep defects of graph plots of electrons as shown in Figure 1.6.

Then from the right side of the screen window of the energy bands one can select choices (Gen-Rec, I-V). The behavior of the solar cell like the short circuit current (J_{SC}) may be derived from the I-V curve. The results can be saved for further editing or use in other programs as ASCII files (e.g., Excel). It is broadly used for the simulation and analysis of different types of solar cells [46–49].

1.7 Conclusion

From the study of these simulation softwares, it can be definitely said that simulation of perovskite solar cells can be used as a tool for designing and optimizing advanced perovskite solar cell structures and also to interpret the measurements made on different device structures. It can also be conclusively inferred that simulation can bring about better understanding of the detailed specifics of the working of perovskite solar cell structures. However, the simulation of perovskite solar cells requires a lot of input parameters, which makes it a tedious task.

All the softwares discussed here use Graphical User Interface except ASA, which uses ASCII file to input the values of various parameters. The number of layers that can be simulated is fixed in SCAPS and AMPS 1D, whereas there is no such fixation in AFORS-HET and ASA. All softwares studied use SRH recombination mechanism, whereas SCAPS and AMPS 1D also use band-to-band recombination. Auger recombination is considered in SCAPS. All the software studied here give the option to model defects. In SCAPS, up to seven defects can be simulated in any semiconductor layer, and all the parameters of each defect can be modified by "defect properties panel." AMPS 1D allows Discrete and Banded Defect (Structural and Impurity) Levels and Generalized Defect (Structural and Impurity) Level Distributions. In AFORS-HET, defect distribution of states (DOS) has to be quantified for all layers and for the interfaces. In ASA, both the extended and localized states can be modeled using the defect-pool models. Modeling of tunnelling is available in all the softwares except AMPS 1D. All softwares are capable of doing electrical and spectral response measurements.

References

1. Neamen, D.A., Semiconductor physics and devices: basic principles, McGraw-Hill, New York, 2003.
2. Mandadapu, U., Vedanayakam, S.V., Thyagarajan, K., Babu, B.J., Optimisation of high efficiency tin halide perovskite solar cells using SCAPS-1D. *Int. J. Simul. Process Model.*, 13, 3, 221–227, 2018.
3. Zeman, M., van den Heuvel, J., Pieters, B.E., Kroon, M., Willemen, J., *Advanced semiconductor analysis*, TU Delft, Delft, 2003.
4. Pieters, B.E., Krc, J., Zeman, M., May. Advanced numerical simulation tool for solar cells-ASA5, in: *2006 IEEE 4th World Conference on Photovoltaic Energy Conference*, vol. 2, IEEE, pp. 1513–1516, 2006.
5. Pieters, BE., Zeman, M., Metselaar, JW., Extraction of the defect density of states of a-Si:H using Q-DLTS. In s.n. (Ed.), *Proceedings of the STW annual workshop on semiconductor advances for future electronics and sensors (SAFE 2005)*, 38–42, STW, 2005.
6. Zeman, M., Willemen, J.A., Vosteen, L.L.A., Tao, G., Metselaar, J.W., Computer modelling of current matching in a-Si: H/a-Si: H tandem solar cells on textured TCO substrates. *Sol. Energy Mater. Sol. Cells*, 46, 2, 81–99, 1997.
7. Meier, J., Dubail, S., Fluckinger, R., Fisher, D., Keppner, H., Shah, A., Intrinsic microcrystalline silicon(μc-Si:H)—A promising new thin film solar cell

material. In *Proceedings of the 1st World Conference on Photovoltaic energy conversion*, Waikoloa, HI, USA, 5–9, December 1994.

8. Yamamoto, K., Yoshimi, M., Tawada, Y., Fukuda, S., Sawada, T., Meguro, T., Takata, H., Suezaki, T., Koi, Y., Hayashi, K., Suzuki, T., Large area thin film Si module. *Sol. Energy Mater. Sol. Cells*, 74, 1-4, 449–455, 2002.
9. Nádazdy, V., Durný, R., Pincik, E., Evidence for the improved defect-pool model for gap states in amorphous silicon from charge DLTS experiments on undoped a-Si: H. *Phys. Rev. Lett.*, 78, 6, 1102, 1997.
10. Nádaždy, V. and Thurzo, I., A model of the small-signal charge DLTS response of traps distributed in both energy and space. *Phys. Status Solidi (a)*, 127, 1, 167–177, 1991.
11. Springer, J., Poruba, A., Mullerova, L., Vanecek, M., Reetz, W., Muller, J., 3-dimensional optical model for thin film silicon solar cells. *3rd World Conference on Photovoltaic Energy Conversion, Proceedings of, 2003*, vol. 2, pp. 1827–1830, 2003.
12. Krč, J., Smole, F., Topič, M., Potential of light trapping in microcrystalline silicon solar cells with textured substrates. *Prog. Photovolt: Res. Appl.*, 11, 429–436, 2003, https://doi.org/10.1002/pip.506.
13. Fonash, S., Arch, J., Cuiffi, J., Hou, J., Howland, W., McElheny, P., Moquin, A., Rogosky, M., Tran, T., Zhu, H., *A manual for AMPS-1D for windows 95/NT a one-dimensional device simulation program for the analysis of microelectronic and photonic structures*, The Pennsylvania State University, USA, 1997.
14. Zhu, H., Kalkan, A.K., Hou, J., Fonash, S.J., Applications of AMPS-1D for solar cell simulation. *AIP Conference Proceedings*, vol. 462, pp. 309–314, 1999, https://doi.org/10.1063/1.57978.
15. Dennai Benmoussa, M. and Boukais, H.B., Simulation of hetero-junction (GaInP/GaAs) solar cell using AMPS-1D. *J. Nano- Electron. Phys.*, 8, 1, 01009, 2016.
16. Hossain, E.S., Chelvanathan, P., Shahahmadi, S.A., Sopian, K., Bais, B., Amin, N., Performance assessment of Cu2SnS3 (CTS) based thin film solar cells by AMPS-1D. *Curr. Appl. Phys.*, 18, 1, 79–89, 2018, https://doi.org/10.1016/j.cap.2017.10.009.
17. Stangl, R., Kriegel, M., Schmidt, M., AFORS-HET, Version 2.2, a numerical computer program for simulation of heterojunction solar cells and measurements, in: *2006 IEEE 4th World Conference on Photovoltaic Energy Conference*, vol. 2, IEEE, pp. 1350–1353, 2006, May.
18. Stangl, R., Kriegel, M., Maydell, K.V., Korte, L., Schmidt, M., Fuhs, W., AFORS-HET, an open-source on demand numerical PC program for simulation of (thin film) heterojunction solar cells, version 1.2, in: *Conference Record of the Thirty-first IEEE Photovoltaic Specialists Conference*, IEEE, pp. 1556–1559, 2005, January2005.
19. Wurfel, P., The chemical potential of radiation. *J. Phys. C: Solid State Phys.*, 15, 3967, 1982.

20. Fuhs, W., Korte, L., Schmidt, M., Heterojunctions of hydrogenated amorphous silicon and monocrystalline silicon. *J. Optoelectron. Adv. Mat.*, 8, 6, 1989–1995, 2006.
21. Gudovskikh, A.S., Kleider, J.P., Stangl, R., New approach to capacitance spectroscopy for interface characterization of a-Si: H/c-Si heterojunctions. *J. Non-Cryst. Solids*, 352, 9-20, 1213–1216, 2006.
22. Gudovskikh, A.S., Kleider, J.P., Froitzheim, A., Fuhs, W., Terukov, E.I., Investigation of a-Si: H/c-Si heterojunction solar cells interface properties. *Thin Solid Films*, 451, 345–349, 2004.
23. Schaffarzik, D., Stangl, R., Laades, A., Schubert, C., Schmidt, M., Recombination analysis at the n-doped a-Si: H (n)/c-Si (p) heterojunction by means of time and intensity dependent surface photovoltage, in: *Proc. PVSEC-20, 20th European Photovoltaic Conference*, 2005, June.
24. Maydell, K.V., Conrad, E., Schmidt, M., Efficient silicon heterojunction solar cells based on p- and n-type substrates processed at temperatures <220 C. *Prog. Photovolt: Res. Appl.*, 14, 289–295, 2006, https://doi.org/10.1002/pip.668.
25. Stangl, R., Kriegel, M., Schaffarzik, D., Schmidt, M., AFORS-HET, Version 2.1, a numerical computer program for simulation of (thin film) heterojunction solar cells, in: *Proc. 15th Int. Photovoltaic Sci. Eng. Conf*, pp. 985–986, 2005, October.
26. Stangl, R., Froitzheim, A., Kriegel, M., Brammer, T., Kirste, S., Elstner, L., ... Fuhs, W., AFORS-HET, a numerical PC-program for simulation of heterojunction solar cells, Version 1.1 (open-source on demand), to be distributed for public use. *Proc. 19th PVSEC*, Paris, France, 1497, 2004.
27. Burgelman, M., Nollet, P., Degrave, S., Modelling polycrystalline semiconductor solar cells. *Thin solid films*, 361, 527–532, 2000.
28. Decock, K., Khelifi, S., Burgelman, M., Modelling multivalent defects in thin film solar cells. *Thin Solid Films*, 519, 21, 7481–7484, 2011.
29. Burgelman, M. and Marlein, J., Analysis of graded band gap solar cells with SCAPS, in: *Proceedings of the 23rd European Photovoltaic Solar Energy Conference*, Valencia, pp. 2151–2155, 2008, September.
30. Verschraegen, J. and Burgelman, M., Numerical modeling of intra-band tunneling for heterojunction solar cells in SCAPS. *Thin Solid Films*, 515, 15, 6276–6279, 2007.
31. Degrave, S., Burgelman, M., Nollet, P., Modelling of polycrystalline thin film solar cells: new features in SCAPS version 2.3, in: *3rd World Conference on Photovoltaic Energy Conversion, Proceedings of 2003*, vol. 1, IEEE, pp. 487–490, 2003, May.
32. Niemegeers, A. and Burgelman, M., Numerical modelling of ac-characteristics of CdTe and CIS solar cells, in: *Conference Record of the Twenty Fifth IEEE Photovoltaic Specialists Conference-1996*, IEEE, pp. 901–904, 1996, May.

33. Press, W.H., Teukolsky, S.A., Flannery, B.P., Vetterling, W.T., *Numerical recipes in Fortran 77: of Fortran numerical recipes: the art of scientific computing*, vol. 1, Cambridge university press, Cambridge, 1992.
34. Verschraegen, J., Khelifi, S., Burgelman, M., Belghachi, A., Numerical modeling of the impurity photovoltaic effect (IPV) in SCAPS, in: *21st European Photovoltaic Solar Energy Conference*, vol. 396, WIP, 2006, September.
35. Khelifi, S., Burgelman, M., Verschraegen, J., Belghachi, A., Impurity photovoltaic effect in GaAs solar cell with two deep impurity levels. *Sol. Energy Mater. Sol. Cells*, 92, 12, 1559–1565, 2008.
36. Khelifi, S., Verschraegen, J., Burgelman, M., Belghachi, A., Numerical simulation of the impurity photovoltaic effect in silicon solar cells. *Renew. Energy*, 33, 2, 293–298, 2008.
37. Decock, K., Zabierowski, P., Burgelman, M., Modeling metastabilities in chalcopyrite-based thin film solar cells. *J. Appl. Phys.*, 111, 4, 043703, 2012.
38. Burgelman, M., Decock, K., Khelifi, S., Abass, A., Advanced electrical simulation of thin film solar cells. *Thin Solid Films*, 535, 296–301, 2013.
39. Niemegeers, A., Gillis, S., Burgelman, M., A user program for realistic simulation of polycrystalline heterojunction solar cells: SCAPS-1D. *Proceedings of the 2nd World Conference on Photovoltaic Energy Conversion*, JRC, European Commission, juli, pp. 672–675, 1998.
40. Pauwels, H.J. and Vanhoutte, G., The influence of interface state and energy barriers on the efficiency of heterojunction solar cells. *J. Phys. D: Appl. Phys.*, 11, 5, 649, 1978.
41. Verschraegen, J., Karakterisering en modellering met SCAPS van de CISCuT dunne-filmzonnecel, dissertation Universiteit Gent. Faculteit Ingenieurswetenschappen, 2006
42. Selberherr, S., Analysis and simulation of semiconductor devices, in: *Springer Science & Business Media*, 1984.
43. Marlein, J. and Burgelman, M., Empirical JV modelling of CIGS solar cells. In Proceedings of NUMOS (Int. Workshop on Numerical Modelling of Thin Film Solar Cells, Gent (B), 28-30 March 2007). 227-233, 227–233, 20072007. AU: Please provide journal title.
44. Walter, T., Herberholz, R., Müller, C., Schock, H.W., Determination of defect distributions from admittance measurements and application to Cu (In, Ga) Se2 based heterojunctions. *J. Appl. Phys.*, 80, 8, 4411–4420, 1996.
45. Decock, K., Khelifi, S., Buecheler, S., Pianezzi, F., Tiwari, A.N., Burgelman, M., Defect distributions in thin film solar cells deduced from admittance measurements under different bias voltages. *J. Appl. Phys.*, 110, 6, 063722, 2011.
46. Sharma, B., Mathur, A.S., Rajput, V.K., Singh, I.K., Singh, B.P., Device modeling of non-fullerene organic solar cell by incorporating CuSCN as a hole transport layer using SCAPS. *Optik*, 251, 168457, 2022.

47. Mathur, A.S., Upadhyay, S., Singh, P.P., Sharma, B., Arora, P., Rajput, V.K., Kumar, P., Singh, D., Singh, B.P., Role of defect density in absorber layer of ternary chalcogenide Cu2SnS3 solar cell. *Optic. Mater.*, 119, 111314, 2021.
48. Mathur, A.S. and Singh, B.P., Study of effect of defects on CdS/CdTe heterojunction solar cell. *Optik*, 212, 164717, 2020.
49. Mathur, A.S., Dubey, S., Singh, B.P., Study of role of different defects on the performance of CZTSe solar cells using SCAPS. *Optik*, 206, 163245, 2020.

2

Fundamentals of Perovskite Solar Cells

Neha Patni*, **Rokadia Zulfiqar** and **Krishna Patel**

Chemical Engineering Department, Institute of Technology, Nirma University, Ahmedabad, Gujarat, India

Abstract
Perovskite solar cell (PSC) is a type of third-generation hybrid solar cell based on organic-inorganic metal halide material, having the molecular formula of the type ABX_3. High efficiency, flexibility, cell architecture, and low-cost production of the PSC have caught the attention of researchers and technologists in the field. There is a tremendous growth of efficiency in a short period, i.e., 3.8% in 2009 to nearly 25% in 2020. This chapter discusses the fundamental principle used in PSCs. Its structure, various layers of the cell, and their significance will be conferred. The chapter will also cover the working of the solar cell and discuss the various parameters on which the efficiency of the solar cell will depend. Fundamental properties of PSCs will be examined like high optical absorption, tunable bandgap, high open-circuit voltage, which is one of the most promising aspects of the perovskite technology that the cells can generate under full sun illumination with low loss-in-potential. In the end, drawbacks of PSCs, like stability and toxicity, will be analyzed and its reasons will be examined in brief.

Keywords: Perovskite, solar cell, open-circuit voltage, power conversion efficiency

2.1 Introduction

Our planet's energy consumption has increased because of rapid industrialization and urbanization. To meet this energy requirement, the use of fossil fuels like coal, natural gas, and oil has increased tremendously. As a result, there is depletion in fossil fuel. Fossil fuels have advantages, like production

**Corresponding author*: neha.patni@nirmauni.ac.in

Khursheed Ahmad and Waseem Raza (eds.) Perovskite Materials for Energy and Environmental Applications, (37–58) © 2022 Scrivener Publishing LLC

of enormous energy, easy availability, cost effectiveness, easy transportation, but has several environmental issues. The use of fossil fuel is the main source of local air pollution, escalating carbon emissions, which leads to increase in the greenhouse gases, and their combustion makes the atmosphere more acidic. Moreover, fossil fuel is a nonrenewable source of energy which means in near future, these resources are going to vanish. As per the review of literature, from last six years the oil is left for around fifty years, natural gas for around fifty two years and coal for around one hundred ten years. The solution to this energy crisis is the renewable sources of energy like solar energy, hydrogen energy, wind energy, tidal energy, ocean thermal energy, biogas, biofuels, and geothermal energy.

The flow of research in renewable energy technology has increased tremendously. Out of all the sources of renewable sources of energy, solar energy is the most encouraging and applied approach. Solar energy comes with long-time warranty and low maintenance costs, environment friendly, and easy to adapt. Solar energy can lead us to sustainable development.

Photovoltaic cells are classified into first, second, and third generations as depicted in Figure 2.1. First-generation cells are silicon-based crystalline cells, which are commercially used. Si-based crystalline cells are widely applied cells because of their high stability and high efficiency (nearly 45%). Second-generation cells are thin film–based cells. Third-generation cells are new emerged technology-based cells. Perovskite is the third generation cell. An ideal solar cell should be highly efficient, stable, less costly, simple to fabricate, and environment friendly. None of the existing cells is ideal. Si-based cells have a high cost of production and complex fabrication process. Cadmium tellurium (CdTe) solar cell and copper indium gallium silicon (CIGS) solar cell are second-generation cells that have achieved high efficiency but have some disadvantages. Cadmium tellurium faces the problem of a low abundance of tellurium and toxicity of cadmium, whereas CIGS has the main problem of scarcity of indium in its industrial applications. These limitations lead to the emergence of the third-generation solar cells.

Dye-sensitized solar cell (DSSC) is the first third-generation solar cell and perovskite solar cell (PSC) has evolved from it, as shown in Figure 2.2. The DSSC consists of layers of glass/FTO (fluorine-doped tin oxide)/mesoporous layer of titanium dioxide/light-absorbing dye/electrolyte/FTO/glass. Despite the impressive research work, DSSC was able to achieve max 13% PCE [2]. The major advantages of DSSC include low fabrication costs and a simple manufacturing process, although disadvantages include the absorption layer should be more than 10 μm to ensure complete light absorption and the phenomenon of light bleaching occurs due

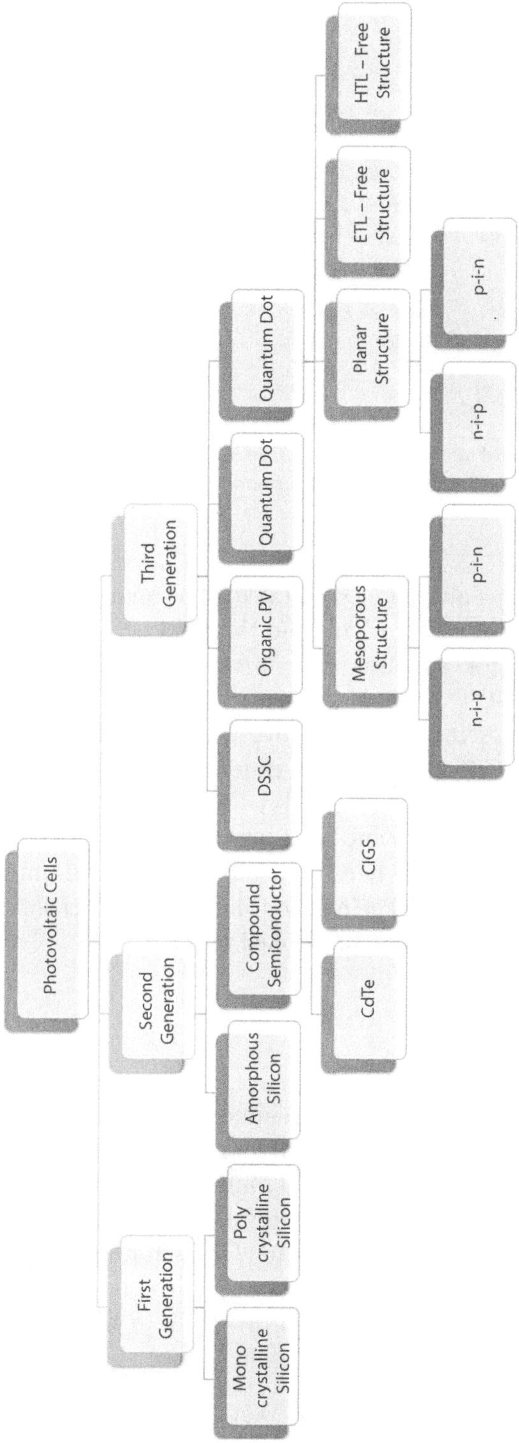

Figure 2.1 Generations of the photovoltaic cells.

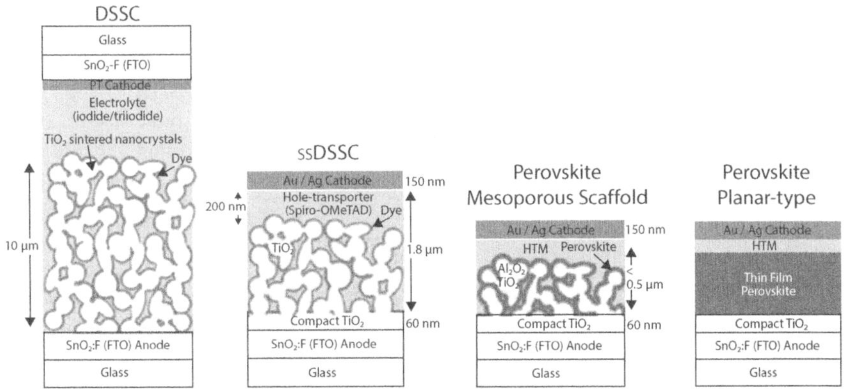

Figure 2.2 Evolution of perovskite solar cell from reduction of size of light absorbing layer from 10 μm (DSSC) to less than 0.5 μm (perovskite solar cell). Reprinted with permission. Copyright {2013} American Chemical Society [1].

to organic dyes [1]. Solid state DSSC (SSDSSC) contains solid hole conductors instead of liquid electrolytes and the thickness is reduced to 2 μm. The hole conductor contains wide bandgap material, such as Spiro-OMeTAD or PEDOT or P3HT. In 2009, Kojima *et al.* [3] utilized organolead halide as a sensitizer to achieve the power conversion efficiency (PCE) of 3.8%. In 2012, for the first time, an all-solid-state PSC was prepared and achieved a PCE of 9.7% [4]. This showed the potential of PSC toward future cost-effective and high-performance solar cells. A considerable amount of research has been done on PSC in a short time. The highest efficiency achieved was 22.1% in 2016 [5]. For commercialization, stability and toxicity of the PSC should be looked upon.

2.2 Structure

Perovskite is the material having the same compound formula as $CaTiO_3$. Perovskite materials have the general formula of ABX_3. Gustav Rose discovered this compound from the Ural Mountains of Russia in 1839 and was named after the mineralogist Lev Perovskite (1792-1856). The Perovskite structure was first described by Victor Goldschmidt in 1926. Perovskite can be divided into two types—inorganic oxide perovskite and halide perovskite. Halide perovskite are further divided into alkali halide perovskite and organo-metal halide perovskite where in A, B, and X vary as mentioned in Table 2.1 [6].

Table 2.1 Types of perovskite structures.

Types of perovskite structure	A	B	X
Inorganic oxide perovskite	Divalent ions—Pb^{2+}, Sn^{2+}, Ca^{2+}, Mg^{2+}, etc	Tetravalent metal ions—Ti^{4+}, Sn^{4+}, etc	O
Alkali halide perovskite	Monovalent ions—Na^{+}, Li^{+}, etc	Divalent ions—Pb^{2+}, Sn^{2+}, Ca^{2+}, Mg^{2+}, etc	Halogen—Cl^{-}, Br^{-}, I^{-}
Organo-metal halide perovskite	Organic cations—methylammonium (CH_3NH_3) and formamidinium ($HC(NH_2)_2$)	Divalent ions—Pb^{2+}, Sn^{2+}, Ca^{2+}, Mg^{2+}, etc	Halogen—Cl^{-}, Br^{-}, I^{-}

Perovskite has a cubic structure as shown in Figure 2.3. A is situated on the vertex, B is at the center, whereas X is situated at the face center. It is equivalent to an octahedral structure.

A small defect in the lattice decreases the symmetry of a crystal [7]. The perovskite mineral is itself distorted, and because of this distortion, many physical properties, like electronic, magnetic, and dielectric behavior, are impacted. The ideal cubic perovskite structure is not common. It has been reported that if the grain surface of perovskite ($CH_3NH_3PbI_3$) is chemically modified by spin-coating of 4-ABPACl, butylphosphonic acid 4-ammonium chloride then the efficiency and stability both improve. 4-ABPACl forms crosslink with perovskite by the formation of hydrogen bond between perovskite ($CH_3NH_3PbI_3$) iodine and $-PO(OH)_2$, $-NH_3^+$,

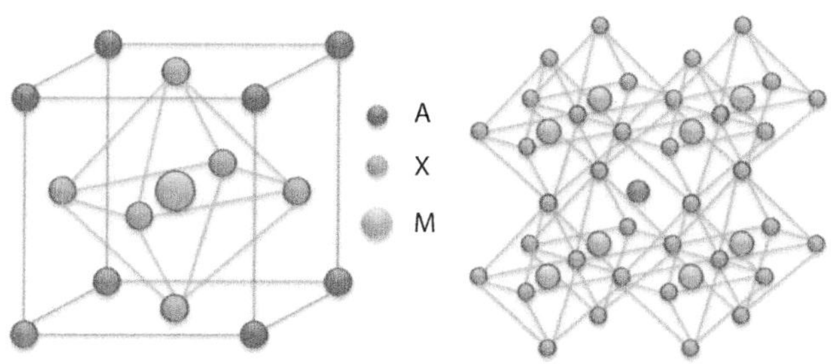

Figure 2.3 The crystal structure of the perovskite [6].

as a result, the cohesion force increases between neighboring perovskite grains [8].

2.3 Working Mechanism of PSC

The working principle of the PSC can be understood by the working of the DSSC and organic photovoltaic. In PSC, the perovskite is used as a light sensitizer, the same as dye in DSSC. The three functions performed by PSC are formation of the exciton (electron-hole pair) by the adsorption of the photon, charge transport, and charge extraction. When light falls on the perovskite, excitons are formed. Because of the difference in the binding energy of exciton of perovskite material, free carriers (free electron and hole) are formed from exciton, which can either generate current or can recombine. Because of the low recombination rate of free carriers in perovskite material and high charge mobility, the diffusion distance and lifetime of the carrier are long. And as a result, PSC is able to achieve such a high efficiency. Highly crystalline $CH_3NH_3PbI_3$ perovskite material is able to achieve charge carrier diffusion length as long as 10 micrometers [9]. As shown in Figure 2.4, the electron is generated in perovskite and then moves to the electron transporting layer (ETL) and finally to the anode. Simultaneously, holes generated in perovskite, move to the hole transporting layer (HTL) and finally to the cathode. The electron migrates to the anode which can be FTO or ITO. The hole migrates to the cathode, which is either Ag or Au [10].

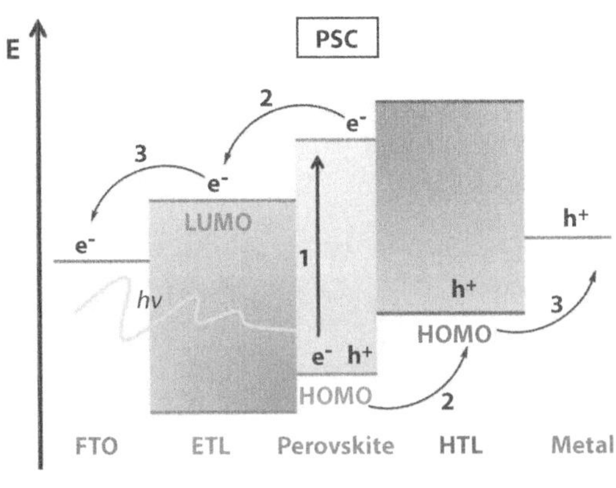

Figure 2.4 Working of the perovskite solar cell [11].

2.4 Device Architecture

Device arrangement is definitely one of the most significant aspects for the performance of the PSC. Depending on the position of the transport layer in the exterior portion, PSC can be classified into two configurations like mesoporous structure and planar structure as depicted in Figure 2.5. These two configurations are further classified into regular (n-i-p) and inverted (p-i-n) as shown in Figure 2.6.

Other two configurations are also possible like ETM-free structure and HTM-free structure, where P-contact is ETM, whereas N-contact is HTM. The brief overview of mesoporous structure and planar structure is discussed further in the chapter.

2.4.1 Mesoporous Structure

A typical n-i-p mesoporous structure is shown in Figure 2.6.a, ETM is on the exterior portion. The first arrangement of the PSC to be tested was a conventional n-i-p mesoporous structure in which the light-harvesting dye was replaced by the lead halide perovskite materials ($CH_3NH_3PbBr_3$ and $CH_3NH_3PbI_3$) in a traditional DSSC [3]. Mesoporous structure of PSC is widely used because of its high porosity and high surface area. As shown in Figure 2.6.a, the mesoporous n-i-p PSC consists of FTO, ETM, a mesoporous layer of metal oxide, perovskite layer, HTM, and an electrode layer. In mesoporous p-i-n PSC, the HTM layer is on the exterior portion as shown in Figure 2.6.b. The perovskite material adheres to the mesoporous structure to increase the light-absorbing area of perovskite, which results in increased efficiency. The mesoporous structure results in efficient crystallization of perovskite and electron injection [9]. There is

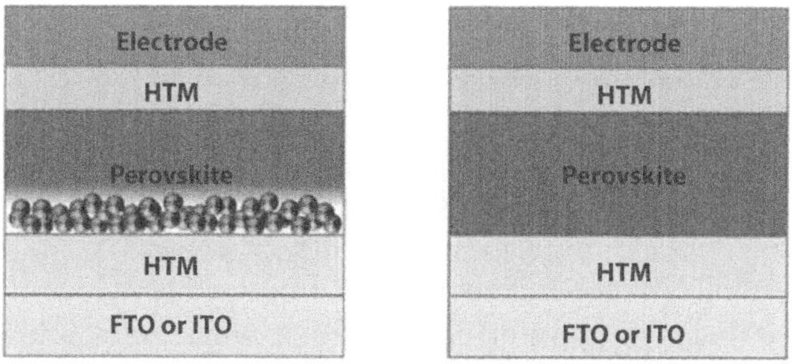

(a) Mesoscopic Device (b) Planar Device

Figure 2.5 Depiction of two main types of PSC [12].

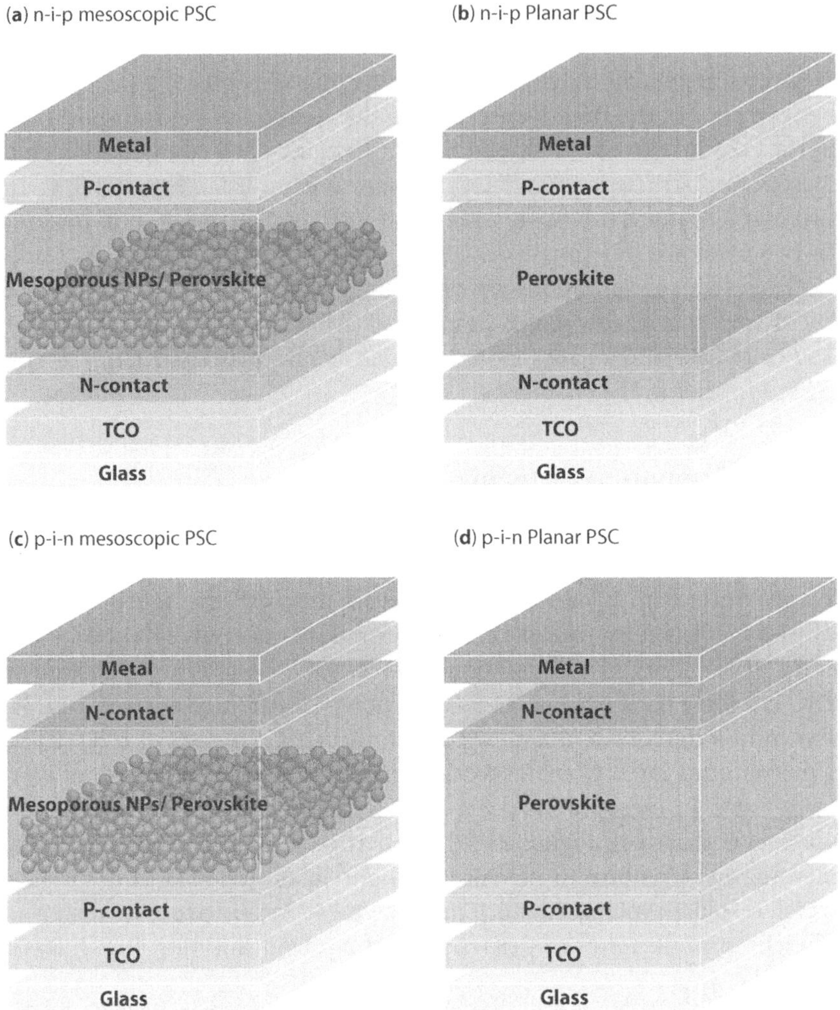

Figure 2.6 Schematic diagram of four types of perovskite solar cell. (a) n-i-p mesoporous structure, (b) p-i-n mesoporous structure, (c) n-i-p planar structure, (d) p-i-n planar structure [13].

a depletion region at the TiO_2 side of the $TiO_2/MAPbI_3$ interface, which provides contribution in separation of photogenerated charge carriers and hence it results in high performance of TiO_2-based cells [14]. TiO_2 mesoporous layer helps to transport the electrons, block holes, and inhibit the recombination of the electron and hole pair. The electron diffusion length and recombination lifetime is governed by the mesoporous layer of TiO_2 and hence is not varied by perovskite composition [15].

The most typical mesoporous framework material is TiO_2. Bi *et al.* achieved PCE of 20.8%, which contains TiO_2 as the mesoporous layer and Spiro-OMeTAD as the electron and hole specified contacts [16]. The limitation of using TiO_2 as a mesoporous layer is comparatively low electron mobility [17]. In place of TiO_2, Al_2O_3, and ZnO have also been used as a mesoporous layer [18]. ZnO has two orders of magnitude of electron mobility higher than TiO_2 [19]. The recombination rate of charge is significantly slower in TiO_2 as compared with Al_2O_3 [20]. Also, Zn_2SnO_4, $BaSnO_3$, and $SrTiO_3$ are tested as mesoporous layers. Zn_2SnO_4-based PSC showed fast electron transport (up to 10 times) and greater charge collection capability as compared with TiO_2-based PSC with the same physical and chemical conditions [21]. $BaSnO_3$ has a large bandgap and relatively higher electron mobility than TiO_2. $BaSnO_3$ was able to achieve 12.3% but has a recombination rate greater than TiO_2 [22]. $SrTiO_3$ had achieved 1.01 eV open-circuit voltage, which is 25% higher than TiO_2 [23]. The most common hole transport material is Spiro-OMeTAD(2,2′,7,7′-Tetrakis [N,N-di(4-methoxyphenyl)amino]-9,9′-spirobifluorene). An alternative to Spiro-OMeTAD, spiro[fluorene-9,9′-xanthene], achieved the PCE of 19.8% with low fabrication cost [16]. KTM3 showed higher V_{oc} (1.08V) and FF (78.3%) than Spiro-OMeTAD [24]. The commonly used counter electrode materials are Ag, Au, and Pt. For large-scale and continuous production of PSC containing the mesoporous layer of TiO_2, the electrospray deposition method is more viable [25].

2.4.2 Planar Heterostructures

Planar heterostructure is an evolution of the mesoporous structure. The main difference in the planar heterostructure is the absence of the mesoporous layer. The perovskite material is sandwiched between ETM and HTM. Hence, it results in a simpler structure. The planar structure can achieve high PCE even in the absence of a mesoporous layer by carefully controlling the interfaces between layers. There are two interfaces present. First is in between ETM and perovskite and second in between perovskite and HTM. Therefore, electron and hole pairs are separated quickly and efficiently by these two interfaces. The planar structure gives better insight in the understanding of the mechanism of photon absorption and charge separation, which will be useful for better optimization of the PSC. The configuration of planar heterostructure is FTO, ETM, perovskite layer, HTM, electrode layer. $FAPbI_3$ solution-processed planar heterostructure PSC showed PCE of 14.2% with large diffusion length and large tunable bandgap [26]. The planar heterostructure PSC constructed with vapor deposition showed PCE of 15% [27]. J-V hysteresis significantly depends on HTM in planar heterostructures. However, it became negligible when Spiro-OMeTAD was replaced with poly(3,4-ethylene dioxythiophene)-poly(styrene sulfonate) (PEDOT: PSS) or any other inorganic HTM [28]. Moreover, J-V hysteresis also depends on the voltage scan direction, scan rate, and range [29]. In the inverted planar heterostructure (p-i-n), HTM is deposited before ETM. The holes travel themselves in an inverted planar heterostructure [29]. It has been stated that if inverted planar heterostructure PSC is made by mixing methylammonium bromide as an additive in lead acetate and methylammonium iodide solution, maximum PCE of 18.3% is achieved [30].

2.5 Properties

2.5.1 High Optical Absorption

The absorber layer is generally referred to as the heart of the solar cell. The material of this absorber is made of organic inorganic hybrid perovskite. This absorber material has a direct band gap with high absorption coefficient. These properties have permitted to be used as very thin compact absorber materials for gathering light which have thickness in the range of hundreds of nanometers. Whereas the thickness of the film for the traditional solar cell (silicon and germanium) is the range of one micrometer to hundreds of micrometers.

Out of the three generations of solar cell, the first-generation absorber is an indirect bandgap semiconductor material. In this type of semiconductor, the minimum value of conduction band and maximum value of valence band does not lie for the same value of k. Because of this, the momentum of the electron in conduction and valence band is different.

In indirect bandgap semiconductor, transition of photon is horizontal in nature; hence, the momentum of photon is comparatively larger. So a large change of momentum is required for electron-hole pair recombination, hence momentum is not conserved here as k value for both the bands is different. Hence, a large change in momentum is required for recombination of the electron-hole pair as compared with direct bandgap semiconductors. So there is a prohibition for transition of electrons between these two levels and the transition becomes weak because of the participation of lattice vibrations during excitation. Hence, the probability of transition is lower in indirect band gap material than those in direct band gap material. Thus, the thickness of the absorber material of indirect

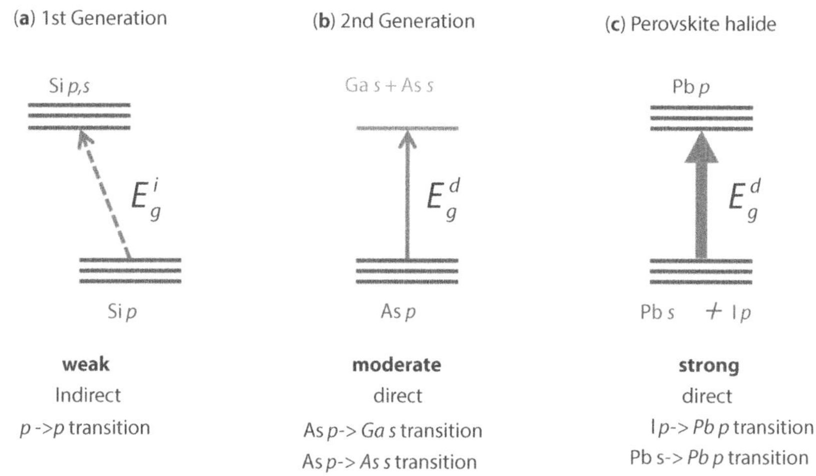

Figure 2.7 Transitions in direct and indirect semiconductor [33].

bandgap semiconductor is way more than that of direct band gap semiconductor material for absorbing identical number of photons per unit area [32]. Because of this, for achieving alike efficiency, there is a rise in cost of production of indirect bandgap semiconductor material solar cells as compared with direct band gap semiconductor material solar cells.

Second-generation absorber material (GaAs) and perovskite ($CH_3NH_3PBI_3$) have direct band gap; hence, they have high optical absorption properties compared to silicon. However, their electronic configuration of both the absorbers is not the same as shown in Figure 2.7. The lower part of conduction of gallium arsenide is obtained from the delocalized s orbitals, whereas the lower part of $CH_3NH_3PBI_3$ mainly consists of Pb p bands. The p orbital is less dispersed as compared to s orbitals [33]. Hence, in GaAs, the transition efficiency is moderate. There is a high probability of transition from intra-atomic Pb s to Pb p, because of this, in $CH_3NH_3PBI_3$, the transition probability between the two bands is almost equivalent to GaAs [33]. Therefore, the optical absorptions are stronger in halide perovskites than GaAs.

2.5.2 High Open-Circuit Voltage

The most distinct property of the PSC technology is their high open circuit voltage, which is the maximum voltage that a solar cell can generate [1]. As from Figure 2.8, we can observe that the best PSC has open circuit voltage greater than 1.1V. The loss-in-potential is around 450 meV, which is

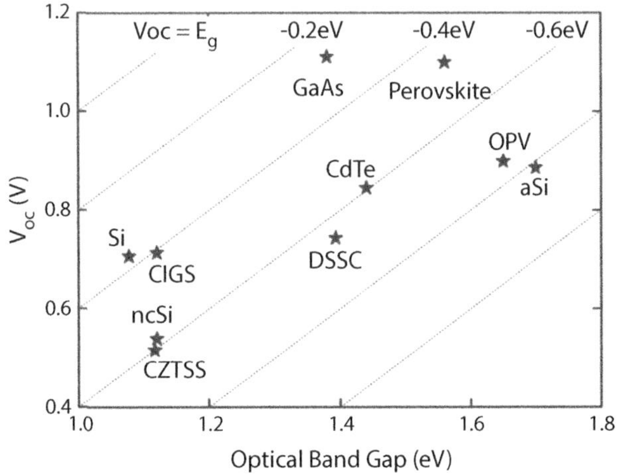

Figure 2.8 Open-circuit voltage (Voc) versus optical band gap (Eg). Reprinted with permission from [1]. Copyright {2013} American Chemical Society.

quite low. There is energy loss of 250 to 300 meV due to thermodynamic constraints, depending on the bandgap) [34].

Every data related to GaAs, Si, CIGS, CdTe, nanocrystalline Si (nc-Si), amorphous Si (a-Si), copper zinc tin sulfide/selenide (CZTSSe), organic photovoltaics (OPVs), and DSSCs were provided by the solar cell efficiency tables of Green *et al.* [35].

CdTe solar cell was the most successful solar cell commercially till 2015 with 19.6% efficiency with losses of approximately 0.59eV. The basic losses of CdTe solar cells are high as compared with perovskites cells [35]. The recombination rate (nonradiative) in perovskite absorber material is lower than polycrystalline film semiconductor [32]. High output voltage is one of the important factors which is responsible for high power conversion efficiency. We know that perovskite is placed in a very good position compared with the other materials for making solar cells. Voltage at maximum power point (V_{mpp}) is to be considered as the most relevant voltage for considering maximum efficiency [36]. Because of the manufacturing defects, there is high series resistance, which greatly affects the V_{mpp}. On optimization of solar cells the V_{mpp} will improve significantly [1].

2.5.3 Low Recombinations

In the recombination process, the excited electrons in the conduction band deexcite in the valence band and occupy the hole. This process can also be termed as electron hole pair recombination. This process competes with the process of separation of the excitation to electrons and holes and their collection at the electrodes. Because of this, there is a decline in the efficiency and current of the solar cell [32].

Recombinations are of two types: radiative and nonradiative recombination. When an electron deexcites from conduction band to valence band it releases a photon to release energy called radiative recombination. Whereas in nonradiative recombination, when an electron deexcites the energy is released in the form of heat. It causes harm to the performance of the cell. The device gets heated up during nonradiative recombinations due to which efficiency gets decreased.

Diffusion length is defined as the average length a carrier moves between its generation/formation and its recombination. On the basis of diffusion length parameters, the semiconductor material can be assessed for solar cell applications. Semiconductor materials have a shorter diffusion length and higher recombinations because they are heavily doped. If the diffusion length is higher, then the longer will be the lifetime of recombinations, the better the collection of carriers at the electrode. $CH_3NH_3PBI_{3-x}Cl_x$ has a diffusion length of more than 1 micron. This diffusion length is almost three times the thickness of the film in solar cells [37].

For constructing planar heterojunction solar devices this characteristic is very important. As the diffusion length of perovskite($CH_3NH_3PBI_3$) is only a hundred nanometer, so for transportation of charges to terminals, a nanoparticle system of a mesoporous TiO_2 is required [38].

2.5.4 Tunable Bandgap

For designing the solar cell, it is necessary that the light absorber is absorbing the maximum amount of the sunlight. To achieve this, we have to tune the band gap of the absorber. So here perovskite has been greatly advantageous as a light absorber because its bandgap is tunable/controllable. The perovskite has a structure of ABX_3. So the bandgap of the perovskite material (absorber) can easily be regulated by altering the organic cation (A) or the metal atom (B) or the halide (X).

2.5.4.1 Organic Cation (A)

In perovskite, structure changes in A do not remarkably alter the structures, but it can alter it to some extent [38]. The structure of the band of $CH_3NH_3PBI_3$ is near to its optimum value for solar applications, so the small change/alteration will help tune the band gap to its optimum value.

Table 2.2 Band gap of different ABX_3 materials [32].

A substitution		**M substitution**		**X substitution**	
Material	Band gap (eV)	Material	Band gap (eV)	Material	Band gap (eV)
$EAPbI_3$	2.2	$MaPbI_3$	1.5	$MAPbI_3$	1.5
$MAPbI_3$	1.5	$MASn_{0.3}Pb_{0.7}I_3$	1.31	$MAPbI_2Br$	1.8
$FAPbI_3$	1.4	$MASn_{0.5}Pb_{0.5}I_3$	1.28	$MAPbBr_3$	2.20
$CsPbI_3$	1.67	$MASn_{0.9}Pb_{0.1}I_3$	1.18	$MAPbCl_3$	3.11
		$MASnI_3$	1.10		

NH_2-$CH=NH_2^+$ (A) is replaced by $CH_3NH_3^+$ in $CH_3NH_3PbI_3$ to narrow the band gap. The band gap of perovskite formed after replacing A (NH_2-$CH=NH_2PbI_3$) is 1.4 eV [39].

The bandgap of $CsPbI_3$ is 1.67 eV [40]. The trend that is observed here is increasing the size of "A| cation, i.e., from Cs^+ to CH_3NH^{+3} and NH_2-$CH=NH^{+2}$, the value of the bandgap steadily declines [32] as shown in Table 2.2. However, on further increasing the size of "A" to $CH_3CH_2NH_2$ (EA), the perovskite $[CH_3CH_2NH_3]^+$ PbI_3 structure turns into orthorhombic symmetry with a bandgap of 2.2 eV [38].

2.5.4.2 *Metal Cation (M)*

Lead (Pb) was majorly used as a metal cation. European Union has restricted the use of Pb as it is toxic to the environment (Restriction of the use of certain hazardous substances (RoHS), Directive 2011/65/EU.). These reasons have led to synthesize leadless perovskite for light absorbing. As an alternative of Pb, we can use tin because tin and Pb are in the same group, so obviously tin would be the first choice. After Sn and Pb ratio was optimized in $CH_3NH_3Sn_{1-x}Pb_xI_3$, it was observed that bandgap was tunable in the range of 1.17 to 1.55 eV. Because of this, it was found that the absorption capacity of light could be increased from visible region to the near-infrared region (1060 nm) [41].

The problem in $CH_3NH_3Sn_{1-x}Pb_xI_3$ solar cells is that it is hard to make a consistent and thick absorber layer(film) on the area of the device, and because of these reasons, the efficiency drops down to 3% [41]. Tin-based solar perovskite cell ($CH_3NH_3SnI_3$) has 1.3eV bandgap. This perovskite can also absorb the light of the near-infrared region. The final power conversion efficiency of this device obtained was 5.37% [42]. Open circuit voltage of tin PSC always gets decreased because of downshifting of the conduction bandgap [38].

2.5.4.3 *Halide Anion (X)*

X position in perovskite can be replaced with elements chlorine, bromine, and iodine. There is a considerable proportion of change in bandgap when there is alternation in the element. Altering the position of X from chlorine to iodine the band gap value ranges from 3.11eV to 1.51eV. The smallest band gap of 1.51eV is obtained by tuning the position of X in $CH_3NH_3PbX_3$ [38].

M and X elements should be changed simultaneously so that the appropriate bandgap can be obtained. The bandgap of $CH_3NH_3SnI_{3-x}Br$ can be modulated from 1.30 to 2.15 eV by substitution of Pb with tin and partially substitution of I (iodine) with Br $(bromine)_x$ [42]. A linearity behavior is observed between bandgap and ratio of I/Br. Efficiency is achieved up to 5.23% when $CH_3NH_3SnI_3$ is used as a light absorber in the solar cell [42]. On optimizing iodine and bromine ratio in $CH_3NH_3SnIBr_2$, a way better PSC can be produced.

2.5.5 Rapidly Increasing Efficiency

The power conversion efficiency of PSCs has raised from 3.8% [3] in 2009 to 22.10% in 2016. For cells having square area aperture greater than 1 cm^2 has shown a verified efficiency of 19.6%. This efficiency is almost equal to efficiency obtained in 2017 by thin film CIGS and CdTe which is 21% [43]. Because of tunable band gaps in the solar cell, it can be optimized accordingly to the solar spectrum. Shockley–Queisser efficiency limit can be approached by this solar cell which is about 31% for a 1.55-eV band gap [44].

Perovskite solar cells have enormous potential because of its low manufacturing cost, it has low-energy payback time when compared with other solar cells and high efficiency. The PSC can be made easily and at a very low-cost [32]. Because of this, solar cell can be printed using inkjet printers. Perovskite solar cell can also be used in tandem configuration. In tandem cell configuration with the top layer of perovskite cell and bottom layer of crystalline silicon cell has achieved an efficiency of 29.6% [27].

2.6 Drawbacks and Ongoing Challenges of PSCs

1. Pb-free alternatives/toxicity of Pb
As we know, Pb is highly toxic and harmful to nature and the environment. Lead harms the microorganisms, which are present in the soil; hence, it is a potential threat to the ecosystem. Lead-based PSCs give high light power conversion efficiencies but to prevent the toxicity issues, we have to find alternatives to Pb. We can use tin as it is safe to replace Pb. In 2014, Pb-free perovskite cells were reported. Here, tin was used ($MASnI_3$) in replacement of Pb, and it achieved power conversion efficiency of 5% to 6% when made with mesoporous TiO_2 scaffolds [45, 46]. Cons of tin-based PSCs is that they are not stable in air, and even if they are prepared in an inert environment, they quickly degrade. The main problem is oxidation of tin from +2 to +4 because of self-doping. So the metal which can be stabilized in +2 oxidation state, then that could be used as an alternative to Pb [47].
2. Stability
The power conversion efficiency of PSCs has risen due to considerable amount of research and focus on architectures, compositions, and manufacturing techniques. For commercialization of perovskite cell technology, it should be able to survive at minimum 25 years without degradation in yield in an external environment. In a controlled environment, high-efficiency solar cells are synthesized but when they are exposed to the ambient atmosphere, efficiency drops rapidly. Light absorbers are very fragile/sensitive to water, because of this, solar cells are unstable [38]. The degradation also depends on oxygen, UV radiation, and temperatures.

The rate of degradation of perovskite is very rapid (few minutes) when there is presence of water. The products formed after the degradation of perovskite are CH_3NH_3I and PbI_2. So to increase the lifetime of PSCs it should be kept away from moisture. An HTM with long alkyl chain was used, which would help to prevent water from passing from HTM to perovskite [48]. A carbon material layer of 10 μm thick was used as back contact to retain water. After these arrangements, the stability got increased and power conversion efficiency remained as it is for 1008 hours [49].

The formation of perovskite $CH_3NH_3PbI_3$ film is carried (annealing) at 100°C for 10 minutes [38]. During its annealiation process when the temperature crosses 120°C, the material($CH_3NH_3PbI_3$) starts to degrade [50].

When temperature crosses 300°C, the perovskite splits into PbI_2, CH_3NH_2, and HI. Sometimes, TiO_2 can also bring instability because it can trap electrons in the crystals as it is nonstoichiometric, and this generally happens in an inert environment. So in seal-packed solar cells, there is a huge problem to avoid perovskite to get oxidized [38]. One solution is to dope TiO_2 with aluminium, which can greatly reduce the oxidation [51].

To some extent, the stability of the PSC has increased but it is still far away from long-term use and commercialization. For improving the stability the stability of PSCs, it is necessary to understand the degradation mechanisms and find out the state at which degradation reactions originated or accelerated [52].

2.7 Conclusion

The efficiency of the PSC in the laboratory has reached 23.1%, which is comparable to the single crystalline solar cell. Because of high achievable efficiency, the researchers of different backgrounds of science have contributed in the investigation of PSCs. The higher efficiency of PSCs has been achieved due to high coefficient of absorption and long diffusion lengths. Perovskite solar cell has desired properties, enormous potential, low cost, and high efficiency. Despite achieving high efficiency, there is much difference between the stability of PSCs (1000 hours) and silicon solar cells (20–25 years). The future challenge is to increase stability. Another drawback is it has toxic material, like Pb, which is very harmful for human beings, as well as the environment. There are some problems that are not solved yet, such as the materials, to increase the scalability, processing of materials, and so on [38].

In total, there are three major concerns, which should be overcome before its commercialization. First, the degradation mechanism should be explored as humidity, heat, and UV light are the main reasons for degradation and brings instability in perovskite. Second, Pb is very harmful to nature, hence it should be replaced with some safer option. Third, PSCs has very high efficiency in small areas of 0.1 cm^2, so this area must be increased such that efficiency and stability remains the same so that we can perform high scale [53]. Researches need to focus and improve on the perovskite and electron-hole transport materials in the future, so that we can use PSCs in place of the conventional solar cell. If these problems are solved, surely perovskite material is projected to have a significant part of the solar cell industry.

Acknowledgment

The authors acknowledge the Chemical Engineering Department of Nirma University for providing the necessary help, infrastructure, and support for the preparation of this chapter/work.

References

1. Snaith, H.J., Perovskites: The emergence of a new era for low-cost, high-efficiency solar cells. *J. Phys. Chem. Lett.*, 4, 21, 3623–3630, 2013.
2. Mathew, S. *et al.*, Dye-sensitized solar cells with 13% efficiency achieved through the molecular engineering of porphyrin sensitizers. *Nat. Chem.*, 6, 3, 242–247, 2014.
3. Kojima, A., Teshima, K., Shirai, Y., Miyasaka, T., Organometal halide perovskites as visible-light sensitizers for photovoltaic cells. *J. Am. Chem. Soc*, 131, 17, 6050–6051, 2009.
4. Kim, H. *et al.*, All-Solid-State Submicron Thin Film. *Sci. Rep.*, 2, 1, 1–7, 2012.
5. Yang, W.S. *et al.*, High-performance photovoltaic perovskite layers fabricated through intramolecular exchange. *Science*, 348, 6240, 1234–1237, 2015.
6. Ma, Y. *et al.*, Recent Research Developments of Perovskite Solar Cells. *Chin. J. Chem.*, 32, 10, 957–963, 2014.
7. Cheng, Z. and Lin, J., Layered organic-inorganic hybrid perovskites: Structure, optical properties, film preparation, patterning and templating engineering. *CrystEngComm*, 12, 10, 2646–2662, 2010.
8. Li, X. *et al.*, Improved performance and stability of perovskite solar cells by crystal crosslinking with alkylphosphonic acid ω-ammonium chlorides. *Nat. Chem.*, 7, 9, 703–711, 2015.
9. Shi, Y. *et al.*, CH3NH3PbI3 and CH3NH3PbI3-xClx in planar or mesoporous perovskite solar cells: Comprehensive insight into the dependence of performance on architecture. *J. Phys. Chem. C*, 119, 28, 15868–15873, 2015.
10. Hussain, I. *et al.*, Functional materials, device architecture, and flexibility of perovskite solar cell. *Emergent Mater.*, 1, 3, 133–154, 2018.
11. Marinova, N., Valero, S., Delgado, J.L., Organic and perovskite solar cells: Working principles, materials and interfaces. *J. Colloid Interface Sci.*, 488, 373–389, 2017.
12. Chen, L.-C. and Tseng, Z.-L., ZnO-Based Electron Transporting Layer for Perovskite Solar Cells, in: *Nanostructured Solar Cells*. 1, 203–215, 2017
13. Li, M.H., Shen, P.S., Wang, K.C., Guo, T.F., Chen, P., Inorganic p-type contact materials for perovskite-based solar cells. *J. Mater. Chem. A*, 3, 17, 9318–9319, 2015.

14. Dymshits, A., Henning, A., Segev, G., Rosenwaks, Y., Etgar, L., The electronic structure of metal oxide/organo metal halide perovskite junctions in perovskite based solar cells. *Sci. Rep.*, 5, 1, 1–6, 2015.
15. Zhao, Y., Nardes, A.M., Zhu, K., Mesoporous perovskite solar cells: Material composition, charge-carrier dynamics, and device characteristics. *Faraday Discuss.*, 176, 301–312, 2014.
16. Bi, D. *et al.*, Facile synthesized organic hole transporting material for perovskite solar cell with efficiency of 19.8%. *Nano Energy*, 23, 138–144, 2016.
17. Jung, H.S. and Park, N.-G., Solar Cells: Perovskite Solar Cells: From Materials to Devices (Small 1/2015). *Small*, 11, 1, 2–2, 2015.
18. Carnie, M.J. *et al.*, A one-step low temperature processing route for organolead halide perovskite solar cells. *Chem. Commun.*, 49, 72, 7893–7895, 2013.
19. Zhang, Q., Dandeneau, C.S., Zhou, X., Cao, C., ZnO nanostructures for dye-sensitized solar cells. *Adv. Mater.*, 21, 41, 4087–4108, 2009.
20. Marchioro, A. *et al.*, Unravelling the mechanism of photoinduced charge transfer processes in lead iodide perovskite solar cells. *Nat. Photonics*, 8, 250, 2014.
21. Oh, L.S. *et al.*, Zn2SnO4-based photoelectrodes for organolead halide perovskite solar cells. *J. Phys. Chem. C*, 118, 40, 22991–22994, 2014.
22. Zhu, L. *et al.*, Mesoporous BaSnO3 layer based perovskite solar cells. *Chem. Commun.*, 52, 5, 970–973, 2016.
23. Bera, A. *et al.*, Perovskite oxide SrTiO3 as an efficient electron transporter for hybrid perovskite solar cells. *J. Phys. Chem. C*, 118, 49, 28494–28501, 2014.
24. Krishnamoorthy, T. *et al.*, A swivel-cruciform thiophene based hole-transporting material for efficient perovskite solar cells. *J. Mater. Chem. A*, 2, 18, 6305–6309, 2014.
25. Kim, M.C. *et al.*, Electro-spray deposition of a mesoporous TiO2 charge collection layer: Toward large scale and continuous production of high efficiency perovskite solar cells. *Nanoscale*, 7, 48, 20725–20733, 2015.
26. Eperon, G.E. *et al.*, Formamidinium lead trihalide: A broadly tunable perovskite for efficient planar heterojunction solar cells. *Energy Environ. Sci.*, 7, 3, 982, 2014.
27. Liu, M., Johnston, M.B., Snaith, H.J., Efficient planar heterojunction perovskite solar cells by vapour deposition. *Nature*, 501, 7467, 395–398, 2013.
28. Kim, H.S. *et al.*, Control of I-V Hysteresis in CH3NH3PbI3 Perovskite Solar Cell. *J. Phys. Chem. Lett.*, 6, 22, 4633–4639, 2015.
29. Chen, B., Yang, M., Priya, S., Zhu, K., Origin of J-V Hysteresis in Perovskite Solar Cells. *J. Phys. Chem. Lett.*, 7, 5, 905–917, 2016.
30. Zhao, L. *et al.*, High-Performance Inverted Planar Heterojunction Perovskite Solar Cells Based on Lead Acetate Precursor with Efficiency Exceeding 18%. *Adv. Funct. Mater.*, 26, 20, 3508–3514, 2016.
31. Sun, S. *et al.*, The origin of high efficiency in low-temperature solution-processable bilayer organometal halide hybrid solar cells. *Energy Environ. Sci.*, 7, 1, 399–407, 2014.

32. Mutalikdesai, A. and Ramasesha, S.K., Emerging solar technologies: Perovskite solar cell. *Resonance*, 22, 11, 1061–1083, 2017.
33. Yin, W. *et al.*, Halide Perovskite Materials for Solar Cells: A Theoretical Review Received. *J. Mater. Chem. A*, 2, 1, 1–7, 2014.
34. Queisser, H.J., Slip patterns on boron-doped silicon surfaces. *J. Appl. Phys.*, 32, 3, 510–519, 1961.
35. Green, M.A., Emery, K., Hishikawa, Y., Warta, W., Dunlop, E.D., Solar cell efficiency tables (version 46). *Prog. Photovolt. Res. Appl.*, 29, 1, 3–15, 2015.
36. Nayak, P.K., Garcia-Belmonte, G., Kahn, A., Bisquert, J., Cahen, D., Photovoltaic efficiency limits and material disorder. *Energy Environ. Sci.*, 5, 3, 6022–6039, 2012.
37. Stranks, S.D. *et al.*, Electron-hole diffusion lengths exceeding 1 micrometer in an organometal trihalide perovskite absorber. *Science*, 342, 341–344, 80, 2013.
38. Wang, B., Xiao, X., Chen, T., Perovskite photovoltaics: A high-efficiency newcomer to the solar cell family. *Nanoscale*, 6, 21, 12287–12297, 2014.
39. Pang, S. *et al.*, NH2CH=NH2PbI3: An alternative organolead iodide perovskite sensitizer for mesoscopic solar cells. *Chem. Mater.*, 26, 3, 1485–1491, 2014.
40. Stoumpos, C.C., Malliakas, C.D., Kanatzidis, M.G., Semiconducting tin and lead iodide perovskites with organic cations: Phase transitions, high mobilities, and near-infrared photoluminescent properties. *Inorg. Chem.*, 52, 9019–9038, 2013.
41. Ogomi, Y. *et al.*, CH3NH3SnxPb(1-x)I3 perovskite solar cells covering up to 1060 nm. *J. Phys. Chem. Lett.*, 5, 6, 1004–1011, 2014.
42. Hao, F., Stoumpos, C.C., Chang, R.P.H., Kanatzidis, M.G., Anomalous band gap behavior in mixed Sn and Pb perovskites enables broadening of absorption spectrum in solar cells. *J. Am. Chem. Soc*, 5, 6, 1004–1011, 2014.
43. Green, M.A. *et al.*, Solar cell efficiency tables (version 51). *Prog. Photovolt. Res. Appl.*, 26, 3–12, 2018.
44. Sha, W.E., II, Ren, X., Chen, L., Choy, W.C.H., The efficiency limit of methylammonium lead iodide perovskite solar cells. *Appl. Phys. Lett.*, 106, 22, 221104, 2015.
45. Noel, N.K. *et al.*, Lead-Free Organic-inorganic Tin Halide Perovskites for Photovoltaic Applications. *Energy Environ. Sci.*, 7, 9, 3061–3068, 2014.
46. Hao, F., Stoumpos, C.C., Cao, D.H., Chang, R.P.H., Kanatzidis, M.G., Lead-free solid-state organic–inorganic halide perovskite solar cells. *Nat. Photonics*, 8, 6, 489–494, 2014.
47. Mitzi, D.B., Chondroudis, K., Kagan, C.R., Organic-inorganic electronics. *IBM J. Res. Dev.*, 45, 1, 29–45, 2001.
48. Liu, J. *et al.*, A dopant-free hole-transporting material for efficient and stable perovskite solar cells † Broader context Energy & Environmental Science COMMUNICATION. *Energy Environ. Sci.*, 7, 9, 2963–2967, 2014.

49. Mei, A. *et al.*, A hole-conductor-free, fully printable mesoscopic perovskite solar cell with high stability. *Science*, 80, 345, 6194, 295–298, 2014.
50. Dualeh, A. *et al.*, Effect of annealing temperature on film morphology of organic-inorganic hybrid pervoskite solid-state solar cells. *Adv. Funct. Mater.*, 4, 17, 2880–2884, 2014.
51. Pathak, S.K. *et al.*, Performance and Stability Enhancement of Dye-Sensitized and Perovskite Solar Cells by Al Doping of TiO2. *Adv. Funct. Mater.*, 24, 38, 6046–6055, 2014.
52. Asghar, M., II, Zhang, J., Wang, H., Lund, P.D., Device stability of perovskite solar cells - A review. *Renew. Sust. Energ. Rev.*, 77, 131–146, 2017.
53. Liu, P., Wang, W., Liu, S., Yang, H., Shao, Z., Fundamental Understanding of Photocurrent Hysteresis in Perovskite Solar Cells. *Adv. Energy Mater.*, 9, 13, 1803017, 2019.

3

Surface Morphological Effects on the Performance of Perovskite Solar Cells

Srinivasa Rao Pathipati

Vignan's Foundation for Science Technology and Research Division of Physics, Department of Science and Humanities, Vadlamudi, Guntur, Andhra Pradesh, India

Abstract
In this chapter, we have discussed the effect of the morphology of the perovskite layer on the performance of the perovskite layers. We have discussed the various methods for the deposition of perovskite layers and how the morphology is controlled by the various processing conditions. This includes one-step, two-step, and dual-source precursor approach methods for the deposition of the perovskite layers. Besides, we have correlated the relation between the morphology of the perovskite layers and photovoltaic parameters of the perovskite-based solar cells. In each deposition technique, we have discussed the factors that govern the crystallization of the perovskite, such as material composition, additives, and post-deposition treatment.

Keywords: Morphology, solvent engineering, crystallization, grain growth, solvent annealing, seed mediated growth, cap-induced crystallization

3.1 Introduction

Organic–inorganic metal halide perovskite materials have attracted a lot of attention from scientists due to its high-performance and its easy processability [1, 2]. Among these perovskite materials$CH_3NH_3PbI_3$ has attracted considerable attention due to its direct bandgap of 1.55 eV, which determines its absorption offset at 800 nm and the weak exciton binding

Email: srinivasphm@gmail.com

Khursheed Ahmad and Waseem Raza (eds.) Perovskite Materials for Energy and Environmental Applications, (59–90) © 2022 Scrivener Publishing LLC

energy of merely 0.03 eV facilitates easy charge separation and long diffusion lengths [3, 4]. The performance and stability of the perovskite solar cells (PSCs) depends on the morphology of the perovskite layer. The morphology of the perovskite layer mainly depends upon the processing methods, the morphology of the underlying layer and post-deposition annealing [2]. Among the solution processing methods, one-step and two-step deposition techniques are widely used in solar cell applications [5, 6]. In the one-step deposition method, the precursors CH_3NH_3I (MAI) and PbI_2 are dissolved in the mixed solvent γ-butyrolactone(GBL), N,N-dimethylformamide (DMF), or dimethylsulfoxide (DMSO) [6, 7]. The two-step method, named sequential deposition method in which PbI_2 is deposited first and MAI dissolved in propanol either spin-coated on top it or PbI_2-coated substrate dipped in it for few seconds [5, 8]. Because of the diffusion of MAI into PbI_2, perovskite layer will be formed with subsequent annealing. Other novel synthesis methods include the vacuum deposition is employed to improve the perovskite purity and uniformity [9]. Vacuum deposition method involves the evaporation of the precursor materials onto the desired substrate under a vacuum environment.

3.2 Morphology Control

The light-harvesting capability of the perovskite layer depends upon its quality. Improving the thickness of the perovskite layer maximizes the photon absorption; however, it causes the formation of pinholes which act as shunting pathways. The absorption quality of the perovskite layer depends on the crystallization nature of the perovskite grains [10]. Since the photovoltaic parameters are mainly dependent on the quality of the perovskite layer, fabrication of high-quality perovskite films without pinholes is required [11]. The crystallinity of the perovskite films has decisive significance on the factors such as charge separation, charge transport and diffusion length [10]. The crystallinity of the perovskite layer thus depends on the deposition method and deposition conditions, interface between charge transport layer and perovskite, hydrophobicity and morphology of the substrate, composition and solvents/additives.

3.2.1 The Effect of Device Architecture on the Morphology and the Device Performance

Initially, these perovskite materials are used in the dye-sensitized solar cells as a sensitizer, but the stability is poor due to the solubility of the electrolyte. Hence perovskite layer was used as an active layer in the mesoscopic solar cells. The mesoscopic device performance is necessary for better performance [6]. Device architecture plays an important role in the morphology of the perovskite layer and hence the device performance. The schematic diagram of the mesostructured PSCs is shown in Figure 3.1 (left). In this type of solar cells, the mesoporous layer is deposited on top of a compact layer. If this scaffold layer participates in the charge transport process, it is called as "active scaffold" and if it does not participate in carrier transport then it is called as "inactive scaffold" or "passive scaffold." The thicker and more uniform perovskite layers can be obtained by using scaffold layer irrespective of whether they are active or passive. The planar heterojunction solar device architecture is shown in Figure 3.1 (right). In this structure, the perovskite layer is sandwiched between the ETL (electron-transport layer) and HTLs (hole-transport layers) without the scaffold layer. This scaffold layer consists of mesopores of TiO_2, Al_2O_3, NiOx, or SnO_2 having nanocrystals size of 50 nm. This results in a thick film of the good quality perovskite layer. However, in the case of planar heterojunction solar cells, either HTL or ETL must be compact and it must be free from pinholes to avoid short-circuiting. The controlled crystallization is an essential parameter in the film deposition due to the sensitivity of the solar cell parameters on the thin-film morphology. Crystallization involves mainly two steps: nucleation and growth. Nucleation starts from a continuous liquid phase either homogeneously or heterogeneously. Homogeneous nucleation occurs when there are no foreign species and it is uniform throughout the phase. Heterogeneous nucleation occurs in the presence of foreign species and it is non-uniform and it enhances the crystal growth rate. Therefore the surface chemistry and surface treatment, morphology of the electrode material on which the perovskite is deposited governs the growth of the perovskite layer. The Gibb's free energy for the heterogeneous nucleation is related to Gibb's free energy for the homogeneous nucleation by the relation

$$\Delta G_{Heterogeneous} = \Delta G_{Homogeneous} \times \left(\frac{2 - 3\cos\theta + \cos^3\theta}{4} \right),$$ where θ is the contact angle

Therefore Gibb's energy for the heterogeneous nucleation is less compared to Gibb's free energy for homogeneous nucleation. Hence the nucleation of the perovskite layer depends on the contact angle of the bottom layer and its interaction with the perovskite layer.

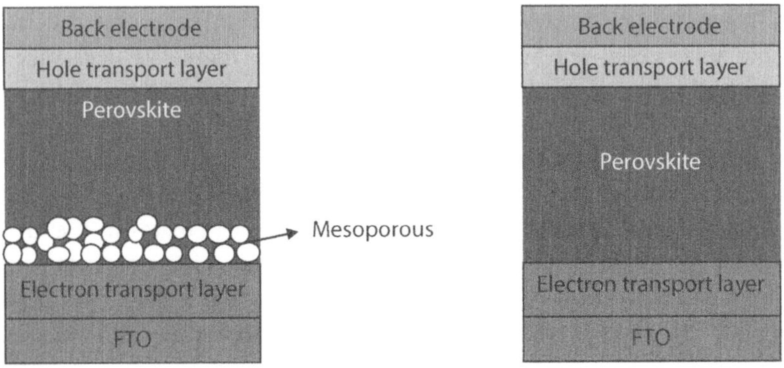

Figure 3.1 Schematic diagrams of mesoscopic heterojunction solar cell structure (left) and planar heterojunction solar cell structure (right).

Therefore, better wetting will occur on the non-wetting surfaces due to low nucleation barrier. Hence nucleation will occur on the surfaces containing larger surface area such as mesoporous areas than the planar structures, due to its lower contact angle and a large radius of curvature. For both mesoporous and planar heterojunction PSCs, several methods have been developed to control the nucleation and growth kinetics. The planar heterojunction PSCs has the advantage of low-temperature processing which is useful for fabrication on flexible substrates and the fabrication of tandem solar cell devices.

3.2.2 Effect of Deposition Technique on the Morphology of the Perovskite Layer

The morphology of the perovskite layer depends on the deposition method since the precursor consists of two different materials. Since the solubility of these components in the common solvent is different, it leads to uncontrollable crystallization during the film deposition. The perovskite active layer can be deposited by several methods. This includes one-step deposition, two-step deposition, dual-source precursor approach and vacuum deposition. In this section, we will discuss each deposition method briefly.

3.2.2.1 One-Step Deposition Method

In the one-step deposition method, MAI and PbI_2 were dissolved in a common polar solvent such as DMF and spin-coated on a substrate. This approach is easy to implement and cost-effective. But the film quality of the

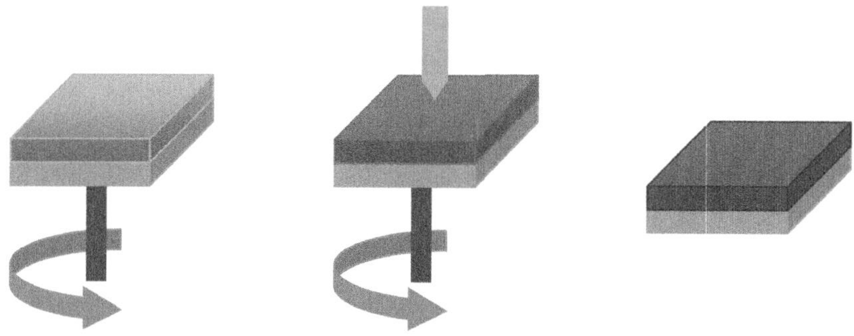

Figure 3.2 Schematic of the solvent engineering process in the one-step deposition.

resulting layers will be very poor and causes incomplete surface coverage and a lot of pinholes. The solvent which can dissolve both the materials in the same proportion is also limited. The efficiency of the devices reported using this method was only about 3-4% only [1].

Solvent engineering approach: The solvent engineering approach was adopted by Jeon *et al.* to produce a homogenous film. The schematic view of solvent engineering method is as exemplified in Figure 3.2. In this method, the precursor was dissolved in a mixed solvent of GBL: DMSO (7:3) ratio, followed by toluene drip during spin coating process[6]. DMSO will form a complex with PbI_2 and improves the morphology dramatically. The anti-solvent must be miscible with the precursor solvent and removes the excess DMSO. It forms an intermediate complex PbI_2-DMSO, and coordination between them leads to uniform film. The formation of the intermediate complex MAI·PbI_2·DMSO reduces the interaction between Pb atoms and halide ions. This intermediate complex inhibits the crystallization of the perovskite layer. During the annealing process, DMSO escapes from the film and this intermediate complex species controls the rate of crystallization of the perovskite. A planar device with the configuration of FTO/TiO_2/$CH_3NH_3Pb(I_{1-x}Br_x)_3$/PTAA/Au exhibited a PCE of 9.1% under forward scan and 14.4% under reverse scan, which indicates a large hysteresis. On the other hand, the mesoporous structure having a thickness of 200 nm with mixed halide perovskite as the active layer showed a PCE of 15.8% and 15.9%, for forward and reverse scans, respectively.

A similar approach followed by Xiao *et al.* where he has used chlorobenzene drip for fast deposition crystallization and recorded efficiency of 13.9%. In this method, only DMF is used as the solvent and chlorobenzene was dropped on perovskite precursor after 6 s, which removes the excess DMF and induces fast crystallization. This can be observed from the change of the color immediately after dropping chlorobenzene onto the spinning substrate. This immediate change of color implies the instantaneous formation of perovskite.

3.2.2.2 Two-Step Deposition Technique

In this section, we have discussed the two-step deposition technique, which involves the deposition of the different precursors sequentially and conversion into perovskite and grain growth mechanism.

Precursors: Since the organometallic halide perovskite consists of two essential precursors, the morphology of the perovskite layer depends on the deposition route. The precursors can be simultaneously deposited in one-step deposition or they can be deposited sequentially. Since the solubility of the perovskite precursors is not the same in the common solvent it is quite difficult to control the morphology. Although this one-step

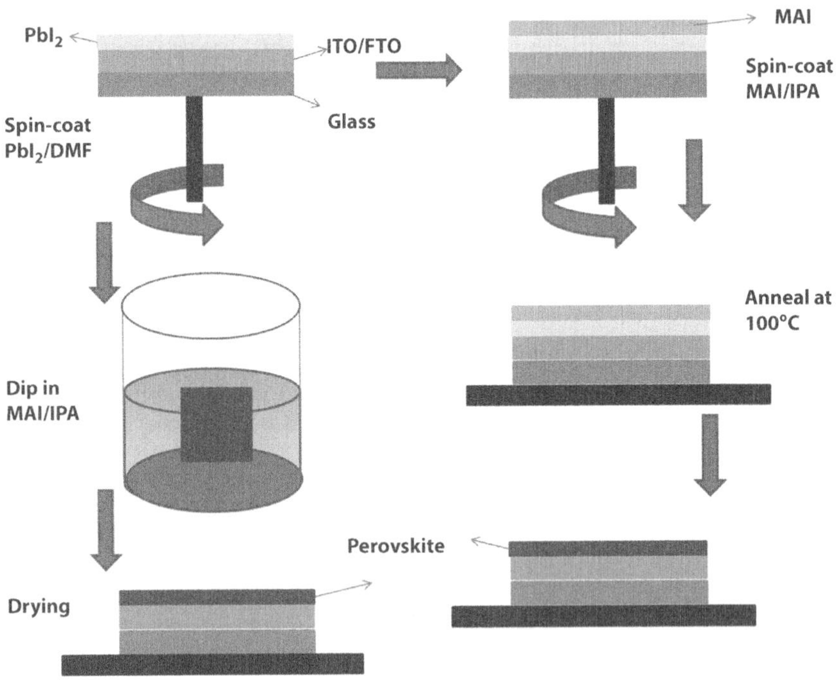

Figure 3.3 Schematic diagram of the sequential-deposition technique: Dipping in MAI/IPA solvent (left) and spin-coating of MAI/IPA solution (right).

deposition is easy to implement, the efficiencies of planar devices prepared using this method are less due to poor surface coverage. The two-step sequential deposition method has offers the possibility of controlled crystallization. In this method, the PbI_2 layer is deposited on to the HTL/ETL and then it is subsequently exposed to the MAI solution either by dip coating or spin coating and the schematic procedure for the two-step deposition is shown in Figure 3.3 [5]. The morphology of the perovskite layer depends on the morphology of the PbI_2 layer. Upon annealing, the MAI diffuses into the PbI_2 film and complete conversion into perovskite takes place. But the main problem in this technique is that the incomplete conversion of PbI_2 into perovskite. The top portion of PbI_2 when exposed to MAI converts into perovskite and inhibits the further growth of perovskite in the bottom layers. To fully convert a thick layer of PbI_2, exposing the substrate to and dipping in the MAI solution might cause perovskite soluble in MAI vapor and it can be peeled off from the substrate. To overcome this problem, making porous PbI_2 might increase the diffusion of MAI into it for conversion into perovskite. There are several strategies to obtain porous PbI_2 films. This includes 1) pre and post-treatment 2) intermediate phase formation 3) additive.

In the pre-treatment strategy, the additive is added in the precursor solution before deposition. Zhang *et al.* incorporated 4-tert-butylpyridine (tBP) into the PbI_2 precursor solution, which will form a PbI_2-*x*tBP after deposition. These tBP molecules retard the crystallization of PbI_2 [12]. These tBP molecules will slowly evaporate after annealing at 70°C and forms a porous structure. In the post-treatment strategy, the porous PbI_2 film will be formed after the deposition. In this strategy, after the PbI_2 film was deposited by spin-coating, dried very slowly, to obtain a porous PbI_2. The transport of MAI via the mesoporous channels results in complete conversion of the perovskite layer and improves the quality of the

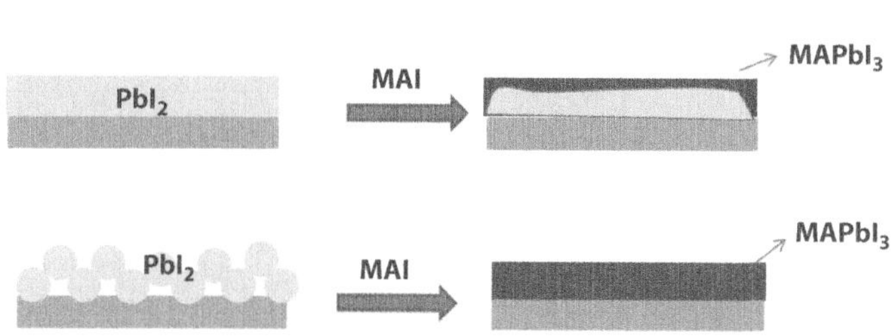

Figure 3.4 Schematic diagram of perovskite layer from compact PbI_2 (top) and porous PbI_2 (bottom).

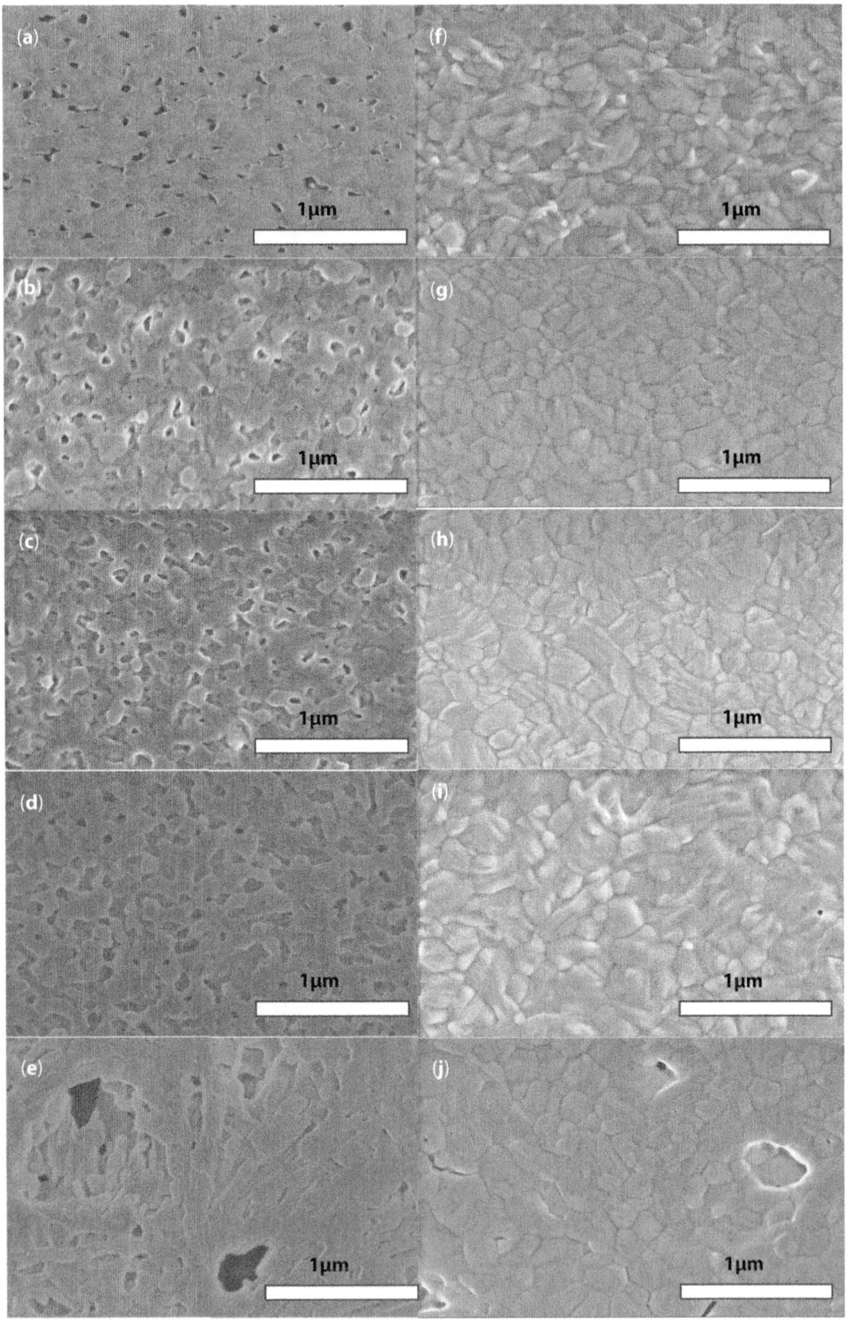

Figure 3.5 Scanning electron microscope images of PbI_2 films (a–e) and perovskite films (f–j) that are formed with varying concentration of NH_4I.

absorbing layer. The schematic view of the conversion from compact PbI_2 film and porous film is as exemplified in Figure 3.4.

To accelerate the formation of porous structure of PbI_2 from the wet film, El-Henawey *et al.* have treated it by toluene and chlorobenzene during spin-coating to mediate the nucleation and growth of PbI_2 crystals [13]. The porous structure of PbI_2 allows MAI more easy penetration into it for complete conversion into perovskite. In a similar fashion, Pathipati *et al.* have added NH_4I halogen additive, which can react with PbI_2 and can form the porous structure. This halogen additive reacts with PbI_2 and forms the porous structure, which allows the diffusion of MAI into the underlying layers [14]. In Figure 3.5, we have presented the PbI_2 films with the addition of different amounts of NH_4I (a–e), and morphology of the perovskite layers (f–j). It is observed that the porosity of the PbI_2 films increases with increasing the addition of NH_4I concentration and also there is a noticeable improvement in the grain size of the perovskite layer. The better performance of the devices and the corresponding larger grain size of the perovskite layer were observed at a concentration of 5 wt% NH_4I. The device efficiency has increased from 15.2% to 17.4%.

Another strategy to promote the conversion reaction is employing of inhibiting the crystallization of PbI_2. The mostly used solvent for the retardation of the crystallization of PbI_2 is DMSO, which can strongly coordinate with Pb^{2+} atoms. When the DMSO solvent additive is added in DMF, the PbI_2 molecules prefers to coordinate with the DMSO molecules, and a complex $PbI_2(DMSO)_x$ will be homogeneously distributed in the DMF solution. During the spin-coating process, these molecules will form a self-assembled order. After the MAI deposition, DMSO will be exchanged by MAI, due to its small affinity towards PbI_2. The concentration of DMSO is an essential parameter, for the formation of highly dense and enhanced coverage of the film for the exchange interaction of DMSO and MAI. Li *et al.* have achieved an efficiency of 17% by using controllable grain morphology with DMSO additive. Wu *et al.* have replaced the DMF solvent with DMSO for the preparation of highly reproducible and uniform perovskite films with a narrow crystallite size distribution [15]. A highest PCE of 13.5% was achieved with a standard deviation of < 0.6% for 120 samples compared to ~ 2.5 % deviation for DMF-based PSCs.

N-Methyl-2-pyrrolidone (NMP) is an additive to control the morphology of the perovskite layer. Jo *et al.* have used the NMP as an additive for the formation of a smooth and homogeneous PbI_2/NMP precursor layer. In this method, NMP plays the same role as DMSO but produces a smoother film than DMSO. The NMP molecules which are attached to PbI_2 molecules are exchanged from PbI_2 to FAI molecules after its deposition.

Compared to the PCE 17.6% obtained with PbI_2/DMSO precursor layers, a PCE of 19.5% was obtained using NMP additive [16]. Zhang *et al.* have used MAI as an additive in the precursor solution of PbI_2. They have prepared PbI_2(+MAI)/DMF precursor solution and obtained very smooth layers by avoiding the full PbI_2 crystallization. The addition of MAI into the precursor reduces the crystallization of PbI_2 and expands the morphology of PbI_2 film. By using a precursor ratioPbI_2:0.15 MAI the best PCE of 17.22% was obtained, which is significantly higher than the conventional two-step method 6.11% [17]. Zhang *et al.* have proposed a new scheme using pyridine for the room temperature processing of perovskite films. In this method, they have used the pyridine vapor to increase the porosity into PbI_2 films [18]. The $PbI_2 \cdot (Py)_2$ intermediate precursor retard the crystallization of PbI_2 films and helps in forming a highly crystalline and pure phase perovskite after immersion into MAI/IPA and the PCE boosts up to 17.1%. In this approach, after the spin-coating of PbI_2 films they were exposed to pyridine vapor. Due to this treatment the platelet structure of PbI_2 is changes to porous $PbI_2.(Py)_2.$ These films are dipped in a solution of MAI for 20 s, such that PbI_2 molecules react with MAI which can form the perovskite.

$$PbI_2.\ (Py)_2 + MAI \rightarrow MAPbI_3 + 2Py(Aq)$$

On the other hand it is evident from the DFT calculations that the activation energy for the formation of perovskite with pyridine treated films is less compared to the pristine PbI_2 films. In the conventional pathway, the formation of perovskite needs to overcome a large energy barrier. The activation energy for the formation of $PbI_2.(Py)_2$ is 0.095 eV. The energy barrier for the formation of perovskite using pyridine is only 0.741 eV, which is lower than that of the conventional two-step deposition (E_a = 1.004 eV). The pyridine assisted deposition of PSC device exhibited a J_{SC} of 22.60 mA cm^{-2}, V_{OC} = 1.05 V, and FF of 0.72, which leads to a PCE of 17.1%. This low-temperature processed synthesis of high-quality perovskite films allows them to fabricate on flexible substrates, which can reduce the weight and as well as the cost.

Conversion: The spin-coated layer of PbI_2 is immersed in the MAI/IPA solution for 1-3 minutes. However, in this method, the morphology of the perovskite layer is difficult to control and perovskite layer can be dissolved in the solvent due to long exposure time [2, 19–25]. Besides the excess MAI on the surface of the perovskite layer causes rough and even the deposition of subsequent layers is quiet difficult. This leads to uncontrollable morphology and difficult to reproduce the photovoltaic parameters. An alternative method for the deposition of this problem is that spin-coating a MAI film on top of the PbI_2 precursor followed by annealing [26]. This method is known two-step interdiffusion method or sequential deposition method. The serious limitation in this method is that the thickness of the film is not uniform through the surface and it leads to wide distribution of photovoltaic parameters. To overcome this problem Chen *et al.* have adopted the vapor assisted solution process in which the PbI_2 deposited substrate is exposed to the MAI vapor at 150°C in N_2 atmosphere for 2 hours to form perovskite layer [27]. Still this process also requires high temperature which might decompose MAI into its constituents and the conversion rate is also low.

Grain growth: After the deposition of the MAI on top of PbI_2 layer the interdiffusion of MAI into PbI_2 lattice will takes place due to temperature. Due to the temperature, the intercalation of MAI molecules into PbI_2 lattice will takes place and complete conversion into perovskite take place. The higher annealing temperatures are required to grow large crystals, and the final grain size of the perovskite layer depends upon the size of the PbI_2 crystals.

3.2.2.3 Dual-Source Precursor Approach

The main challenge with the solvent engineering method is that the reproducibility of the device parameters. The antisolvent is usually dropped on the center of the substrate, which might leads to non-uniform crystallization on the surface of the thin film. In addition to that the antisolvent that is used in this method such as chlorobenzene and toluene are toxic and hazardous. Hence the attention is paved towards the antisolvent free methods for the fabrication of PSCs. On the other hand the mixed cation (MA and FA)-based perovskites have attracted a lot of attention due to its lower bandgap. However α-$FAPbI_3$ that is formed at higher temperature changes to non-perovskite phase at room temperature [28–30]. Zhang *et al.* have developed an antisolvent free method using Lead acetate precursor in the ratio of $PbAc_2$: MAI (1:3) [31]. However the same method we cannot apply to mixed cations of FA^+ and MA^+, since the undesired non-perovskite is always formed at annealing temperature greater than 100°C. To overcome this issue Luo *et al.* have developed a dual source precursor approach for the fabrication of mixed cation perovskite. In this method dual precursors $PbAc_2$: MAI (1:3) and PbI_2: FAI (1:1) are dissolved in the solvent. During the annealing process at appropriate high temperature MA·Ac and DMF evaporated and yields rapid crystallization and pure α-phase perovskite.

Due to the formation of pure α-phase, the thin films exhibit enhanced coverage and bulk charge transport. The thin films formed at room temperature to 120°C contain non-perovskite δ-phase and intermediate adducts. The films annealed at 140°C, form stable and pure α-phase. These results indicate that phase transition takes place in mixed cation perovskite due to inclusion MA^+ ions compared to pure α-phase $FAPbI_3$ formed at 175°C. Hence by using this method temperature of the phase transition temperature reduces to 120°C. By using this method, a champion PCE of 20.15% was achieved with a J_{SC} of 23.11 mA cm^{-2}, a V_{OC} of 1.09 V and a high FF of 0.80.

3.2.2.4 *Vacuum Deposition Technique*

The highest reported efficiencies of PSCs are small active area devices of <0.1 cm^2 [32–37]. When PSCs are fabricated in large areas the PCE losses are significant due non-uniformity of the thin film over a large area. The widely adopted technique for large area deposition in the semiconductor industry is vacuum-based physical deposition method. The main advantage of this physical deposition is that the high purity of the sublimed materials and precise control over the deposition and film thickness. Besides, this method is also free from toxic solvents and we can control the stoichiometric composition of the thin film. In the case of perovskite since there are two raw materials, multiple sources are required and precise control over the deposition rate must be tuned precisely. This method also has an advantage of low-substrate temperature fabrication, which enables us to fabricate the devices on polymer substrates and also on flexible-plastic substrates. In this method, the perovskite precursors are loaded in separate crucibles and heated to their corresponding sublimation temperatures in a high vacuum chamber. The stoichiometry and the thickness of the film are controlled by the evaporation rate of the individual precursor and overall formation of the perovskite material. If the deposition rate of the organic and inorganic precursor is carefully monitored to maintain the stoichiometry the perovskite will be formed without any thermal annealing. The parameters that govern the growth of the film are substrate temperature and its morphology and evaporation rate of the precursor. However, the deposition rate of the organic compound is difficult to control due to its high vapor pressure, which can complicate the co-deposition of the precursors [38]. Recently, Li *et al.* have reported a PCE of 20.28% for small area devices and 18.13% for 21 cm^2 active devices [9].

3.3 Effect of Various Parameters on Growth of Perovskite

3.3.1 Effect of Solvent Additive

One of the strategies to optimize the crystallinity of the perovskite films is that the addition of the solvent in the precursor solvent. The growth kinetics of the perovskite film depends on the solution concentration, interface and temperature [11, 39–41]. To control the morphology of the perovskite layer Liang *et al.* have incorporated 1,8-diiodooctane (DIO) into perovskite precursor solution to modulate the morphology of the perovskite layer. These DIO molecules can coordinate with Pb^{+2} ions during the crystal growth and improves the homogeneous nucleation. Since this coordination alters the interfacial energy and kinetics of crystal growth, the morphology can be controlled. Liang *et al.* have added the optimized ratio of 1 wt% of DIO into the perovskite precursor. They have used $PbCl_2$:MAI (1:3) as the precursor in the DMF solvent [42]. The control device exhibit a PCE of 7.8% with a V_{OC} of 0.90 V, a J_{SC} of 15 mA cm^{-2}, and a FF of 0.58 while the devices fabricated using 1% DIO in precursor solution exhibited a significantly enhanced PCE of 10.3% with a V_{OC} of 0.92 V, a J_{SC} of 15.6 mA cm^{-2}, and a FF of 0.71. The addition of DIO not only improves the morphology and it can also improve the solubility of $PbCl_2$ in DMF solvent. The improvement in the solubility results from the chelation of Pb^{2+} with DIO as Cl^- ligands. This will reduce the rational entropy of the configuration during formation and reduces the Gibb's free energy. Consequently, the incorporation of additive influences the kinetics of the growth of the perovskite film. The temporary chelation of Pb^{2+} with DIO prolongs the growth of perovskite films and thus more regular faceting and enhanced coverage.

Gong *et al.* have used water as an additive in the perovskite precursor solution. They have added 2 wt% water in the precursor solution to improve the crystallinity of the deposited perovskite layers [43]. Adding a small amount of water into DMF enhances the solubility of the perovskite which can enhance the crystallinity of the perovskite. Compared to DMF, water has low boiling point and high vapor pressure which leads to rapid crystallization. By using this method, they have obtained rather compact thin films of perovskite without any pin holes. By using 2 wt% additive, the PCE of PSCs has increased from 12.13% to 16.06%. Wu *et al.* have added 2 wt% of water in PbI_2 precursor solution in two-step deposition and obtained a high quality dense film of PbI_2 [44]. The PCE of PSCs increased to 18% compared to pristine films. Similarly, Nirmal *et al.* have added the

5 vol% of water in MAI/IPA solution, which helps for perovskite crystallization to form larger grains in the sequential deposition method. The PCE of the PSCs has improved from 9.04%to 12.42% [45].

3.3.2 Effect of Solid Additive

The solar cell parameters are determined from the scan direction and scan rate. These PSCs often exhibit hysteresis due to the ion migration, charge carrier trapping at the grain boundaries and interfaces [46–49]. Therefore improving the grain size and crystallinity of the perovskite film reduce the hysteresis in J-V characteristics. To reduce the interface trap density, the interface must be passivated. Some reports reveal the metal oxide layers coated with fullerenes will passivate the traps and reduce the hysteresis and improve the performance. Ke *et al.* have added the solid additive such as $Pb(SCN)_2$ in the perovskite precursor, to control the grain growth and crystallinity of the perovskite films [50]. When this solid additive is added in the precursor solution $CH_3NH_3^+$ cations and SCN^- anions react and form HSCN and CH_3NH_2 gases will be formed and they escape during the film deposition and no SCN ions will appear in the final perovskite, which can be an added advantage. The PSC fabricated using the precursor containing 5% $Pb(SCN)_2$ exhibit a PCE of 18.15% with a V_{OC} of 1.11 V, a J_{SC} of 21.43 mA cm^{-2} and a FF of 76.35% with minimal hysteresis. It is revealed from the SEM images, that the pristine perovskite layer exhibits a grain size of 100–400 nm. However, the perovskite layer with 5% $Pb(SCN)_2$ in the precursor exhibits grain size of the order of micron size.

$$CH_3NH_3I + PbI_2 + x\, Pb\, (SCN)_2 \rightarrow (1 - 2x)\, MAPbI_3 + 3x\, PbI_2 + 2x\, (HSCN \uparrow + CH_3NH_2 \uparrow)$$

The enlarged size of grains and improved crystallinity is due to the formation of CH_3NH_2 gas accompanied by the formation of HSCN gas and results in the formation of excess PbI_2 at the grain boundaries.

Tai *et al.* have used the $Pb(SCN)_2$ as the precursor in the two-step deposition, which is prepared under relatively high humid conditions [51]. In this process, first $Pb(SCN)_2$ layer was deposited on TiO_2 on which MAI layer was deposited on top of it. The resulting films are $CH_3NH_3PbI_{3-x}(SCN)_x$. The resulting films exhibit a maximum efficiency of 15% and relatively high stability under humid conditions. Yu *et al.* have added the $Pb(SCN)_2$ additive into $FA_{1-x}Cs_xPbI_3$ films to increase the grain size [52]. $FA_{1-x}Cs_xPbI_3$ films exhibit a maximum PCE of 17.6% and these films exhibit a grain size of 120 nm, which can limit the PCEs. By the addition of 0.5 mol% of $Pb(SCN)_2$, the average grain size has improved to 19.57% under reverse bias and 18.57% under forward bias.

3.3.3 Seed-Induced Growth of Perovskites

The PCE of the PSCs is strongly depended upon the morphology, surface coverage, grain size and the degree of crystallinity. The morphology of the perovskite layers is governed by the nucleation and kinetics of growth. The incorporation of seed crystals in the perovskite precursor layer or it can be introduced through the antisolvent process is proved to be an effective way to control the morphology of the perovskite layer. Sian Li *et al.* have introduced MAI capped PbS nanoparticles [53]. Through this intermixing seed growth the domain size, orientation, film coverage and uniformity of the film can be substantially improved. The intermixed crystals function like seed-like crystals to manipulate the nucleation and growth of perovskite crystals. The schematic view of the perovskite crystal grown from a solution is depicted in Figure 3.6. The MAI capped PbS nanocrystals will initiate the nucleation, and act as centers for the growth of perovskite.

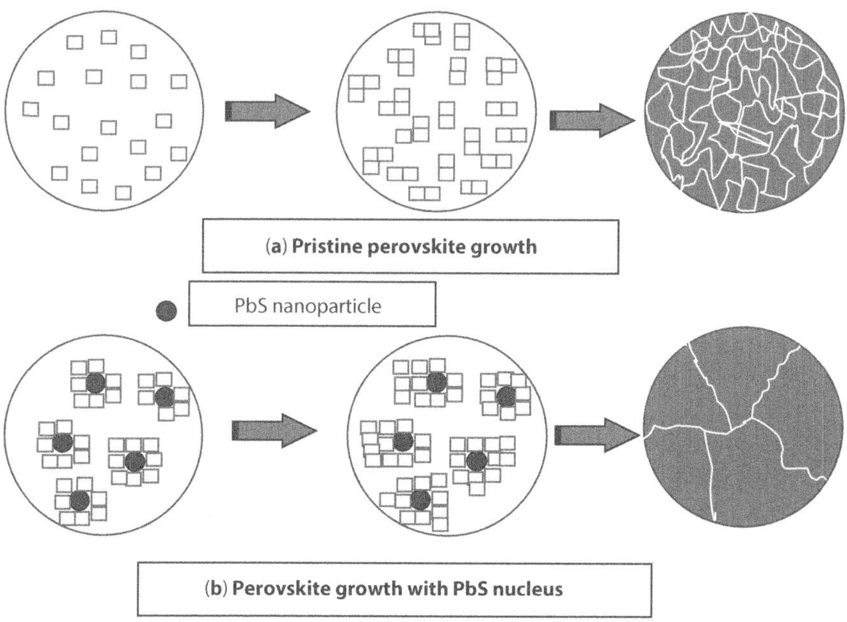

Figure 3.6 Nucleation and growth mechanism (a) without and (b) with MAI-capped PbS nanoparticles.

Sian *et al.* have used the precursor $PbAc_2$: MAI (1:3 molar ratio) for device fabrication and MAI capped PbS nanoparticles were intermixed with different weight ratios of 0, 0.5,1, 1.5 wt%. The incorporation of these nanoparticles into the precursor solution results in substantial morphological changes and crystallinity. It can be observed that in case of pristine layers pinholes and voids appear, whereas in case of samples deposited by using nanoparticles (0.5 and 1.0 wt %) in the perovskite solution exhibit enhanced coverage and larger grains.

The grain size of the perovskite crystals deposited by using 1.0 wt% MAI-capped PbS nanoparticles is of the order of 2 to 5 μm with enhanced coverage. The pristine perovskite solar cell exhibited a PCE of 14.11%, with a V_{OC} of 0.99 V, a J_{SC} of 21.01 mA cm^{-2} and a FF of 0.68. The PSCs fabricated by using 1.0 wt% MAI-capped PbS nanoparticles dispersed perovskite precursor exhibits a PCE of 17.40%, with a V_{OC} of 1.07 V, a J_{SC} of 21.12 mA cm^{-2}, and a FF of 0.77. The improvement in the device performance is due to the improvement in the grain size of the perovskite layer. Sakai *et al.* have demonstrated that a fractional substitution of PbI_2 in$PbAc_2$: MAI has a profound influence on the thin-film morphology. They found that replacing a 2 mol% of $PbCl_2$ with PbI_2 changes the nucleation and growth kinetics. The presence of PbI_2 results in the formation of $MAPbI_3$ crystals which will act as nucleation centers for the growth of precursor crystals. The PbI_2 content is varied from 2 mol% to 100%. The pristine films of the perovskite film grains are from the 0.5 to 1 μm with few pinholes. The crystals grow with the 2 mol% additive in the precursor solution impinge on each other to form pinhole free films. These films exhibit rectangular shape crystallites with enhanced surface coverage. With the addition of more PbI_2 content, the surface results in poor surface coverage and pinholes and cracks. This nucleation and growth of the films were studied by the optical microscopy method. From these images, it is evident that the nucleation starts from the $MAPbI_3$ crystals and the growth of the films takes place subsequently. There are certain limitations for the fabrication of cesium-based PSCs in two-step deposition. These crystals will grow in quadrilateral shape at a concentration of 2 mol%. With further increase in concentration, small crystals are formed and coalesce together and pinholes will be formed. The PCE estimated from J-V curves is 19.1% (Jsc = 23.1 mA cm^{-2}, V_{OC} = 1.08 V, and FF = 0.77) for the cell fabricated with 2 mol% PbI_2 ($x = 0.02$), compared to PCE of 16.8% (J_{SC} = 20.7 mA cm^{-2}, V_{OC}=1.07 V, and FF = 0.77) for the pristine devices.

There is a serious limitation for the fabrication of mixed halide PSCs by sequential deposition technique due to limited solubility of cesium halides. The incorporation of CsI into PbI_2 leads to the formation of unstable δ-phase $CsPbI_3$. On the other hand, the solubility of CsI in alcohols is limited. To overcome this issue, Zhao *et al.* have incorporated cesium-based perovskite colloids into PbI_2 precursor solution in two-step deposition [54]. When the alkylammonium halide layer is deposited on top of PbI_2 layer, these cesium perovskite nanoparticles will act as a seed for the growth of perovskite layers. Using this method, we can incorporate cesium into high-quality perovskite films. The V_{OC} of the devices fabricated in this method increases considerably by 60 mV and the PCE increases from 19.4 (with V_{OC} =1.07 V, J_{SC} =23.7 mA cm^{-2}, and FF = 0.75) to 21.7% (with V_{OC}=1.13 V, J_{SC}=24.1 mA cm^{-2}, and FF = 0.81) for the reverse scans, respectively.

Bi *et al.* have introduced a polymer-assisted growth of perovskites. They have used a polymer PMMA (poly (methyl methacrylate) as an additive as the antisolvent in one-step deposition [55]. This polymer will interact with PbI_2 and forms an intermediate adduct and reduces the crystallization of the perovskite. The incorporation of these PMMA molecules will form the randomly oriented nuclei to reduce the formation of Gibbs free energy and grow in the thermodynamically favored orientation. The nonhomogenous nucleation coupled with the lower free energy results in the formation of smooth perovskite with few pinholes and defects. In this method, during the deposition of perovskite layer by spin-coating, antisolvent (mixed solvent chlorobenzene/toluene) containing different concentration of PMMA from 0 to 4 mg/mL was dropped on it. The PMMA induce nucleation centers and directs the growth of these crystals. The photovoltaic parameters of the devices fabricated using this method is as follows: J_{SC}= 23.7 mA cm^{-2}, V_{OC} = 1.14 V, and FF = 0.78. From these images, it is evident that the grain size of the perovskite layer gradually increases with PMMA concentration to a certain extent and then decreases. At PMMA concentration more than 0.6 mg/mL, the PMMA starts to deposit on the grain boundaries. In addition, the grains are growing perpendicular to the substrate in a preferred direction, consequently surface area reduces.

Hence, the PMMA molecules will acts as centers for heterogeneous nucleation, improve the grain size and favors them to grow in a preferred direction.

3.3.4 Homogenous Cap-Induced Crystallization

In this method, a homogeneous capping perovskite film is placed over the precursor film during the solid-phase crystallization (Figure 3.7a). By using this method the grain size, uniformity, composition, crystallinity and defect density of thin films. Zhu *et al.* have adopted this method for the perovskite film to control the growth dynamics such as nucleation rate, film composition and surface roughness [56]. The perovskite layer was first synthesized by the solvent engineering method and then cap layer was placed on top of it annealed at 100°C. In the conventional annealing process, no cap layer was introduced on top of it. The schematic images of the grain growth of the perovskite layers annealed by the homogeneous cap mediated (HCM) crystallization produces high-quality films with larger grain size compared to controlled crystallization (Figures 3.7c and 3.7d). The average grain size is estimated as 1.1 μm for the cap-induced crystallized films and 0.17 μm for conventional crystallized films. In this method, the precursor solution was made by the mixed solvent of DMF/DMSO. During the spin–coating of the precursor solution, some of the DMSO molecules are left in the film due to its high boiling point. In the conventional crystallization, these DMSO molecules evaporate very fast and induce rapid nucleation. But in the case of HCM crystallization, the evaporation rate of DMSO molecules is significantly reduced, which can suppress the nucleation density which

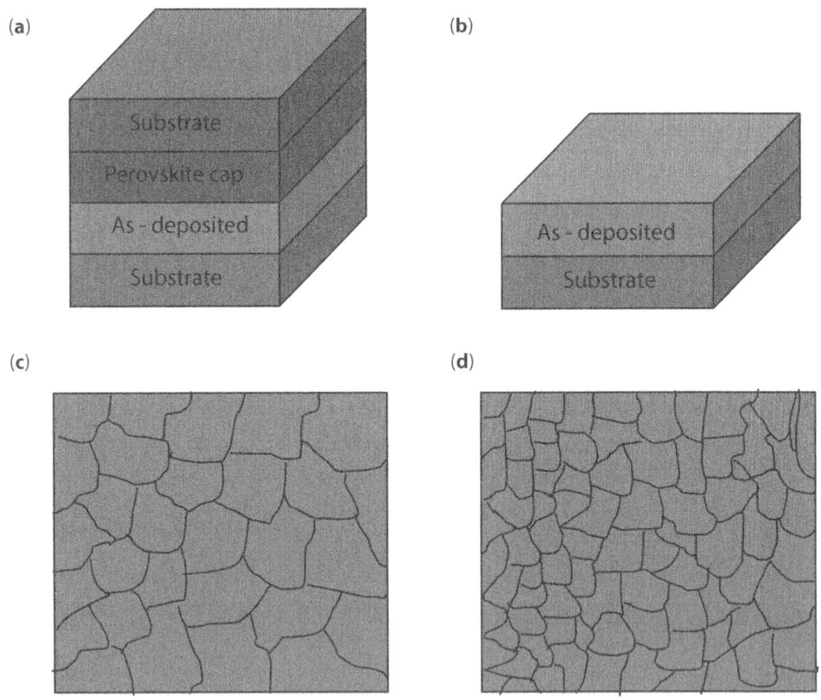

Figure 3.7 Schematic view of the (a) HCM crystallization and (b) conventional crystallization (c) surface morphology of perovskite layer by grown by HCM crystallization (d) grown by conventional crystallization.

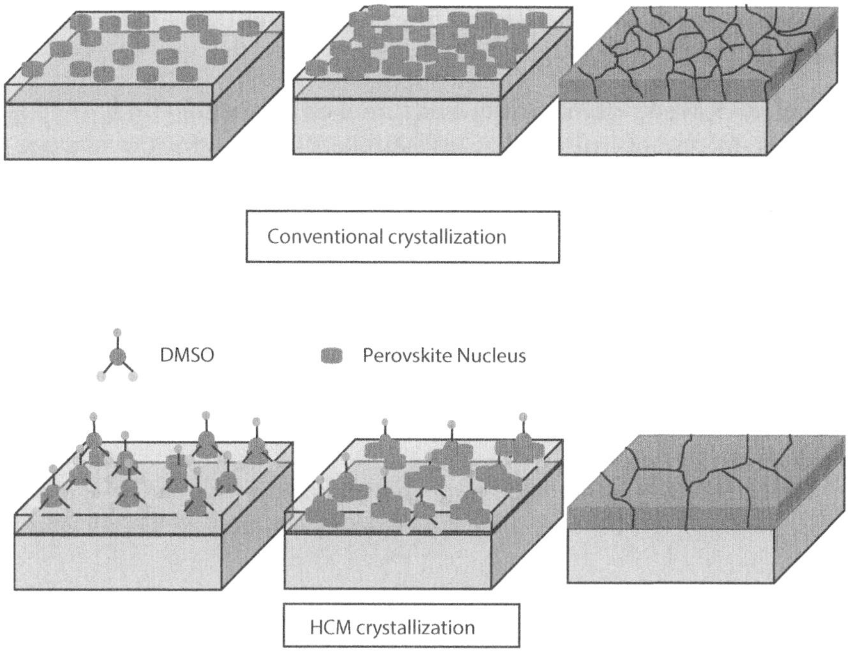

Figure 3.8 Schematic view of conventional crystallization (top) and HCM-crystallization (bottom).

results in the growth of larger grains (Figure 3.8). The average PCE value is 17.87% for the cells fabricated using HCM crystallization and 13.84% for the cells with control samples. The same group has used the PbI_2 to enhance the crystallization of the perovskite and improve the efficiency from 12.99% to 15.77%.

3.3.5 Effect of Hydrophobicity

For better performance of the PSCs, high aspect ratio grains are required. The large grain size of these perovskite films reduces the charge carrier recombination at grain boundaries. Cheng Bi *et al.* have reported the growth of high aspect ratio perovskite films on different hydrophobic layers [57]. These hydrophobic non-wetting HTLs reduce the heterogeneous nucleation and promote the migration of the grain boundaries during grain growth by imposing less drag force. The improvement in the grain size of the perovskite films reduces the grain boundary area and improves the crystallinity.

Growth mechanism of perovskite films on hydrophilic vs hydrophobic surfaces:

In Figure 3.9, we depict the growth mechanism on the wetting vs non-wetting surfaces. In case of hydrophilic surfaces like PEDOT: PSS, the homogeneous nucleation will take place. Due to this homogeneous nucleation, non-uniform grain growth takes place which results in a wide distribution of grain size. These wetting surfaces prevent the migration of the grains during the annealing process and formation of larger grains. In the case of non-wetting surfaces, the surface dragging force of the substrate decreases dramatically and it facilitates the grain boundary mobility for the formation of larger grains. This non-wetting smooth and clean surface polymer allows the formation of highly-dense grains due to heterogeneous nucleation.

Cheng Bi *et al.* have studied this phenomenon on a wide variety of materials, such as PVA, PEDOT: PSS, C-OTPD, PTAA, and PCDTBT whose

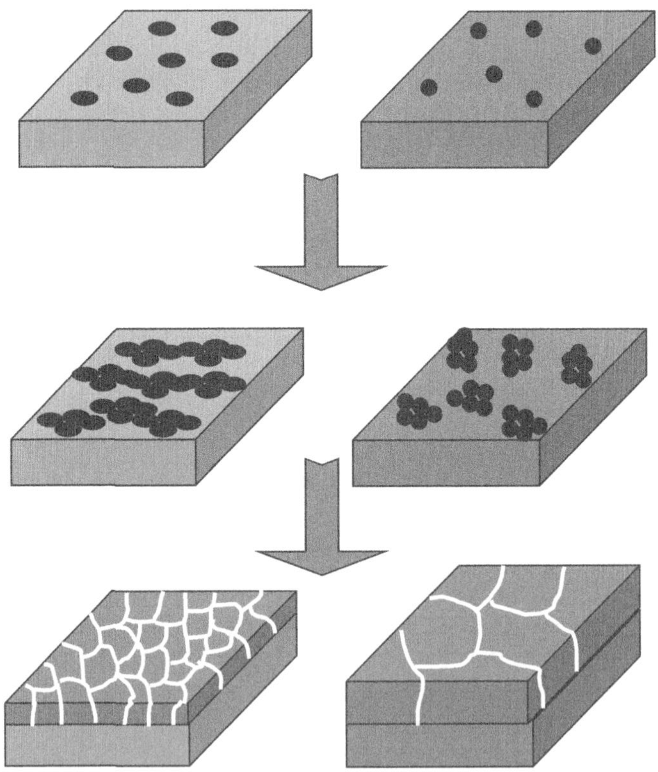

Figure 3.9 Growth mechanism of perovskite layer on hydrophilic vs hydrophobic layer.

contact angles are 10°, 12°, 79°, 105° and 108°. PVA and PEDOT: PSS are less hydrophobic due to the presence of hydroxyl groups on the surface and on the other hand c-OTPD is relatively less hydrophobic due to the presence of oxygen on its terminals compared to PTAA and PCDTBT. On an average the grain size is relatively larger for $MAPbI_3$ films on hydrophobic HTLs. The sizes of $MAPbI_3$ grains on wetting PVA (277 nm) and PEDOT:PSS (301 nm) are smaller than the film thickness (360 nm) after annealing. The grain size on the non-wetting HTLs were larger compared to the film thickness, and the aspect ratio reaching 2.3, 3.2 and 7.9 for c-OTPD, PTAA, and PCDTBT respectively. The grain size of the perovskite layer is about ~5 μm, which is 14 times more than the thickness, and grain boundaries are perpendicular to the substrate. It is noted that PCDTBT is highly hydrophobic so that the perovskite films on PCDTBT were mostly not continuous. Compared to the photovoltaic parameters of the PEDOT: PSS (J_{SC}= 18.6 mA cm^{-2}, V_{OC}=0.92 V, FF=0.72 and PCE=12.3%) based PSCs, c-OTPD-based devices exhibit better photovoltaic parameters(J_{SC}= 21.2 mA cm^{-2}, V_{OC}=1.09 V, FF=0.75 and PCE=17.5%). Hysteresis is also negligible for these devices due to the formation of large grains which is an indication for the less number of traps. The improvement in the performance of the device is due to the improvement in the interface between the c-OPTD and perovskite layer and also formation of fine crystallites and larger grains near the interface. Compared with the device fabricated using c-OTPD, the PSC device with PTAA showed a slightly higher efficiency of 18.1% with comparable V_{OC} of 1.07 V and J_{SC} of 22.0 mA cm^{-2} but a larger FF of 76.8%. The PTAA-based devices exhibit a better performance due to its relatively large contact angle and high work function. Although larger perovskite grains were observed on PCDTBT, most perovskite layers on PCDTBT have low film coverage, and thus not suitable for device optimization.

In summary, the non-wetting HTLs was demonstrated to be effective in enhancing the efficiency of PSC devices to 18.3%. In addition to the higher PCE, these HTLs are expected to be much more stable than the acidic PEDOT:PSS[57]. Yang Li *et al.* have developed a new small molecule 4,4'-Bis(4-(di-p-toyl)aminostyryl)biphenyl, which can lower the density of nuclei, which promotes the growth of perovskite into a dense film with larger grain size [58].

M N Shah *et al.* has fabricated the devices on top of porphyrin-based HTLs [59]. In Figure 3.10, we have presented the contact angles of water on Porphyrin (Po) and cobalt-Porphyrin (CoP) deposited on glass substrates, and it is 34.0° for Po films and for CoP films, it is 98.4°. The hydrophobic nature of the surface on which the perovskite layer is deposited has decisive

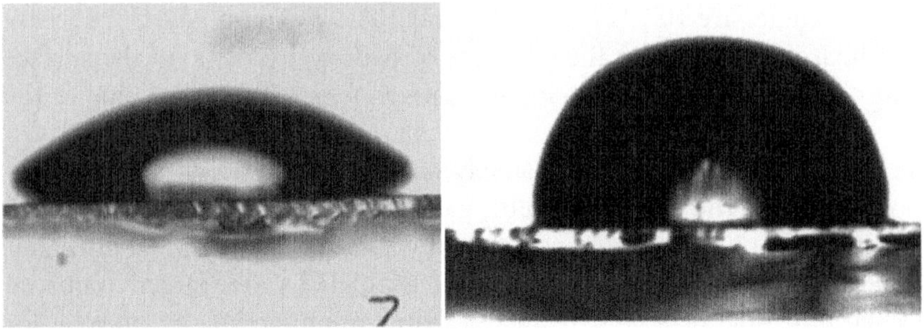

Figure 3.10 Contact angles of the Porphyrin (left) and Cobalt-Porphyrin (right) deposited on glass substrates.

significance on the morphology. The hydrophobic nature of the HTL facilitates the formation of high aspect ratio of the grains and larger grains because of the formation of lower number of nucleation sites. This effect is further confirmed from the SEM analysis, that the perovskite layers that are deposited on top of CoP exhibit better crystalline size compared to the films that are deposited on top of Po. Because of the improved grain size, the less trapping of the charge carriers at the grain boundaries, and hence the recombination reduced. The device efficiency has increased from 14% to 16.9% due to the improvement in the grain size of the perovskite layer. The SEM images of the perovskite layers deposited on top of Po and CoP HTL layers are presented in Figure 3.11.

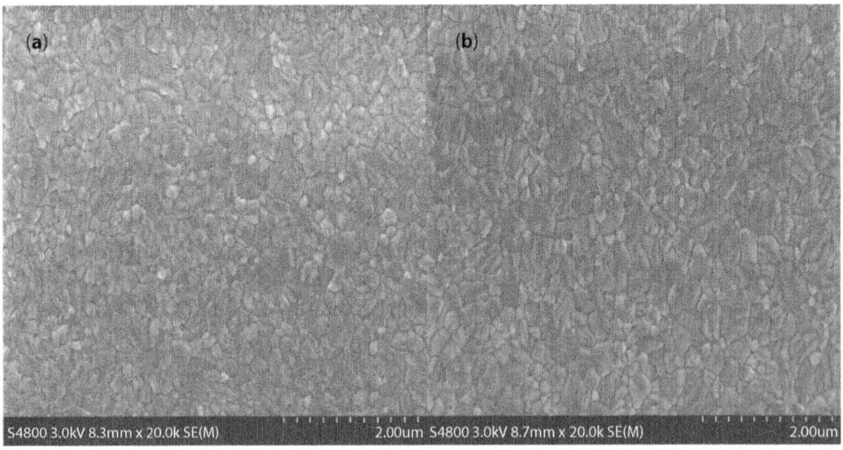

Figure 3.11 SEM images of perovskite layer deposited on (a) Porphyrin and (b) Cobalt-Porphyrin.

3.3.6 Effect of Interface Modification

Low-temperature solution-processed ETL/HTLs are required to be compatible with the flexible substrates and for simple manufacturing and they must be performed with the high-temperature processed mesoporous TiO_2, ZnO, SnO_2 nanoparticles [2, 60, 61]. These thin films contain a large number of defects which are detrimental to the device performance and poor quality perovskite layer [62]. However, to reduce the interfacial recombination and to improve the long-term operational stability, passivation of these ETL/HTLs is required [62, 63]. In addition to that, this charge transport layer surface may be rough to grow a good quality perovskite. Surface engineering has been demonstrated to be a promising way to simultaneously improve the surface morphology and electrical properties of metal oxide thin films in optoelectronic applications. Tan *et al.* have demonstrated that chlorine-modification is an effective for passivation of the surface defects of TiO_2 colloidal nanocrystals films. They found that the interfacial Cl atoms on the surface of TiO_2 passivate the traps at the interface and minimize the charge carrier recombination [63]. In addition, these Cl atoms make a strong electronic coupling with the perovskite. The devices fabricated using Cl-capped TiO_2 exhibit PCE >20%, and it can retain 90% of initial PCE after 500 hours of operation. With the incorporation of Cl atoms into PSCs, the average V_{OC} has increased from 1.06 V to 1.14 V, and FF increased from 0.69 to 0.77. Planar heterojunction solar cells have simple structure compared with a scaffold layer structure. A cathode layer, such as TiO_2, is used as an electrode, which can form Ohmic contact with the perovskite. Various types of TiO_2 layers doped with different elements are used. The underlying charge extraction layer should have high conductivity, smoothness and energy level alignment with the perovskite [64]. However, because of the low conductivity of TiO_2, charge accumulation will take place because of poor charge extraction capabilities. In addition to exposing the TiO_2 to ultraviolet light causes formation of oxygen vacancies which results in the formation of deep-level trap states in the bandgap. Various materials have been adopted to modify the interface between the TiO_2 and perovskite, such as PCBM and self-assembled monolayers. Yang *et al.* treated the TiO_2 layer with $TiCl_4$ to improve the interface, by suppressing charge carrier recombination and improve charge extraction [65]. The presence of unsaturated bonds on the surface of HTLs or ETLs causes trap states at the interface with the perovskite causes recombination and reduces the PCE. The surface defects act as nucleation sites, which might cause unprecedented growth of perovskite crystals. The serious limiting factor for the synthesis of the interfacial layer is that it should not be soluble in perovskite solvent, and it is compatible with the annealing temperature in further processing steps. Self-assembled monolayers anchor on the surface of TiO_2 reduce the density-of-states resulting in reducing the number of states. Treating the SnO_2 surface with alkali halides such as KCl causes an improvement of 100 mVin V_{OC} and PCEs up to 20.5%, with negligible hysteresis. In the same way, in p-i-n architecture, HTL/perovskite interface is treated to enhance the V_{OC} as well as PCE. The surface modification of PEDOT/PSS has been widely studied to enhance the PCE of the PSCs. NiOx was also found to be an alternative inorganic HTL, but the highly reactive oxygen vacancies require passivation. We note that the V_{OC} represents the most evident parameter that benefits from passivation because it directly relates to the reduction of nonradiative recombination processes. However, the FF of devices is perhaps an even more critical parameter to be considered.

3.3.7 Effect of Solvent Annealing

The grain size of the perovskite has a substantial influence on the photovoltaic parameters. To grow large crystals, Xiao *et al.* have adopted the solvent vapor annealing method which is depicted in Figure 3.12 [66]. In this method, immediately after the deposition of the precursors one by one, the substrate was simply placed on a hotplate under a Petri dish together with a small amount of solvent of approximately 10 µl and annealed to 100°C. The schematic view of the solvent vapor annealing is depicted in Figure 3.13. The DMF vapor penetrates the perovskite and helps for the growth of large crystals. The size of the perovskite crystals has improved up to micron size compared with the thermal annealed films, which is only 260 nm.

To improve the grain size of the perovskite layer, You *et al.* have demonstrated that perovskite films annealed in the air exhibit better crystallinity

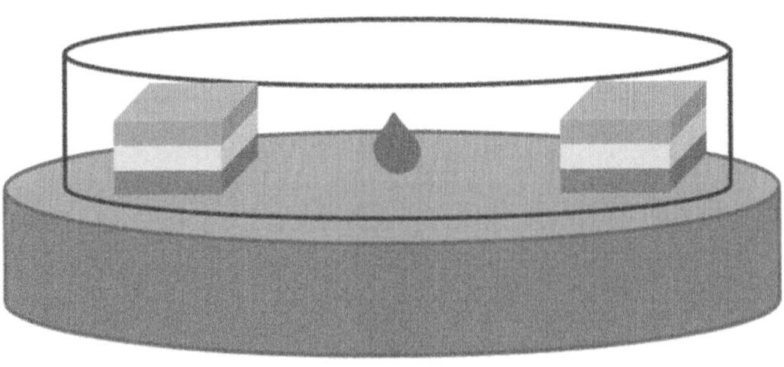

Figure 3.12 Schematic of the solvent vapor annealing method.

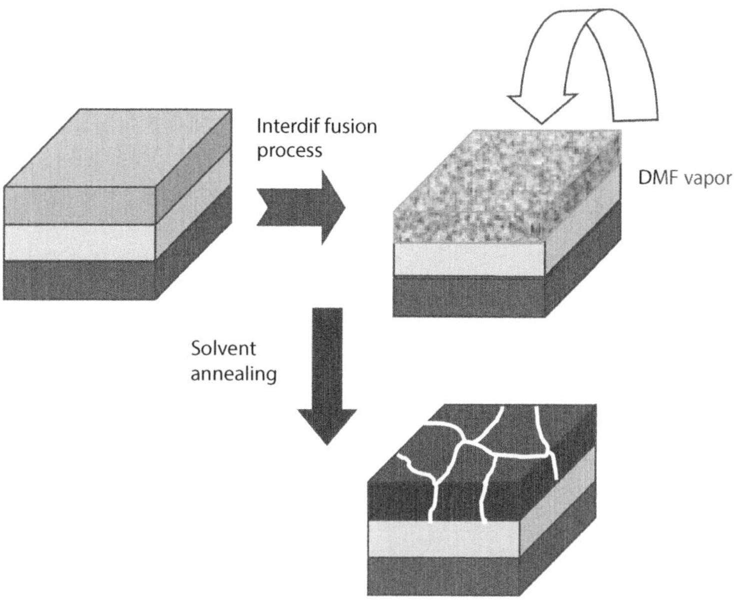

Figure 3.13 Schematics of the interdiffusion process and solvent vapor annealing.

than in the nitrogen ambient. Jiang Liu *et al.* have reported the annealing of perovskite films under different solvent vapors. He found that annealing the perovskite films under different vapors have a pronounced effect on perovskite grain growth. They have tested solvent vapor annealing with nitrogen, H_2O, GBL, DMF, and DMSO. They found that solvent vapor annealing with DMSO gives the best photovoltaic performance. These solvent vapors condense on the surface of the perovskite and dissolve the perovskite. The liquid phase of these films will act as a binder for the grains and bind the adjacent grains, leading to larger grain growth and heals the grain boundary defects. Recrystallization of the perovskite layer take place and densification of the films will takes place (Figure 3.13). This effect was further studied by Zuo *et al.* that exposing PbI_2:MAI films to DMF or DMSO vapor affects the morphology of the perovskite [52]. This is due to DMF or DMSO forms an intermediate phase. Because of the formation of the complex, additional peaks in the X-ray diffraction pattern are appearing. The formation of the intermediate complex helps the diffusion of the MAI into PbI_2. In addition, this solvent vapor treatment helps for intercalation of MAI into PbI_2 lattice. Because of the interaction of DMF and DMSO molecules, it improves the interlayer distance of PbI_2, thus facilitating the MAI molecules into PbI_2 lattice. Because of the interdiffusion between PbI_2 and

MAI in the two-step deposition, the formation of an intermediate phase changes the kinetics of the crystallization. This improves the grain size of the crystals from 300 nm for pristine films to 500 nm for the films with vapor exposure.

References

1. Kojima, A., Teshima, K., Shirai, Y., Miyasaka, T., Organometal Halide Perovskites as Visible-Light Sensitizers for Photovoltaic Cells. *J. Am. Chem. Soc*, 131, 6050, 2009.
2. Wang, J.T.-W., Ball, J.M., Barea, E.M., Abate, A., Alexander-Webber, J.A., Huang, J., Saliba, M., Mora-Sero, I., Bisquert, J., Snaith, H.J., Nicholas, R.J., Low-Temperature Processed Electron Collection Layers of Graphene/TiO2 Nanocomposites in Thin Film Perovskite Solar Cells. *Nano Lett.*, 14, 724, 2014.
3. Park, N.-G., Perovskite solar cells: an emerging photovoltaic technology. *Mater. Today*, 18, 65, 2015.
4. Leijtens, T., Eperon, G.E., Pathak, S., Abate, A., Lee, M.M., Snaith, H.J., Overcoming ultraviolet light instability of sensitized TiO2 with meso-superstructured organometal tri-halide perovskite solar cells. *Nat. Commun.*, 4, 2885, 2013.
5. Burschka, J., Pellet, N., Moon, S.-J., Humphry-Baker, R., Gao, P., Nazeeruddin, M.K., Grätzel, M., Sequential deposition as a route to high-performance perovskite-sensitized solar cells. *Nature*, 499, 316, 2013.
6. Jeon, N.J., Noh, J.H., Kim, Y.C., Yang, W.S., Ryu, S., Seok, S., II, Solvent engineering for high-performance inorganic–organic hybrid perovskite solar cells. *Nat. Mater.*, 13, 897, 2014.
7. Xiao, M., Huang, F., Huang, W., Dkhissi, Y., Zhu, Y., Etheridge, J., Gray-Weale, A., Bach, U., Cheng, Y.-B., Spiccia, L., A Fast Deposition-Crystallization Procedure for Highly Efficient Lead Iodide Perovskite Thin-Film Solar Cells. *Angew. Chem. Int. Ed.*, 53, 9898, 2014.
8. Li, W., Fan, J., Li, J., Mai, Y., Wang, L., Controllable Grain Morphology of Perovskite Absorber Film by Molecular Self-Assembly toward Efficient Solar Cell Exceeding 17%. *J. Am. Chem. Soc*, 137, 10399, 2015.
9. Li, J., Wang, H., Chin, X.Y., Dewi, H.A., Vergeer, K., Goh, T.W., Lim, J.W.M., Lew, J.H., Loh, K.P., Soci, C., Sum, T.C., Bolink, H.J., Mathews, N., Mhaisalkar, S., Bruno, A., Highly Efficient Thermally Co-evaporated Perovskite Solar Cells and Mini-modules. *Joule*, 4, 1035, 2020.
10. Xing, G., Mathews, N., Sun, S., Lim, S.S., Lam, Y.M., Grätzel, M., Mhaisalkar, S., Sum, T.C., Long-Range Balanced Electron- and Hole-Transport Lengths in Organic-Inorganic CH3NH3PbI3. *Science*, 342, 344, 2013.

11. Eperon, G.E., Burlakov, V.M., Docampo, P., Goriely, A., Snaith, H.J., Morphological Control for High Performance, Solution-Processed Planar Heterojunction Perovskite Solar Cells. *Adv. Funct. Mater.*, 24, 151, 2014.
12. Zhang, H., Mao, J., He, H., Zhang, D., Zhu, H.L., Xie, F., Wong, K.S., Grätzel, M., Choy, W.C.H., A Smooth CH3NH3PbI3 Film via a New Approach for Forming the PbI2 Nanostructure Together with Strategically High CH3NH3I Concentration for High Efficient Planar-Heterojunction Solar Cells. *Adv. Energy Mater.*, 5, 1501354, 2015.
13. El-Henawey, M., II, Gebhardt, R.S., El-Tonsy, M.M., Chaudhary, S., Organic solvent vapor treatment of lead iodide layers in the two-step sequential deposition of CH3NH3PbI3-based perovskite solar cells. *J. Mater. Chem. A*, 4, 1947, 2016.
14. Pathipati, S. and Shah, M., Performance enhancement of perovskite solar cells using NH4I additive in a solution processing method. *Sol. Energy*, 162, 8, 2018.
15. Wu, Y., Islam, A., Yang, X., Qin, C., Liu, J., Zhang, K., Peng, W., Han, L., Retarding the crystallization of PbI2 for highly reproducible planar-structured perovskite solar cells via sequential deposition. *Energy Environ. Sci.*, 7, 2934, 2014.
16. Jo, Y., Oh, K.S., Kim, M., Kim, K.-H., Lee, H., Lee, C.-W., Kim, D.S., High Performance of Planar Perovskite Solar Cells Produced from PbI2(DMSO) and PbI2(NMP) Complexes by Intramolecular Exchange. *Adv. Mater. Interfaces*, 3, 1500768, 2016.
17. Zhang, T., Yang, M., Zhao, Y., Zhu, K., Controllable Sequential Deposition of Planar CH3NH3PbI3 Perovskite Films via Adjustable Volume Expansion. *Nano Lett.*, 15, 3959, 2015.
18. Zhang, H., Cheng, J., Li, D., Lin, F., Mao, J., Liang, C., Jen, A.K.-Y., Grätzel, M., Choy, W.C.H., Toward All Room-Temperature, Solution-Processed, High-Performance Planar Perovskite Solar Cells: A New Scheme of Pyridine-Promoted Perovskite Formation. *Adv. Mater.*, 29, 1604695, 2017.
19. Docampo, P., Hanusch, F.C., Stranks, S.D., Döblinger, M., Feckl, J.M., Ehrensperger, M., Minar, N.K., Johnston, M.B., Snaith, H.J., Bein, T., Solution Deposition-Conversion for Planar Heterojunction Mixed Halide Perovskite Solar Cells. *Adv. Energy Mater.*, 4, 1400355, 2014.
20. Cohen, B.-E., Gamliel, S., Etgar, L., Parameters influencing the deposition of methylammonium lead halide iodide in hole conductor free perovskite-based solar cells. *APL Mater.*, 2, 081502, 2014.
21. Ito, S., Tanaka, S., Nishino, H., Lead-Halide Perovskite Solar Cells by CH3NH3I Dripping on PbI2–CH3NH3I–DMSO Precursor Layer for Planar and Porous Structures Using CuSCN Hole-Transporting Material. *J. Phys. Chem. Lett.*, 6, 881, 2015.
22. Zheng, L., Chung, Y.-H., Ma, Y., Zhang, L., Xiao, L., Chen, Z., Wang, S., Qu, B., Gong, Q., A hydrophobic hole transporting oligothiophene for planar

perovskite solar cells with improved stability. *Chem. Commun.*, 50, 11196, 2014.

23. Zhang, C., Luan, W., Yin, Y., High Efficient Planar-heterojunction Perovskite Solar Cell Based on Two-step Deposition Process. *Energy Proc.*, 105, 793, 2017.
24. Sun, W., Li, Y., Yan, W., Peng, H., Ye, S., Rao, H., Zhao, Z., Liu, Z., Bian, Z., Huang, C., Rapid and Complete Conversion of CH3NH3PbI3 for Perovskite/C 60 Planar-Heterojunction Solar Cells by Two-Step Deposition. *Chin. J. Chem.*, 35, 5, 2017.
25. Yella, A., Heiniger, L.-P., Gao, P., Nazeeruddin, M.K., Grätzel, M., Nanocrystalline Rutile Electron Extraction Layer Enables Low-Temperature Solution Processed Perovskite Photovoltaics with 13.7% Efficiency. *Nano Lett.*, 14, 2591, 2014.
26. Gao, K., Li, L., Lai, T., Xiao, L., Huang, Y., Huang, F., Peng, J., Cao, Y., Liu, F., Russell, T.P., Janssen, R.A.J., Peng, X., Deep Absorbing Porphyrin Small Molecule for High-Performance Organic Solar Cells with Very Low Energy Losses. *J. Am. Chem. Soc*, 137, 7282, 2015.
27. Chen, Q., Zhou, H., Hong, Z., Luo, S., Duan, H.-S., Wang, H.-H., Liu, Y., Li, G., Yang, Y., Planar Heterojunction Perovskite Solar Cells via Vapor-Assisted Solution Process. *J. Am. Chem. Soc*, 136, 622, 2014.
28. Binek, A., Hanusch, F.C., Docampo, P., Bein, T., Stabilization of the Trigonal High-Temperature Phase of Formamidinium Lead Iodide. *J. Phys. Chem. Lett.*, 6, 1249, 2015.
29. Zhou, Y., Zhou, Z., Chen, M., Zong, Y., Huang, J., Pang, S., Padture, N.P., Doping and alloying for improved perovskite solar cells. *J. Mater. Chem. A*, 4, 17623, 2016.
30. Aguiar, J.A., Wozny, S., Holesinger, T.G., Aoki, T., Patel, M.K., Yang, M., Berry, J.J., Al-Jassim, M., Zhou, W., Zhu, K., In situ investigation of the formation and metastability of formamidinium lead tri-iodide perovskite solar cells. *Energy Environ. Sci.*, 9, 2372, 2016.
31. Luo, D., Zhao, L., Wu, J., Hu, Q., Zhang, Y., Xu, Z., Liu, Y., Liu, T., Chen, K., Yang, W., Zhang, W., Zhu, R., Gong, Q., Dual-Source Precursor Approach for Highly Efficient Inverted Planar Heterojunction Perovskite Solar Cells. *Adv. Mater.*, 29, 1604758, 2017.
32. Saliba, M., Matsui, T., Seo, J.-Y., Domanski, K., Correa-Baena, J.-P., Nazeeruddin, M.K., Zakeeruddin, S.M., Tress, W., Abate, A., Hagfeldt, A., Grätzel, M., Cesium-containing triple cation perovskite solar cells: improved stability, reproducibility and high efficiency. *Energy Environ. Sci.*, 9, 1989, 2016.
33. Yang, W.S., Noh, J.H., Jeon, N.J., Kim, Y.C., Ryu, S., Seo, J., Seok, S., II, High-performance photovoltaic perovskite layers fabricated through intramolecular exchange. *Science*, 348, 1234, 2015.
34. Saliba, M., Matsui, T., Domanski, K., Seo, J.-Y., Ummadisingu, A., Zakeeruddin, S.M., Correa-Baena, J.-P., Tress, W.R., Abate, A., Hagfeldt, A.,

Grätzel, M., Incorporation of rubidium cations into perovskite solar cells improves photovoltaic performance. *Science*, 354, 206, 2016.
35. McMeekin, D.P., Sadoughi, G., Rehman, W., Eperon, G.E., Saliba, M., Hörantner, M.T., Haghighirad, A., Sakai, N., Korte, L., Rech, B., Johnston, M.B., Herz, L.M., Snaith, H.J., A mixed-cation lead mixed-halide perovskite absorber for tandem solar cells. *Science*, 351, 151, 2016.
36. Bi, D., Tress, W., Dar, M., II, Gao, P., Luo, J., Renevier, C., Schenk, K., Abate, A., Giordano, F., Baena, J.-P.C., Decoppet, J.-D., Zakeeruddin, S.M., Nazeeruddin, M.K., Grätzel, M., Hagfeldt, A., Efficient luminescent solar cells based on tailored mixed-cation perovskites. *Sci. Adv.*, 2, e1501170, 2016.
37. Zheng, X., Hou, Y., Bao, C., Yin, J., Yuan, F., Huang, Z., Song, K., Liu, J., Troughton, J., Gasparini, N., Zhou, C., Lin, Y., Xue, D.-J., Chen, B., Johnston, A.K., Wei, N., Hedhili, M.N., Wei, M., Alsalloum, A.Y., Maity, P., Turedi, B., Yang, C., Baran, D., Anthopoulos, T.D., Han, Y., Lu, Z.-H., Mohammed, O.F., Gao, F., Sargent, E.H., Bakr, O.M., Managing grains and interfaces via ligand anchoring enables 22.3%-efficiency inverted perovskite solar cells. *Nat. Energy*, 5, 131, 2020.
38. Ávila, J., Momblona, C., Boix, P.P., Sessolo, M., Bolink, H.J., Vapor-Deposited Perovskites: The Route to High-Performance Solar Cell Production? *Joule*, 1, 431, 2017.
39. Docampo, P., Ball, J.M., Darwich, M., Eperon, G.E., Snaith, H.J., Efficient organometal trihalide perovskite planar-heterojunction solar cells on flexible polymer substrates. *Nat. Commun.*, 4, 2761, 2013.
40. Jeng, J.-Y., Chiang, Y.-F., Lee, M.-H., Peng, S.-R., Guo, T.-F., Chen, P., Wen, T.-C., CH3NH3PbI3 Perovskite/Fullerene Planar-Heterojunction Hybrid Solar Cells. *Adv. Mater.*, 25, 3727, 2013.
41. Conings, B., Baeten, L., De Dobbelaere, C., D'Haen, J., Manca, J., Boyen, H.-G., Perovskite-Based Hybrid Solar Cells Exceeding 10% Efficiency with High Reproducibility Using a Thin Film Sandwich Approach. *Adv. Mater.*, 26, 2041, 2014.
42. Liang, P.-W., Liao, C.-Y., Chueh, C.-C., Zuo, F., Williams, S.T., Xin, X.-K., Lin, J., Jen, A.K.-Y., Additive Enhanced Crystallization of Solution-Processed Perovskite for Highly Efficient Planar-Heterojunction Solar Cells. *Adv. Mater.*, 26, 3748, 2014.
43. Gong, X., Li, M., Shi, X.-B., Ma, H., Wang, Z.-K., Liao, L.-S., Controllable Perovskite Crystallization by Water Additive for High-Performance Solar Cells. *Adv. Funct. Mater.*, 25, 6671, 2015.
44. Wu, C.-G., Chiang, C.-H., Tseng, Z.-L., Nazeeruddin, M.K., Hagfeldt, A., Grätzel, M., High efficiency stable inverted perovskite solar cells without current hysteresis. *Energy Environ. Sci.*, 8, 2725, 2015.
45. Adhikari, N., Dubey, A., Gaml, E.A., Vaagensmith, B., Reza, K.M., Mabrouk, S.A.A., Gu, S., Zai, J., Qian, X., Qiao, Q., Crystallization of a perovskite film for higher performance solar cells by controlling water concentration in methyl ammonium iodide precursor solution. *Nanoscale*, 8, 2693, 2016.

46. Kang, D.-H. and Park, N.-G., On the Current–Voltage Hysteresis in Perovskite Solar Cells: Dependence on Perovskite Composition and Methods to Remove Hysteresis. *Adv. Mater.*, 31, 1805214, 2019.
47. Liu, P., Wang, W., Liu, S., Yang, H., Shao, Z., Fundamental Understanding of Photocurrent Hysteresis in Perovskite Solar Cells. *Adv. Energy Mater.*, 9, 1803017, 2019.
48. Elumalai, N.K. and Uddin, A., Hysteresis in organic-inorganic hybrid perovskite solar cells. *Sol. Energy Mater. Sol. Cells*, 157, 476, 2016.
49. Chen, B., Yang, M., Priya, S., Zhu, K., Origin of J–V Hysteresis in Perovskite Solar Cells. *J. Phys. Chem. Lett.*, 7, 905, 2016.
50. Ke, W., Xiao, C., Wang, C., Saparov, B., Duan, H.-S., Zhao, D., Xiao, Z., Schulz, P., Harvey, S.P., Liao, W., Meng, W., Yu, Y., Cimaroli, A.J., Jiang, C.-S., Zhu, K., Al-Jassim, M., Fang, G., Mitzi, D.B., Yan, Y., Employing Lead Thiocyanate Additive to Reduce the Hysteresis and Boost the Fill Factor of Planar Perovskite Solar Cells. *Adv. Mater.*, 28, 5214, 2016.
51. Tai, Q., You, P., Sang, H., Liu, Z., Hu, C., Chan, H.L.W., Yan, F., Efficient and stable perovskite solar cells prepared in ambient air irrespective of the humidity. *Nat. Commun.*, 7, 11105, 2016.
52. Zuo, L., Dong, S., De Marco, N., Hsieh, Y.-T., Bae, S.-H., Sun, P., Yang, Y., Morphology Evolution of High Efficiency Perovskite Solar Cells via Vapor Induced Intermediate Phases. *J. Am. Chem. Soc*, 138, 15710, 2016.
53. Li, S.-S., Chang, C.-H., Wang, Y.-C., Lin, C.-W., Wang, D.-Y., Lin, J.-C., Chen, C.-C., Sheu, H.-S., Chia, H.-C., Wu, W.-R., Jeng, U.S., Liang, C.-T., Sankar, R., Chou, F.-C., Chen, C.-W., Intermixing-seeded growth for high-performance planar heterojunction perovskite solar cells assisted by precursor-capped nanoparticles. *Energy Environ. Sci.*, 9, 1282, 2016.
54. Zhao, Y., Tan, H., Yuan, H., Yang, Z., Fan, J.Z., Kim, J., Voznyy, O., Gong, X., Quan, L.N., Tan, C.S., Hofkens, J., Yu, D., Zhao, Q., Sargent, E.H., Perovskite seeding growth of formamidinium-lead-iodide-based perovskites for efficient and stable solar cells. *Nat. Commun.*, 9, 1607, 2018.
55. Bi, D., Yi, C., Luo, J., Décoppet, J.-D., Zhang, F., Zakeeruddin, S.M., Li, X., Hagfeldt, A., Grätzel, M., Polymer-templated nucleation and crystal growth of perovskite films for solar cells with efficiency greater than 21%. *Nat. Energy*, 1, 16142, 2016.
56. Zhu, W., Bao, C., Lv, B., Li, F., Yi, Y., Wang, Y., Yang, J., Wang, X., Yu, T., Zou, Z., Dramatically promoted crystallization control of organolead triiodide perovskite film by a homogeneous cap for high efficiency planar-heterojunction solar cells. *J. Mater. Chem. A*, 4, 12535, 2016.
57. Bi, C., Wang, Q., Shao, Y., Yuan, Y., Xiao, Z., Huang, J., Non-wetting surface-driven high-aspect-ratio crystalline grain growth for efficient hybrid perovskite solar cells. *Nat. Commun.*, 6, 7747, 2015.
58. Li, Y., Xu, Z., Zhao, S., Song, D., Qiao, B., Zhu, Y., Meng, J., Benefits of the Hydrophobic Surface for CH3NH3PbI3 Crystalline Growth towards Highly Efficient Inverted Perovskite Solar Cells. *Molecules*, 24, 2027, 2019.

59. Shah, M.N., Pathipati, S.R., Ahmed, N., Porphyrin based hole transport layers for enhanced charge transport and stability in perovskite solar cells. *J. Mater. Sci.: Mater. Electron.*, 30, 7866, 2019.
60. Pinpithak, P., Chen, H.-W., Kulkarni, A., Sanehira, Y., Ikegami, M., Miyasaka, T., Low-temperature and Ambient Air Processes of Amorphous SnOx-based Mixed Halide Perovskite Planar Solar Cell. *Chem. Lett.*, 46, 382, 2017.
61. Kumar, M.H., Yantara, N., Dharani, S., Graetzel, M., Mhaisalkar, S., Boix, P.P., Mathews, N., Flexible, low-temperature, solution processed ZnO-based perovskite solid state solar cells. *Chem. Commun.*, 49, 11089, 2013.
62. Ball, J.M. and Petrozza, A., Defects in perovskite-halides and their effects in solar cells. *Nat. Energy*, 1, 16149, 2016.
63. Hamada, K., Tanaka, R., Kamarudin, M.A., Shen, Q., Iikubo, S., Minemoto, T., Yoshino, K., Toyoda, T., Ma, T., Kang, D.-W., Hayase, S., Enhanced Device Performance with Passivation of the TiO2 Surface Using a Carboxylic Acid Fullerene Monolayer for a SnPb Perovskite Solar Cell with a Normal Planar Structure. *ACS Appl. Mater. Interfaces*, 12, 17776, 2020.
64. Stolterfoht, M., Caprioglio, P., Wolff, C.M., Márquez, J.A., Nordmann, J., Zhang, S., Rothhardt, D., Hörmann, U., Amir, Y., Redinger, A., Kegelmann, L., Zu, F., Albrecht, S., Koch, N., Kirchartz, T., Saliba, M., Unold, T., Neher, D., The impact of energy alignment and interfacial recombination on the internal and external open-circuit voltage of perovskite solar cells. *Energy Environ. Sci.*, 12, 2778, 2019.
65. Liu, Z., Chen, Q., Hong, Z., Zhou, H., Xu, X., De Marco, N., Sun, P., Zhao, Z., Cheng, Y.-B., Yang, Y., Low-Temperature TiOx Compact Layer for Planar Heterojunction Perovskite Solar Cells. *ACS Appl. Mater. Interfaces*, 8, 11076, 2016.
66. Xiao, Z., Dong, Q., Bi, C., Shao, Y., Yuan, Y., Huang, J., Solvent Annealing of Perovskite-Induced Crystal Growth for Photovoltaic-Device Efficiency Enhancement. *Adv. Mater.*, 26, 6503, 2014.

4

Advanced Synthesis Strategies for Single Crystal Perovskite Halides

Prerna and Sandeep Arya*

Department of Physics, University of Jammu, Jammu, Jammu and Kashmir, India

Abstract

Organic-inorganic hybrid halide perovskites, as well as purely organic and purely inorganic halide perovskites, have been grown by several techniques until now. However, rapid crystal growth of perovskites exhibiting high quality is of utmost importance. This is because the quantity and quality of a crystal is the basic feature responsible for industrial and scientific applications. Furthermore, preferable size of the crystal is also a major issue regarding optoelectronic device applications. Hence, for desired applications, the most significant research task is the rapid growth of large size and good quality single crystal perovskites. This chapter will describe some important crystal growth methods.

Keywords: Perovskite halides, synthesis strategies, single-crystal

4.1 Introduction

The successful application of halide perovskites that have been studied for several decades strongly depends upon the growth of high-quality perovskites with well developed qualities like high carrier mobility, low trap density, and longer diffusion lengths etc. These perovskites can be synthesized in various forms like single crystals, nanostructures, and polycrystalline films, etc by adopting different growth methods [1–4]. Single crystals are known to exhibit continuous and unbroken crystal lattices. Prominently, they display no grain boundaries leading to fewer defects in them. Due to their monocrystalline nature, they show exceptional optical

**Corresponding author*: snp09arya@gmail.com

Khursheed Ahmad and Waseem Raza (eds.) Perovskite Materials for Energy and Environmental Applications, (91–122) © 2022 Scrivener Publishing LLC

and electrical properties depending upon the shape and size of their crystallites [5]. The current developments in the field of perovskite single-crystals demonstrate that they could effectively conquer the challenges faced by other counterparts because of their outstanding optoelectronic properties. Moreover, the active area thickness in the single crystal perovskites is comparatively much greater leading to a longer carrier diffusion length and therefore the photogenerated charge carriers are successfully extracted [6]. Thus, as a result of the insignificant effect of grain boundary defects and surface defects, single crystal perovskites among all other forms exhibit greatly improved intrinsic semiconductor properties and are therefore superior materials for achieving high performance in optoelectronic device applications [7–10].

The direct application of grown single crystals is confined only to a few areas and therefore, further investigation is required for its successful application in other fields. Almost every technique that has been utilized for the growth of halide based single-crystal perovskites involves either liquid or solid state reactions. Several studies were however limited to the growth of only large sized crystals. But for the stability and toxicity of large scale production, large barriers are yet to overcome and several factors need to be evidently understood. This chapter briefly focuses on the growth techniques of single crystal perovskites. Some of the reported methods for synthesizing single crystal halide perovskites involve the growth of hybrid perovskites only, while others can grow both hybrid as well as purely inorganic halide perovskites. Moreover, a few methods like the Bridgman growth technique can grow only pure inorganic single crystal halide perovskites. Therefore, to have more detailed information, we have focused on each reported growth technique of single crystal halide perovskites. Table 4.1 summarizes various perovskite single crystals with their size and functional properties, formed using different synthesis methods.

4.2 Popular Single Crystal Growth Techniques

Out of the various methods reported, the five most popular techniques that have been utilized several times by the researchers for the growth of single crystal perovskite halides are: (**1**) Anti-solvent vapor-assisted crystallization method; (**2**) Inverse temperature crystallization; (**3**) Modified inverse temperature crystallization; (**4**) Solution temperature lowering method and (**5**) Bridgman method.

Table 4.1 Various perovskite single crystals with their size and functional properties, formed using different synthesis methods [54].

Synthesis method	Perovskite single crystals	Size	Functional properties	Ref.
AVC	$MAPbX_3$	Millimeter-sized single crystals	Low trap-state densities of the order 10^9-10^{10} per cubic centimeter and charge carrier diffusion lengths > 10 μm	[11]
	$MAPbBr_3$	Single crystal of dimensions ~1.4 × 1.4 × 0.7 mm^3	The surface recombination velocity is found to be 3.4±0.1 × 10^3 cm s^{-1} which is ~2–3 orders of magnitude lower than that in many important unpassivated semiconductors employed in solar cells	[12]
	$MAPbBr_3$	Single crystals of dimensions of 1–6 mm	A record μτ product of 1.2 × $10^{-2}cm^2V^{-1}$, lowest surface recombination velocity of 64 cm s^{-1}, high sensitivity of 80 μC $Gy_{air}{}^{-1}$ cm^{-2} and a lowest detectable dose rate of 0.5 μGy_{air} s^{-1}	[13]
	$(NH_4)_3Sb_2I_9$	2.1 mm thick single crystal with a side length of 5.7 mm	A low PCE of 0.51%, good stability in glovebox but poor stability in ambient air	[15]

(Continued)

Table 4.1 Various perovskite single crystals with their size and functional properties, formed using different synthesis methods [54]. (*Continued*)

Synthesis method	Perovskite single crystals	Size	Functional properties	Ref.
ITC	$MAPbI_3$	Large crystals of 1 mm length	After exposure to ambient air, the crystals degraded. The nature of the nucleation process and the kinetic processes remain under investigation	[20]
	$CsPbBr_3$	A several-mm long and flat single crystal	A moderately high carrier mobility and a high gamma photon absorptivity but low resistivity (2 GΩ cm)	[23]
	$(FAPbI_3)_{0.9}(MAPbBr_3)_{0.05}(CsPbBr_3)_{0.05}$	Millimeter-sized single crystals	A band gap of 1.52 eV, long carrier lifetime (up to 16 μs), at least 10000 h water-oxygen stability, 1000 h light stability, and excellent thermal stability	[24]

(*Continued*)

Table 4.1 Various perovskite single crystals with their size and functional properties, formed using different synthesis methods [54]. (*Continued*)

Synthesis method	Perovskite single crystals	Size	Functional properties	Ref.
MITC	$MAPbI_3$	Two-inch-sized single crystals	Low trap density (1.8×10^{9} cm^{-3}), very low hole concentration (8.8×10^{11} cm^{-3}), high carrier mobility (34 cm^2 $V^{-1}s^{-1}$)	[18]
	$FAPbI_3$	20 mm large single crystal	Very low trap state density (1.34×10^{10} cm^{-3}), high carrier mobility and remarkable thermal stability	[26]
	$MAPbBr_3$	Laminar crystal of size 6 × 8 mm and thickness 16 μm	High mobility (23.7 $cm^2s^{-1}V^{-1}$) and low trap density (2.5×10^{10} cm^{-3}), contributing to an intriguing efficiency of 7.11% with impressive stability of maintaining 93% initial PCE after 1000 h aging	[29]
	$MAPbI_3$	Millimeter-sized single crystals	A shift of ASE peak wavelength with temperature at different excitation wavelength led to a phase change in the bulk single crystals from orthorhombic to tetragonal phase	[30]
	$FAPbI_3$	5 mm sized single crystal	Long carrier lifetime (484 ns), high carrier mobility (4.4 $cm^2V^{-1}s^{-1}$ and high conductivity (1.1×10^{-7} Ω^{-1} cm^{-1})	[31]
	$CsPbBr_3$	Millimeter-sized single crystals	ON/OFF ratio as high as 10^5 at zero bias, good EQE (over 6%) and a peak responsivity of 28 mAW^{-1} at 550 nm.	[32]

(*Continued*)

Table 4.1 Various perovskite single crystals with their size and functional properties, formed using different synthesis methods [54]. *(Continued)*

Synthesis method	Perovskite single crystals	Size	Functional properties	Ref.
TSSG	$MASnI_3$	Single crystal of dimensions 2016 × 10 mm^3	A band gap of ~1.15 eV, trap densities of the order of 10^{11} cm^{-3}, thermally stable up to 200°C, stable under an inert (Ar or N_2) atmosphere for several months	[34]
	$FASnI_3$	Single crystal of dimensions 8 × 6 × 5 mm^3	A bandgap of ~1.4 eV, trap densities of the order of 10^{11} cm^{-3}, thermally stable up to 175°C, stable under an inert (Ar or N_2) atmosphere for several months	[34]
	$MAPbI_3$	Average size of 3.3 mm and a largest size of ~10 mm	Electron-hole diffusion lengths >175 μm under 1 sun illumination and >3 mm under weak light	[37]
BSSG	$MAPbI_3$	Single crystal of dimensions 12 × 12 × 7 mm^3	Using 532 nm laser, high responsivity of 953 A W^{-1} and EQE of 2.22 × 10^5% under weakest detectable light of 2.12 $nWcm^{-2}$ while on exposure to relatively strong light at 1 $mWcm^{-2}$, higher responsivity of 2.55 AW^{-1} and EQE of 5.95 × 10^2%.	[35]

(Continued)

Table 4.1 Various perovskite single crystals with their size and functional properties, formed using different synthesis methods [54]. (*Continued*)

Synthesis method	Perovskite single crystals	Size	Functional properties	Ref.
BG	$CsPbBr_3$	Single crystal ingot of diameter 7 mm	High attenuation, significant photoconductivity response and high resistivity (343 GW cm)	[41]
	$CsPbBr_3$	Single crystal ingot of diameter 11 mm	Energy resolution of 3.9% and 3.8% for 122 keV ^{57}Co and 662 keV ^{137}Cs gamma-rays respectively	[43]
	$CsPbBr_3$	Single crystal ingot of diameter 24 mm and length 90 mm	High electron mobility (11.61 $cm^2/(V\ s)$), high responsivity (5.83 A/W) and high EQE (1360%)	[44]
	$CsSnI_3$	-	Conversion efficiencies of up to 10.2%	[45]
SE	$MA_2Pb(SCN)_2I_2$	-	The crystal structure is closely related to the K_2NiF_4-type, a sharp absorption onset at 790 nm	[48]
	BA_2PbCl_4	Large crystals up to 5 × 10 × 2 mm^3	Larger ferroelectric spontaneous polarization (13 μCcm^{-2}) and a higher Curie temperature (438 K) with a band gap of 3.65 eV	[49]

(*Continued*)

Table 4.1 Various perovskite single crystals with their size and functional properties, formed using different synthesis methods [54]. (*Continued*)

Synthesis method	Perovskite single crystals	Size	Functional properties	Ref.
SG	$MAPb(Br_{1-x}Cl_x)_3$	Large crystals up to 5 mm	The band gap of single crystals increased and the unit cell dimensions decreased with increased Cl content x	[50]
	$MAPbBr_{3-x}Cl_x$, $MAPbI_{3-x}Br_x$	0.5–5 mm thick single crystals	An ultra-narrow EQE peak with an FWHM of <20 nm, a high sensitivity with a detection limit down to 80 pW cm^{-2}, and a high off-resonance rejection ratio of >200	[51]
LSG	$(C_6H_5C_2H_4NH_3)_2PbCl_4$	Crystals ranged from small block-like to larger plate-like crystals (dimensions >4 mm)	Well-ordered sheets of corner-sharing distorted $PbCl_6$ octahedra separated by bilayers of phenethylammonium cations	[52]
DPC	$MAPbI_3$	Ribbon-like morphology with a width of tens of μm	A low PCE of 1.73%	[53]

4.2.1 Anti-Solvent Vapor-Assisted Crystallization (AVC) Method

The basic principle behind the AVC technique is based on achieving crystallization because of different solubility of a perovskite in different solvents. Using this method, one can grow a perovskite which is completely soluble in a specific good solvent but poorly soluble in some other solvent. In this process, two or more solvents should be properly chosen. Here, the solvent which is less volatile is a good solvent while the other one with more volatility is a bad solvent. For that reason, a solvent is a crucial component of this process. AVC method is not subjected to the temperature. So, with a variation in temperature, this technique does not involve any change in the phase of the grown single crystal perovskites. Figure 4.1(a) shows the diagrammatic depiction of the AVC method. In this process, the precursors are first dissolved in the appropriately selected good solvent. The contents are then shifted to a small vial which is positioned in a large vial. The large vial is filled with a bad solvent and sealed properly. Within a few hours, the bad solvent in the large vial diffuses slowly into the small vial containing the solution of precursors and a good solvent. Since, the

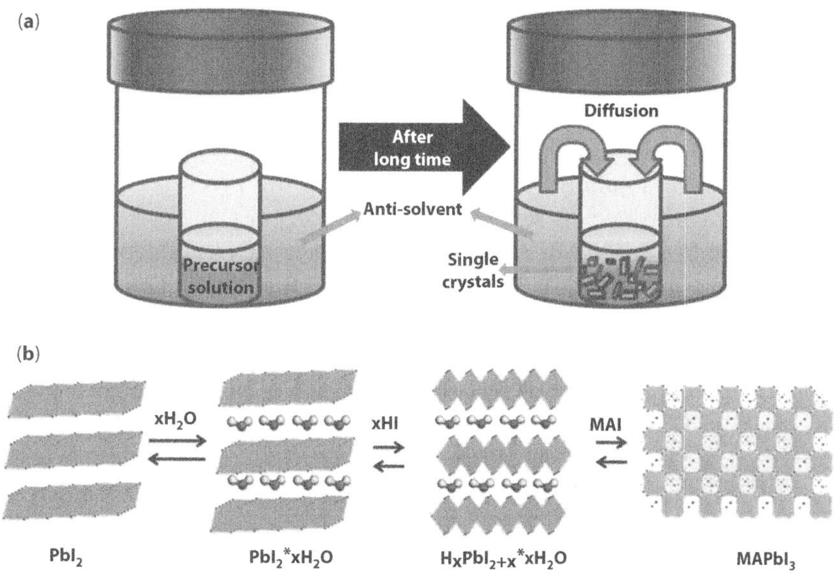

Figure 4.1 (a) Diagrammatic depiction of AVC process for the growth of single crystals; (b) Diagrammatic depiction showing configurations of layered PbI_2 structure, $PbI_2{\cdot}xH_2O$, $H_xPbI_{2+x}{\cdot}xH_2O$, and $MAPbI_3$ single crystal. Reproduced from [14] with permission. Copyright 2015, The Royal Society of Chemistry.

material is insoluble in the bad solvent, therefore, at the small vial bottom, the crystal creation ability increases resulting in the formation of single crystals in the small vial.

This technique was employed at the initial stage for successfully synthesizing single crystals of $MAPbBr_3$ and $MAPbI_3$ where MA = $CH_3NH_3^+$. As a good solvent, Gamma-butyrolactone (GBL) was utilized for $MAPbI_3$ and N,N-dimethylformamide (DMF) was utilized for $MAPbBr_3$. Dichloromethane (DCM) was however taken as a poor solvent for both $MAPbBr_3$ and $MAPbI_3$. For growing $MAPbBr_3$ crystals, $PbBr_2$ and MABr were taken in molar ratio 1:1 and then mixed in DMF forming precursor solution. This was followed by diffusion of anti-solvent DCM to precursor solution forming millimeter-sized $MAPbBr_3$crystals. Similarly, $MAPbI_3$ single crystals were grown by dissolving MAI and PbI_2 precursors in the ratio of 3:1 in GBL solvent. After this, the bad solvent DCM diffused into the precursor solution and formed desired $MAPbI_3$ single crystals. The prepared $MAPbX_3$ single crystals showed higher carrier mobility and lifetime with remarkably low trap density than the corresponding polycrystalline thin-films [11]. $MAPbBr_3$ single crystals were grown in a similar way by using a mixture of MABr and $PbBr_2$ in DMF solvent followed by slow diffusion of anti-solvent toluene and the process resulted in $MAPbBr_3$ crystals with superior features [12]. Large-sized $MAPbBr_3$ single crystals were also prepared by the AVC process [13]. This time, the precursors $PbBr_2$ and MABr were taken in two different ratios of 0.8:1 and 1:1 in DMF solvent. After the diffusion of anti-solvent into the precursor solutions, the grown crystals of $MAPbBr_3$ were of different transparency. As a result of the unbalanced solubility of the precursors in the precursor solution of ratio 1:1, the obtained crystal was opaque. In contrast, the precursor solution with $PbBr_2$ and MABr ratio of 0.8:1 formed a clear crystal of thickness 1-2 mm and side length 6 mm. The device formed using this crystal for detecting X-rays was found to be 70 times efficient than the polycrystalline thin film.

A different procedure was reported for growth of $MAPbI_3$ single crystals using the AVC technique [14]. Here, an intermediate product was formed that was verified by XRD diffraction pattern. A yellow colored solution of PbI_2 in good solvent HI was prepared at 120°C which changes to $H_xPbI_{2+x}{\cdot}xH_2O$ intermediate product. CH_3NHI was then added slowly with continuous stirring that dissolved completely into the solution. When this solution was cooled down, anti-solvent diethyl ether was poured which led to the formation of $MAPbI_3$ single crystals. The diagrammatic depiction of various configurations involved is illustrated in Figure 4.1(b).

Lead-free perovskite single crystals of $(NH_4)_3Sb_2I_9$ were also grown by employing the AVC technique [15]. In this process, the precursors NH_4I and SbI_3 were dissolved in C_2H_5OH. The resulting solution was then placed in the sealed beaker containing anti-solvent $CHCl_3$. Chloroform diffuse into the precursor solution and didn't dissolve which led to the precipitation of $(NH_4)_3Sb_2I_9$ single crystals of dimensions 5.7 mm × 2.1 mm. Recently, mixed-cation lead bromide perovskite single crystals of $Cs_x(MA)_{1-x}PbBr_3$, (x = 0.05–0.3) were also prepared by using AVC technique [16]. Firstly, MABr, $PbBr_2$, and CsBr precursors were kept for drying in vacuum at 120°C overnight. After that, mixing of 0.2 M $PbBr_2$ and 0.2 M MABr was done in DMF solution followed by constant stirring for 12 h at room temperature (RT). CsBr/$PbBr_2$ molar ratio was set at 5%, 9%, 13%, 17%, and 33%. The respective solutions were filtered using a 0.2 μm pore-size syringe PTFE filter. The single crystals were ultimately formed after the slow diffusion of anti-solvent methanol into the precursor solution and exhibited good quality.

4.2.2 Inverse Temperature Crystallization (ITC)

Based on the principle of an unusual retrograde solubility regime, ITC technique involves loss of solubility in particular organic solvent or solvents at elevated temperature [17]. Figure 4.2 illustrates the diagrammatic depiction of ITC method for the synthesis of single crystal perovskites. In view of the fact that the crystallization occurs by inverse solubility in some organic solvents, as a result ITC method strongly depends upon temperature. This synthesis technique is a balance of dissolution and precipitation. The solvent molecules entirely bound the molecules of the perovskite in the complexes at low temperatures. This restricts the unbound molecules to attain the saturation point. Nevertheless, as soon as there is an increase

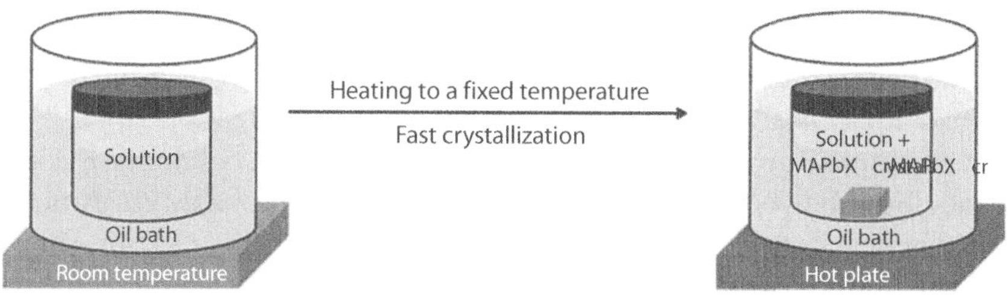

Figure 4.2 Diagrammatic depiction of ITC technique for the formation of $MAPbX_3$ single crystals. Reproduced from [19] with permission. Copyright 2015, Nature Publishing Group.

in temperature, the bonding energy decreases and free perovskite molecules get concentrated in the solution. The moment the solution acquires supersaturation stage, nucleation appears at some point, followed by crystal growth. Just like in AVC method, the solvent must be appropriately chosen in ITC technique also, for growing a good single crystal. As an example, GBL, DMF, and dimethylsulfoxide (DMSO), were found to be the most favorable solvent for $MAPbI_3$, $MAPbBr_3$, and $MAPbCl_3$, respectively [18]. However, it is an exigent task to grow mixed halide perovskite as it is required to choose one solvent for both precursor materials.

Using ITC technique, $MAPbX_3$ single crystals were initially prepared [19, 20], and soon after, many single crystal perovskites were prepared. For the synthesis of $MAPbI_3$ crystals [20], PbI_2 and MAI in the molar ratio of 1:1 were dissolved in GBL. With the purpose of forming a clear solution, it was made to whisk vigorously on a hot plate at 100°C. The temperature was then increased to 190°C and after 25 minutes (the schematic of the experiment examination is outlined in Figure 4.3(a)), rhombic dodecahedral or rhombo-hexagonal dodecahedral $MAPbI_3$ crystals as shown in Figure 4.3(b) were obtained that exhibited space group *I4/mcm*. However, the crystals get dissolved in the solution rapidly upon cooling to RT. The freshly grown crystals were black and the surface was glossy that degrade and form a yellow phase on the surface within some days when kept under ambient air atmosphere. During storage, the crystals get exposed to humidity that formed spike-like elongated structures specifying that the decomposition initiates and the crystal surface become partly rough as shown in Figure 4.3(c).

Single-crystal perovskites of both $MAPbX_3$ and $FAPbX_3$ (where FA = $HC(NH_2)_2^+$) were also prepared by this method [21]. The solubilities of these perovskites were checked in different solvents at various temperatures to confirm with their retrograde solubility behavior confidently. DMF and GBL were found to be appropriate solvents for bromine and iodine based perovskites respectively to persuade the retrograde solubility. The role of surface tension was described in the crystallization mechanism of metal halide perovskites that were grown by ITC method and it paves the way to control the crystallization process [22].

One can also grow completely inorganic single crystal halide perovskites in addition to hybrid halide perovskites. This would probably get through the chemical instability of hybrid perovskites. Good quality $CsPbBr_3$ single crystals were grown under ambient conditions via ITC technique using DMSO as a solvent [23]. CsBr and $PbBr_2$ were first dissolved in a blend of DMSO, DMF, cyclohexanol, and the contents were then transferred to a vial and heated to 90°C. One to three nuclei were formed and when further

Figure 4.3 (a) Schematic of the experiment examination; (b) grown $MAPbI_3$ crystals; (c) Secondary electron image of $MAPbI_3$ crystal. Reproduced from [20] with permission. Copyright 2015, Nature Publishing Group. (d) Obtained $CsPbBr_3$ crystals of lengths 7 and 8 mm. Reproduced from [23] with permission. Copyright 2016, American Chemical Society. (e) Growth process for the formation of $(FAPbI_3)_{0.95}(CsPbBr_3)_{0.05}$ single crystal. Reproduced from [24] with permission. Copyright 2019, American Chemical Society.

heated to 110°C, about 8 mm long, flat and orange-colored $CsPbBr_3$ crystals were formed. Figure 4.3(d) shows the freshly prepared $CsPbBr_3$ single crystals that exhibited good quality. Moreover, a series of highly stable and well-defined mixed cation and mixed halide perovskite single crystals with a new composition, $(FAPbI_3)_{0.9}(MAPbBr_3)_{0.05}(CsPbBr_3)_{0.05}$ was prepared using ITC method [24]. The crystals were prepared using GBL solution, and the process is outlined in Figure 4.3(e). The crystals exhibited a carrier lifetime of 16 μs and excellent stability for 1000 hours under light and for at least 10,000 hours against water and oxygen. Further high-quality mix cation lead iodide single crystals of $FA_{1-x}Cs_xPbI_3$ (x = 0, 0.1) were grown by using ITC method, and the thermodynamic properties as well as their crystal structure was studied. The prepared crystals had a hexagonal pyramidal shape and exhibited a typical size of 1 to 2 mm^3 [25].

4.2.3 Modified Inverse Temperature Crystallization

As the name suggests, modified inverse temperature crystallization (MITC) method is basically a form of ITC method with some modification. In this method, a seed crystal is initially grown followed by its utilization to grow large size single crystals. By employing this method, it is possible to grow hybrid perovskites in an effortless manner. Several organic-inorganic large single crystal perovskites were grown by using MITC method [18, 26, 27]. The schematic illustration of MITC technique that was employed for the growth of large and good quality single crystal perovskites of $MAPbX_3$ and $FAPbI_3$ [28] following a step-by-step procedure is shown in Figure 4.4. MAX or FAX and PBX_2 were first dissolved in an appropriate solution, and the resulting solution was kept at a high temperature for about overnight. It was observed that the precursors lose their solubility and formed a huge number of small seed crystals of size 1 to 2 mm. Out of the formed crystals, a superior seed crystal was appropriately chosen and positioned inside a newly prepared precursor solution. This solution was subsequently subjected to inverse heating, which yielded larger crystals of 7 mm size. Taking the larger size seed crystal as a new one, the repetitive process gives comparatively extremely big single crystals. The obtained crystals of $MAPbCl_3$ and $MAPbBr_3$ were cuboidal. Further, $MAPbBr_3$ were orange colored while $MAPbCl_3$ were transparent.

Big laminar single crystals of $MAPbBr_3$ of about 16 μm thickness were prepared by MITC technique [29]. The process involved the construction of a box using TiO_2 coated FTO glass. In between the box, a space gap was

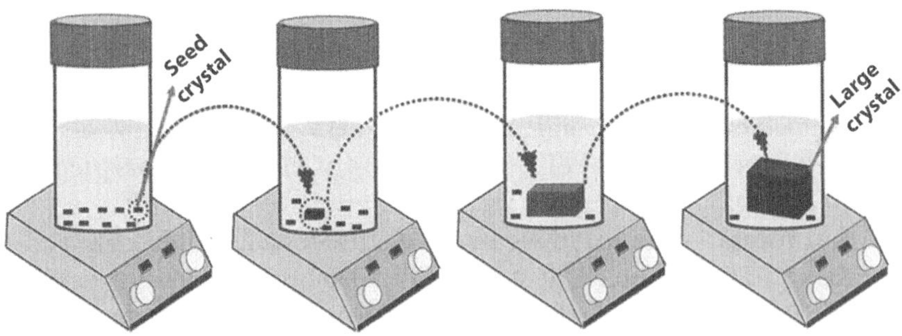

Figure 4.4 Schematic illustration of MITC technique. Reproduced from [28] with permission. Copyright 2018, American Chemical Society.

maintained. Firstly, MABr and $PbBr_2$ precursors were dissolved in DMF solution and the resulting solution was inserted in the provided space gap. The solution was then heated with the help of a homothermal aluminium block at 90°C. Schematic depiction of the heating process is described in Figure 4.5(a). The obtained crystal as shown in Figure 4.5(b) showed high power efficiency, high mobility, and low-trap density in comparison to analogous thin-film polycrystalline sample. Nonlinear optical properties of MITC grown $MAPbI_3$ single crystal were also studied [30]. $FAPbI_3$ single crystals as large as 5 mm were prepared by MITC technique using $Pb(ac)_2.3H_2O$ and FAac. First, a seed crystal was synthesized and subsequently, ITC method was repeated three times to yield a single crystal of $FAPbI_3$ [31].

In contrast, for the growth of completely inorganic bulk perovskites, it is required to avoid the impurities, which are in the form of undesired materials. So basically, the growth of pure and completely inorganic single crystals is comparatively a complex task. For instance, owing to the good solubility of $PbBr_3$ and poor solubility of CsBr, it is difficult to prepare pure $CsPbBr_3$ single crystals because there occurs an unbalanced solubility of the precursors. As a consequence, this process can form a fusion of

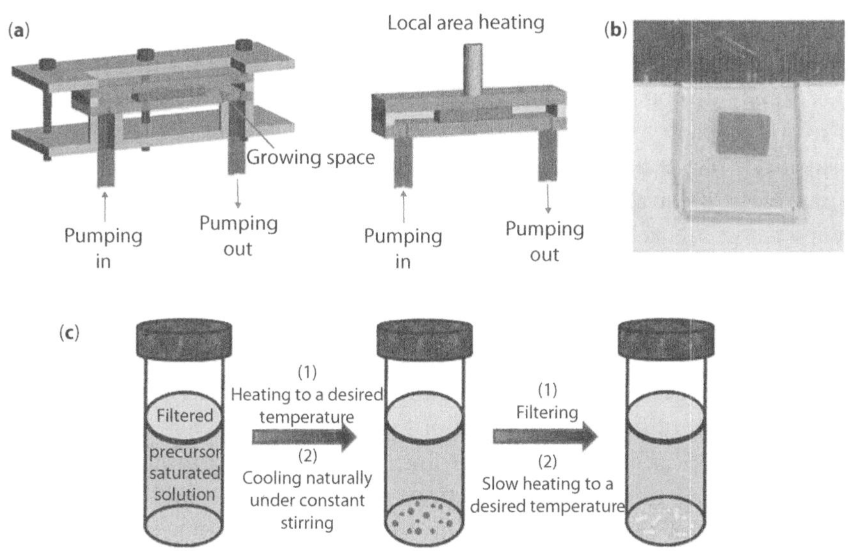

Figure 4.5 (a) Schematic depiction of the heating process utilizing homothermal aluminum block; (b) Grown $MAPbBr_3$ laminar crystal. Reproduced from [28] with permission. Copyright 2018, American Chemical Society. (c) Diagrammatic illustration for the synthesis of pure $CsPbBr_3$ single crystals.

$CsPbBr_3$, Cs_4PbBr_6, and $CsPb_2Br_5$ crystals, where the impurity crystals of Cs_4PbBr_6 and $CsPb_2Br_5$ are superfluous materials and needed to be avoided through MITC process.

Further, this technique was employed by various researchers, and different ways were found to remove the unwanted impurity crystals. In a very effective method, the precursors' molar ratio was varied to observe the effect on undesired materials during the synthesis of $CsPbBr_3$ single crystals [32]. First, the precursors CsBr and $PbBr_2$ were taken in the molar ratio of 1:2 in the solution and filtered at a temperature of 100°C followed by subsequent heating to 120°C for 3 hours. As a result, pure shape-controlled $CsPbBr_3$ single crystals were formed, and there was no impurity crystal present. On the other hand, when CsBr and $PbBr_2$ were taken in the molar ratio of 1:1 and 1:1.5 at a temperature of 120°C, undesired crystals of Cs_4PbBr_6 were formed. In another procedure, nine antisolvents were utilized to prepare saturation solution of CsBr and $PbBr_2$ precursors in DMF [33]. Of them, only three antisolvents, viz. acetonitrile, methanol, and water, were observed to form saturated solution. With acetonitrile and methanol, yellow-orange colored precipitates were formed, whereas with water, white colored precipitates were obtained. Some impurity precipitates were also observed which were removed from MeCN and MeOH saturated solution by filtration. The resultant product was heated at a particular temperature till a mixture of yellow-green and orange colored crystals appeared. It was then cooled to RT under constant stirring and orange-colored crystals finally disappeared. The cooled solution was again filtered to remove the undesired crystals and further heated slowly to a desired temperature. This leads to the formation of desired single crystals of $CsPbBr_3$. The diagrammatic illustration for the synthesis of pure $CsPbBr_3$ single crystals is depicted in Figure 4.5(c).

4.2.4 Solution Temperature Lowering Method

This method is based on the principle of lowering the temperature in a precise way leading to the growth of the crystal on a favored seed that is dipped in a specific solvent. Under this process, two techniques have been developed depending on the position of the seed crystal. One is the top seeded solution growth method and the other is bottom seeded solution growth method. Solution temperature lowering (STL) is a typical method that involves easily controllable and stable crystal growth of good quality large single crystals. It is possible to grow large single crystals of hybrid halide perovskites, as well as 2D perovskites using these seeded growth techniques [34]. The process initiates with the growth of nearly millimeter sized small crystals and then through seeded growth, large single crystals are finally obtained. In equivalent acid halide solvents, hybrid halide perovskites have moderate solubility at RT and the solubility varies with the variation in temperature. As soon as the solute attains the saturation stage, the saturated precursor solution forms large single-crystal perovskites [35]. The main drawback of this technique is that it is not possible to grow those crystals that exhibit low solubility at elevated temperatures. Moreover, the entire growth process involved takes about two to four weeks for suitably growing even about 1-cm size crystal and therefore, this is quite a time-consuming technique [36].

4.2.4.1 Top-Seeded Solution Growth Method

As the name suggests, this process involves the formation of the seed crystal at the top of a vial. In this technique, the saturated precursor solution was firstly poured in a vial and then a silicon substrate was placed over it. The vial was finally sealed. The sealed vial can be viewed as constituting two parts: the bottom part where the precursor solution was poured and the top part where the silicon substrate lies. The bottom of the vial was heated with the help of a hot plate and kept at a temperature of 75°C. Top part was cooled down by airflow, and therefore, a temperature gradient was maintained in the vial. Finally, several seed crystals were formed on the silicon substrate as a consequence of temperature discrepancy. Only one crystal was left there while others were removed. The illustrative diagram of the Top-Seeded Solution Growth (TSSG) technique is shown in Figure 4.6(a). At last, large $MAPbI_3$ single crystals with superior quality were obtained after several hours using TSSG method [37]. The single crystal as shown in Figure 4.6(b) exhibited dimensions of 10 × 3 mm. Similarly, high-quality $MASnI_3$ and $FASnI_3$ single crystals were also prepared, which are shown in Figures 4.6(c, d). In this synthetic procedure, the precursors SnO and MAI or FAI were dissolved in a mixture of HI and H_3PO_2. The resulting solution was made to stir continuously at a temperature of 78°C. This was followed by progressively lowering the temperature from 60°C to 65°C to RT. The process resulted in the formation of black and shiny single crystals of $MASnI_3$ and $FASnI_3$. To check their stability, both the crystals were kept for 1 month. It was observed that $FASnI_3$ crystals were stable, whereas in $MASnI_3$ crystals, oxidation of tin occurred and therefore, $MASnI_3$ crystals were found to be unstable [34].

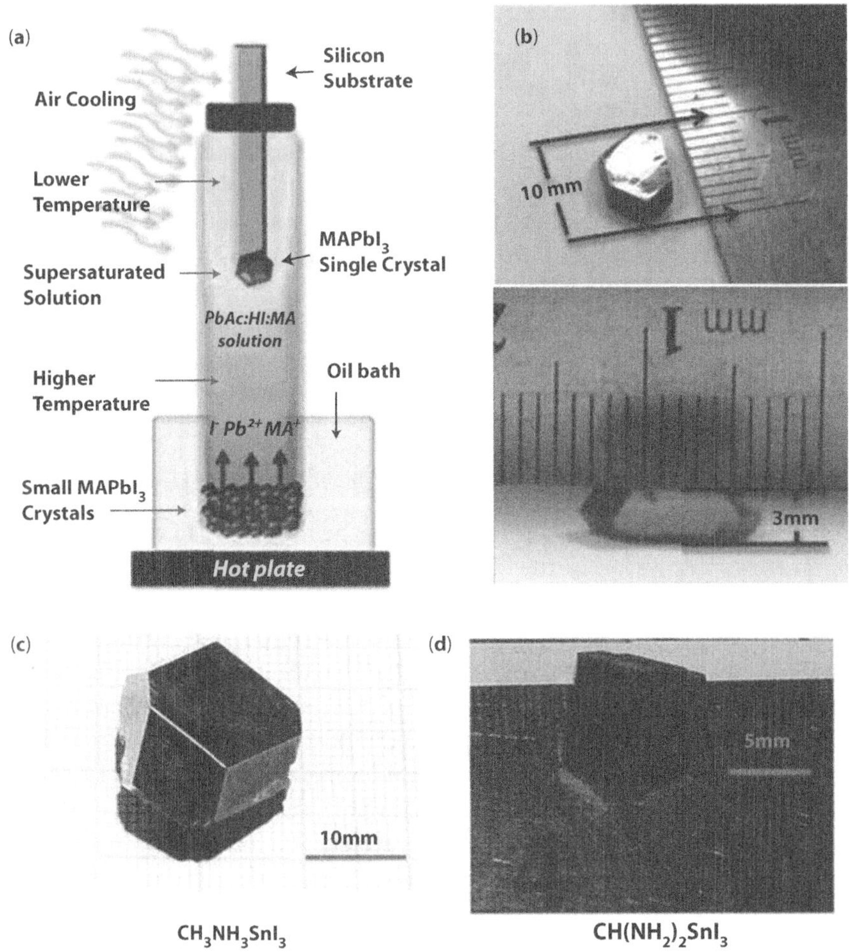

Figure 4.6 (a) Illustrative diagram of TSSG method; (b) Obtained $MAPbI_3$ single crystals; (c) $MASnI_3$ single crystal; and (d) $FASnI_3$ single crystal. Reproduced from [38] with permission. Copyright 2016, The Royal Society of Chemistry.

4.2.4.2 Bottom-Seeded Solution Growth Method

In contrast to TSSG method, Bottom-Seeded Solution Growth (BSSG) process involves the positioning of the seed crystal at the bottom of a vial. In one of the procedure, several seed crystals of $MAPbI_3$ were prepared via spontaneous crystallization. One superior seed crystal was chosen out of them and positioned at center of a tray. The system was heated to a temperature of 65°C and maintained as such. After some time, the temperature was gradually lowered to 40°C, and the tray containing the seeded crystal was simultaneously rotated with the help of an electric motor as shown in Figure 4.7(a). Thus, via BSSG method, $MAPbI_3$ single crystals were finally obtained with dimensions of 10 × 10 × 8 mm [39]. Likewise, $MAPbI_3$ single crystals were prepared by BSSG technique via a slightly different route. Here, the process adopted by Poglitsch and Weber [27] was utilized for the formation of small seed crystals of $MAPbI_3$. First, two separate solutions were prepared in hydroiodic acid, one using methylamine while other with lead(II) acetate trihydrate. Both the solutions were heated up to 100°C and maintained at that temperature for 2 hours and finally mixed. Afterward, the resulting solution was made to cool down gradually to 57°C within about 5 days. After 5 days, small $MAPbI_3$ crystals of approximately 3 mm were obtained. Before attaining a temperature of 47°C, the seed crystals were removed. Now, a separate system is filled with precursor solution and heated to 100°C with the help of a hot plate. Out of the obtained seed crystals, a good quality crystal was carefully chosen and inserted in the growth solution at the bottom using a platinum wire. Figure 4.7(b) depicts the schematic diagram of the adopted BSSG method. Because of the high

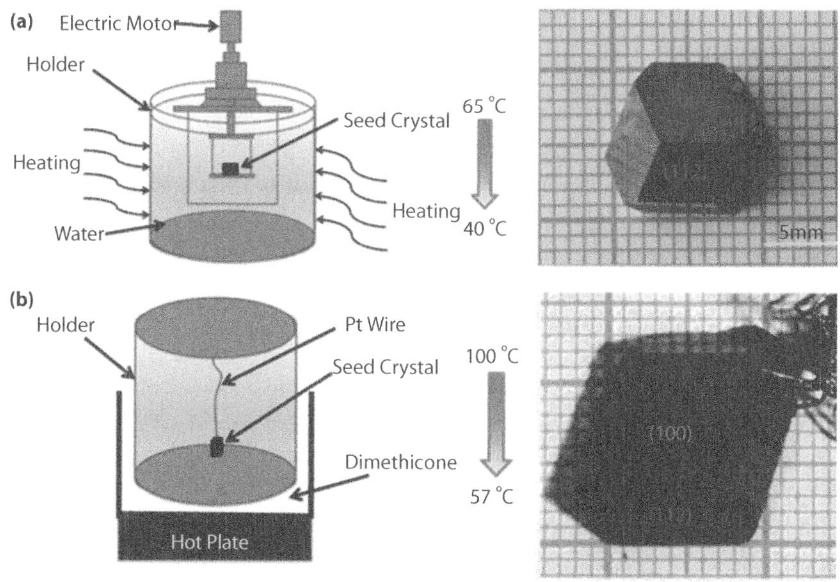

Figure 4.7 (a) Illustrative diagram of BSSG method along with the obtained $MAPbI_3$ single crystal. Here temperature was lowered from 65°C to 40°C; (b) Schematic of the BSSG process where temperature was lowered from 100°C to 57°C. The synthesized $MAPbI_3$ single crystal is also shown. Reproduced from [38] with permission. Copyright 2016, The Royal Society of Chemistry.

temperature of the growth solution, the outer surface of the crystal dissolved after 10 minutes. Afterward, the solution temperature was lowered from 100°C to 57°C within a time span of 15 days. Black $MAPbI_3$ single crystals were finally obtained and drawn from the system before attaining a temperature of 47°C. The obtained crystals were rhombic dodecahedral with dimensions 12 × 12 × 7 mm as shown in Figure 4.7(b) [35].

4.2.5 Bridgman (BG) Method

The Bridgman (BG) method involves the filling of the powdered material and seed inside a quartz ampoule or crucible. The crucible is sealed with inert gas or vacuum and made to move inside a furnace. A temperature gradient is maintained in the furnace, and the crucible is firstly moved into the hot part of the furnace. After sometime, the powder melts, and subsequently, the quartz ampoule is moved toward the cold part of the furnace. At this point of time, crystallization facade starts propagating through melted powder and large perovskite single crystals are obtained inside the sealed ampoules [40]. The illustrative diagram of BG technique is shown in Figure 4.8(a). The BG technique cannot be employed for the synthesis of hybrid perovskites because the organic compounds possess high vapor pressure and show chemical instability at their melting point. In contrast, BG method is successfully utilized for growing completely inorganic single crystal perovskites with a specific melt temperature.

By employing a solid-state reaction at an elevated temperature, $CsPbBr_3$ single crystal perovskites were synthesized by using BG method [41]. Two different samples of $PbBr_2$ were taken, the commercially available 5N grade (A) and the freshly prepared $PbBr_2$ (B). The process initiated by taking CsBr and $PbBr_2$ in equimolar ratio and properly grinding them. The contents were transferred to a fused silica crucible and the ampoule was sealed and evacuated at 10^{-4} mbar. The ampoule was then placed in the furnace and gradually heated up to 600°C for 6 hours. The system was then slowly cooled to RT resulting in the formation of transparent orange colored crystals of $CsPbBr_3$ of diameter 7 mm. Two specimens of $CsPbBr_3$ single crystals were prepared in a three-zone furnace: $CsPbBr_3$-1 using sample A and $CsPbBr_3$-2 using sample B of $PbBr_2$ and are shown in Figure 4.8(b). In another way of obtaining $CsPbBr_3$ crystals, CsBr and $PbBr_2$ were directly mixed in stoichiometric ratio at a very high temperature (>700°C) in a protective environment of Br and HBr. Thus, $CsPbBr_3$ crystals were obtained from the melt by employing the BG method [42].

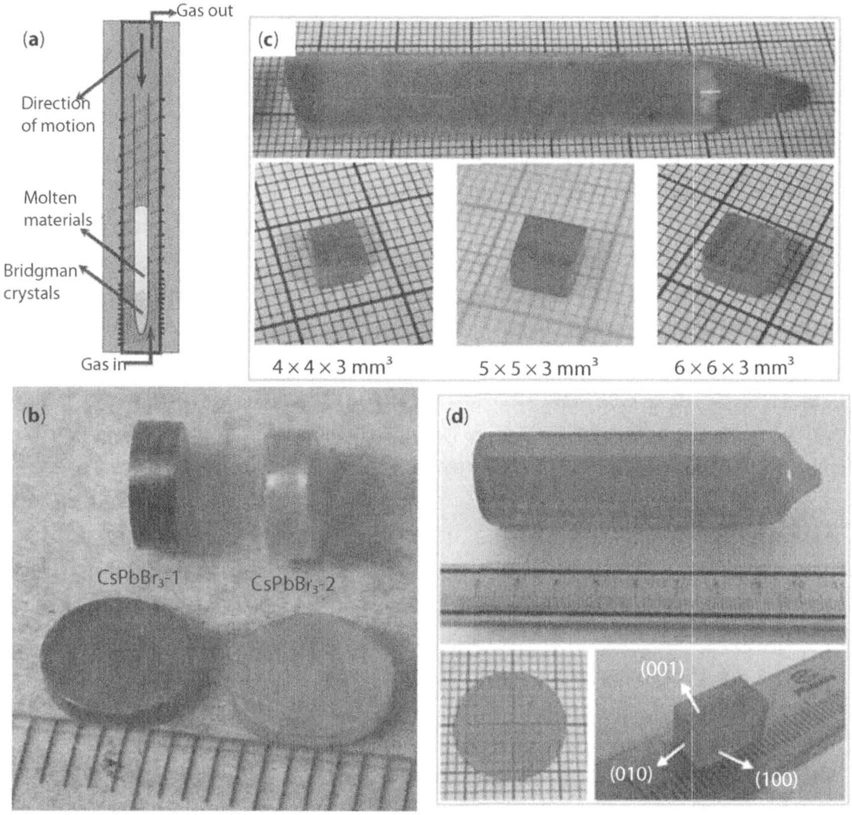

Figure 4.8 (a) Illustrative diagram of BG technique. Reproduced from [40] with permission. Copyright 2013, Materials Research Society. (b) Obtained specimens of $CsPbBr_3$-1 and $CsPbBr_3$-2. Reproduced from [41] with permission. Copyright 2013, American Chemical Society. (c) As-grown $CsPbBr_3$ ingot and the prepared single crystal wafers. Reproduced from [43] with permission. Copyright 2018, Nature Publishing group. (d) As-grown $CsPbBr_3$ ingot and the prepared polished wafer as well as oriented cuboid. Reproduced from [44] with permission. Copyright 2018, American Chemical Society.

Similarly, CsBr and $PbBr_2$ were directly mixed in a stoichiometric ratio of 1:1 at 580 °C to form polycrystalline $CsPbBr_3$. To avoid the chances of unintentional contamination in a multistep process, the synthesis was done in the same fused silica tube having an inner diameter of 11 mm and a conical shaped end. This was done to minimize any possible contamination. The whole growth process was performed in vacuum with an original pressure of $<10^{-4}$ Torr at RT. The silica tube was sealed with polycrystalline

$CsPbBr_3$ and positioned over a support rod. The support rod had a geared translation system that can move downward inside a two-zone vertical BG furnace. The silica tube was heated up to a temperature of 590°C through duration of 10 hours and kept at that temperature for a further 12 hours. This helped in preventing the heterogeneous nucleation of the crystals. It is important to mention that an overheating of the melt even 15 to 30°K helps in minimizing any superfluous nucleation centers. Further, an optimization of the temperature profile assists in preventing any thermally induced cracks. The silica tube was made to pass through the temperature gradient region at a low dropping speed and then the ingot was cooled down to 200°C during 20 hours. It was further cooled to RT at a rate of 2 to 5°K h^{-1} [43]. The desired $CsPbBr_3$ ingot with a diameter of 11 mm was formed and is shown in Figure 4.8(c). A large single crystal ingot of $CsPbBr_3$ of diameter 24 mm and length 90 mm was also prepared by using BG technique [44]. Here, a quartz crucible of inner diameter 24 mm was utilized. The polycrystalline $CsPbBr_3$ was firstly sealed inside the crucible and was then transferred to a two-zone BG furnace. The hot part of the furnace was maintained at a temperature of 600°C while the cold part was kept at a temperature of 200°C. The temperature gradient was nearly 30 to 40°C/cm. The ampoule was kept at high-temperature zone for 12 hours and then dropped slowly at a speed of 1 mm/h to the low-temperature zone. The furnace was made to cool down to RT within 3 days time span and finally large single crystal of $CsPbBr_3$ was obtained as shown in Figure 4.8(d).

The growth of $CsSnI_3$ single crystals via BG technique involves a stoichiometric mixture of CsI, SnI_2 and SnF_2. The mixture was made to heat in evacuated Pyrex at 450°C for 30 minutes. After cooling down to RT, large black colored $CsSnI_3$ perovskite single crystal was formed [45]. In BG method, the purification should be carried out separately before the growth of the crystal. This is because the chemical purification is not that much efficient because of the utilization of a crucible that is totally sealed [40]. The impurities possessing different solubility in the melt than that of the crystal get separated from the crystals during the movement of the crucible inside the furnace. However, for the impurities that have almost same solubility as that of the crystal, BG method is not efficient in removing them [46]. The temperature difference between the top and bottom of the quartz ampoule can be changed and therefore the interface curvature can be easily adjusted. Consequently, we can say that the instrument can be effortlessly operated, but as the grown single crystal touches the quartz wall, strain is produced at the boundary between grown crystal and the crucible wall. This can induce mechanical stresses, cracks, small-angle grain boundaries, and even the impurity level and defect density gets increased. It is, therefore, required for the semiconductors to certainly possess the prescribed characteristics that are impractical without consistent dopant segregation. This is the main limitation for the application of BG grown crystals in the fabrication of high quality devices [47]. Further, it is important to note that the parameters temperature gradient, dropping speed and cooling rate should not be very high as all these parameters in surplus would probably lead to the formation of cracks in the ingots.

4.3 Other Techniques

In addition to the above-described methods, there are other techniques also that have not been utilized much in practice and are described briefly in this section. These include: slow evaporation method, solvothermal growth method, layered solution growth method, and droplet-pinned crystallization method. All these techniques needed to be explored more in future.

By using the slow evaporation (SE) technique, it is possible to grow single crystals from air-stable samples in solution growth processes. It is an easy method where the solution undergoing evaporation can be prepared by mixing two or more reactants in an appropriate solvent. Subsequently, the resulting mixture is transferred to a small vial, which is provided with a perforated top. After sometime, desired crystals are obtained inside the vial following the evaporation of the solution. The illustrative diagram of SE technique is shown in Figure 4.9(a). The most favorable growth of the crystals in SE technique requires the growth of the crystals from only saturated solutions. This is due to the reason that a supersaturated solute would result in extremely small crystalline size of the grown crystals. Therefore, under the crystallization conditions, the solute should be moderately soluble. Moreover, the solutes and the solvent must be suitably chosen for the proper solubility. A variety of organic as well as inorganic single crystals have been prepared by SE technique. $(MA)_2Pb(SCN)_2I_2$ single crystals in which iodide ion was partly substituted by thiocyanide ion, were grown by SE method where initially $Pb(SCN)_2$ and MAI were dissolved in DMF solution in a ratio of 1:2. Afterward, the solution was left for evaporating slowly at RT. Consequently, black color crystals of $(MA)_2Pb(SCN)_2I_2$ were formed [48]. SE technique was also utilized for the growth of BA_2PbX_4 single crystals where BA is benzylammonium [49]. The dichlorides of benzylammonium and lead were first dissolved in conc. HCl in stoichiometric molar ratio. Thus, single crystals of BA_2PbX_4 were formed but they were extremely small. On the other hand, when the solution of the precursors in

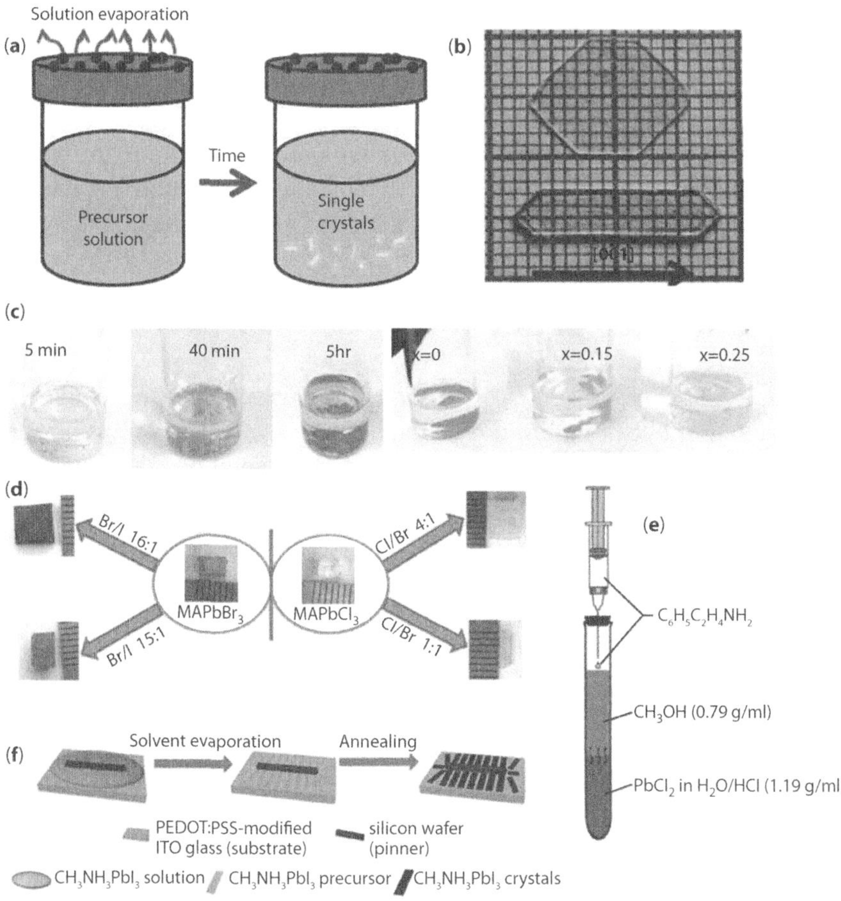

Figure 4.9 (a) Illustrative diagram of SE technique; (b) $(BA)_2PbCl_4$ single crystals. Reproduced from [38] with permission. Copyright 2016, The Royal Society of Chemistry. (c) (left) The three typical growth stages (5 min, 40 min, 5 h) of red single crystals in 35 wt% $MAPbBr_3$ DMF solution at 50°C. (right) Photos of the single crystals of $MAPb(Br_{1-x}Cl_x)_3$; x = 0, 0.15, 0.25 from left to right. Reproduced from [50] with permission. Copyright 2015, The Royal Society of Chemistry. (d) Images of obtained single crystal perovskites (single as well as mixed halide) with different halide ratios. Reproduced from [28] with permission. Copyright 2018, American Chemical Society. (e) Illustrative diagram of LSG technique for the growth of $(C_6H_5C_2H_4NH_3)_2PbCl_4$ plate single crystals. Reproduced from [52] with permission. Copyright 1999, Academic Press. (f) Illustrative diagram of DPC technique for the growth of $MAPbI_3$ single crystals. Reproduced from [53] with permission. Copyright 2016, The Royal Society of Chemistry.

DMF was kept to evaporate at a slow rate at a temperature of 90°C, transparent bulk crystals were formed with dimensions of 5 × 10 × 2 mm^3. The obtained BA_2PbX_4 crystals are shown in Figure 4.9(b) and were further investigated for their crystal structure and properties. The SE method is an effective strategy for synthesizing pure single crystal perovskites where the respective solvents exhibit low boiling points up to 100°C. However, owing to the reduced solubility of reactants of the perovskite precursor, this technique is not that efficient and is therefore less attempted. Hence, it can be said that this method is difficult to precisely control that would sternly restricts its applications.

The solvothermal growth (SG) method, as the name suggests, involves the heating of the precursors in a particular solvent. Three mixed halide single-crystal perovskites of $MAPb(Br_{1-x}Cl_x)_3$ (x = 0, 0.15, 0.25) with different Cl/Br ratios were prepared by SG method [50]. Here, the precursors $PbBr_2$, MABr, and MACl were first dissolved in different molar ratios in DMF solution. The resulting solution was then transferred to a sealed vial and subsequent to the heating at 50°C for up to 5 hours, the desired $MAPb(Br_{1-x}Cl_x)_3$ single crystals were formed as shown in Figure 4.9(c) with different Cl/Br ratios. Similarly, $MAPbBr_{3-x}Cl_x$ and $MAPbI_{3-x}Br_x$ single crystals were also prepared by SG process [51]. This time, Lead(II) acetate along with methylamine and haloid acids were initially mixed. After keeping the resulting solution at a temperature of 100°C, a super-saturated aqueous solution was formed. When the temperature of the precursor solution was slowly decreased, single crystals were observed to be precipitated at the bottom. Figure 4.9(d) shows the obtained single crystal perovskites with thickness of 0.5 to 5 mm. As can be seen from the figure, the color of the obtained single crystals varied from transparent to yellow and finally to orange with increasing Br/(Cl + Br) molar ratio from 0 to 1 in the perovskite $MAPbBr_{3-x}Cl_x$. However, in case of $MAPbI_{3-x}Br_x$, the color of the obtained single crystals varied from orange to red color with an increase in I/(I + Br) ratio.

The single crystal perovskites, in which inorganic and organic parts have incompatible solubility and generally demonstrate unsuitable crystal routine formerly grown by other techniques, are synthesized by layered solution growth (LSG) technique. This technique is suitable for the growth of those perovskites where the components of the system needed to be cooled down so as to avoid any decomposition or polymerization. Therefore, LSG technique is somewhat similar to the AVC technique with the difference that the slow diffusion process in LSG method is controlled by the difference in densities of the involved components. The densities of the components are different essentially due to the dissimilar solubility of the crystallizing material in different solutions. These different solutions form different layers in the LSG apparatus. $(C_6H_5C_2H_4NH_3)_2PbCl_4$ plate single crystals were prepared by utilizing LSG technique, and the illustrative diagram of the process is shown schematically in Figure 4.9(e). Initially, $PbCl_2$ was taken and dissolved in H_2O/HCl solution (1.19 g/ml). The prepared solution was then transferred to a tube for the crystal growth and this solution served as the bottom-most layer of the system. Over this bottom-most layer, methanol (0.791 g/ml) was poured in a dropwise manner. Over methanol, $C_6H_5C_2H_4NH_2$ solution in a stoichiometric ratio was consequently added and this was also performed in a dropwise manner with the help of a syringe. $C_6H_5C_2H_4NH_2$ solution, therefore, served as the top-most layer of the system. Now, because of the difference in densities of the bottom-most layer and the top-most layer, methanol acted as a quite sharp interface between the two layers. Therefore, the crystal growth process was very slow as a consequence of the sharp interface. The actual growth was observed after quite a lot of weeks. Subsequently, small colorless $(C_6H_5C_2H_4NH_3)_2PbCl_4$ plate single crystals were formed at the solution interface. These crystals were removed after one year and investigated for their properties. The obtained crystals demonstrated high quality and for that reason, this method is considered to be highly suitable and proficient for the growth of hybrid halide perovskites. But, this process can yield only small single crystals and it is not possible to synthesize large crystals by LSG technique. For the applications which require large single crystals, this technique cannot be employed. Further, LSG method is very time consuming that limits its applications.

For the growth of single crystals at a microlevel, droplet-pinned crystallization (DPC) method is used. This technique can also be utilized for the growth of single-crystalline perovskite thin films by following a conventional and proficient preparation procedure. The DPC technique has been suitably employed numerous times for the synthesis of aligned single crystals. However, it is an exigent task to fabricate single crystal perovskite based devices through this technique. This is attributed to the complexity in forming dense arrays of single crystals in a controlled alignment over the large-area substrates. This method was employed for the first time and last time by Ye *et al.* [53] for the synthesis of $MAPbI_3$ single crystal perovskites at a micro level. On a cleaned ITO coated glass, a layer of PEDOT/PSS was initially spin-coated. Afterward, precursor solution of $MAPbI_3$ was prepared by taking MAI and PbI_2 in the ratio of 1:1. Nearly, as small as 20 μL of this solution was taken and poured over the spin-coated PEDOT/PSS layer in a dropwise manner. A small silicon wafer of dimensions 0.2

cm ×1 cm was placed over $MAPbI_3$ solution so as to pin the droplet. The entire growth process of $MAPbI_3$ perovskite by DPC method is illustrated in Figure 4.9(f). The ITO with all these layers was then subjected to heating at a temperature of 80°C. Consequently, $MAPbI_3$ crystallize alongside droplet thinning direction after the evaporation of the solvent. Afterward, it was annealed at a temperature of 100°C for 3 minutes and finally arrays of $MAPbI_3$ single crystals were formed. But, the fabricated device showed very low performance. The device resulted in low open circuit voltage because of the large amount of leakage current caused by the direct contact of PCBM and PEDOT: PSS due to the imperfect coverage of the perovskite arrays on the PEDOT: PSS layer. Further, there was transmission of light from the crystal free regions as a consequence of incomplete coverage which resulted in decreased absorption. This was another main reason for the low performance of the device. It can also be attributed to the existence of defects on the surface of the crystal. The obtained low performance of the device might be the definite reason that no other researcher further explored this technique for the growth of perovskite single crystals.

Conclusions

In this chapter, a comprehensive description of different synthesis strategies of single crystal perovskite halides is provided. Some researchers have focused on the examination of crystal quality of the grown perovskites. However, their individual properties like grain boundaries, defects and purity, as well as their interdependence are vague. Consequently, further concern is required to be given to their synthesis strategies so as to minimize the defects and enhance the purity of the crystals. This would certainly assist in facilitating the deep perceptive of the basic characteristics and device physics of perovskites. Therefore, good quality single crystal perovskites will provide a means for the understanding of their fundamental characteristics. Moreover, a few synthesis techniques of single crystal perovskites are carried out in toxic solvents, like DMF, GBL, and DMSO, that can possibly penetrate into the human body and it can be venomous. Further, volatile solvents can diffuse into the atmosphere effortlessly and it might lead to global environmental pollution. This problem is not easy to resolve as volatile materials have poor solubility in less toxic solvents. Hence, it is considerably important to utilize environment-friendly solvents for synthesis strategies, and it is also necessary to grow more crystals in solid state reactions.

References

1. Ye, T., Wang, X., Li, X., Yan, A.Q., Ramakrishna, S., Xu, J., Ultra-high Seebeck coefficient and low thermal conductivity of a centimeter-sized perovskite single crystal acquired by a modified fast growth method. *J. Mater. Chem. C*, 5, 1255, 2017.
2. Stoumpos, C.C. and Kanatzidis, M.G., Halide Perovskites: poor Man's high-performance semiconductors. *Adv. Mater.*, 28, 5778, 2016.
3. Cheng, X., Jing, L., Zhao, Y., Du, S., Ding, J., Zhou, T., Crystal orientation-dependent optoelectronic properties of MAPbCl3 single crystals. *J. Mater. Chem. C*, 6, 1579, 2018.
4. Ding, J., Lian, Z., Li, Y., Wang, S., Yan, Q., The role of surface defects in photoluminescence and decay dynamics of high-quality perovskite MAPbI3 single crystals. *J. Phys. Chem. Lett.*, 9, 4221, 2018.
5. Chen, Y., He, M., Peng, J., Sun, Y., Liang, Z., Structure and growth control of organic–inorganic halide perovskites for optoelectronics: From polycrystalline films to single crystals. *Adv. Sci.*, 3, 1500392, 2016.
6. Murali, B., Kolli, H.K., Yin, J., Ketavath, R., Bakr, O.M., Mohammed, O.F., Single Crystals: The Next Big Wave of Perovskite Optoelectronics. *ACS Mater. Lett.*, 2, 184, 2019.
7. Wehrenfennig, C., Eperon, G.E., Johnston, M.B., Snaith, H.J., Herz, L.M., High charge carrier mobilities and lifetimes in organolead trihalide perovskites. *Adv. Mater.*, 26, 1584, 2014.
8. Liu, Y., Yang, Z., Liu, S., Recent Progress in Single-Crystalline Perovskite Research Including Crystal Preparation, Property Evaluation, and Applications. *Adv. Sci.*, 5, 1700471, 2018.
9. Kour, R., Arya, S., Verma, S., Gupta, J., Bandhoria, P., Bharti, V., Datt, R., Gupta, V., Potential substitutes for replacement of lead in perovskite solar cells: a review. *Global Challenges*, 3, 1900050, 2019.
10. Song, Y., Bi, W., Wang, A., Liu, X., Kang, Y., Dong, Q., Efficient lateral-structure perovskite single crystal solar cells with high operational stability. *Nat. Commun.*, 11, 1, 2020.
11. Shi, D., Adinolfi, V., Comin, R., Yuan, M., Alarousu, E., Buin, A., Chen, Y., Hoogland, S., Rothenberger, A., Katsiev, K., Losovyj, Y., Low trap-state density and long carrier diffusion in organolead trihalide perovskite single crystals. *Science*, 347, 519, 2015.
12. Yang, Y., Yan, Y., Yang, M., Choi, S., Zhu, K., Luther, J.M., Beard, M.C., Low surface recombination velocity in solution-grown CH 3 NH 3 PbBr 3 perovskite single crystal. *Nat. Commun.*, 6, 7961, 2015.
13. Wei, H., Fang, Y., Mulligan, P., Chuirazzi, W., Fang, H.H., Wang, C., Ecker, B.R., Gao, Y., Loi, M.A., Cao, L., Huang, J., Sensitive X-ray detectors made of methylammonium lead tribromide perovskite single crystals. *Nat. Photonics*, 10, 333, 2016.

14. Zhou, H., Nie, Z., Yin, J., Sun, Y., Zhuo, H., Wang, D., Li, D., Dou, J., Zhang, X., Ma, T., Antisolvent diffusion-induced growth, equilibrium behaviours in aqueous solution and optical properties of CH 3 NH 3 PbI 3 single crystals for photovoltaic applications. *RSC Adv.*, 5, 85344, 2015.
15. Zuo, C. and Ding, L., Lead-free Perovskite Materials (NH4) 3Sb2IxBr9– x. *Angew. Chem. Int. Ed.*, 56, 6528, 2017.
16. Liu, F., Wang, F., Hansen, K.R., Zhu, X.Y., Bimodal Bandgaps in Mixed Cesium Methylammonium Lead Bromide Perovskite Single Crystals. *J. Phys. Chem. C*, 123, 14865, 2019.
17. Saidaminov, M.I., Adinolfi, V., Comin, R., Abdelhady, A.L., Peng, W., Dursun, I., Yuan, M., Hoogland, S., Sargent, E.H., Bakr, O.M., Planar-integrated single-crystalline perovskite photodetectors. *Nat. Commun.*, 6, 1, 2015.
18. Liu, Y., Yang, Z., Cui, D., Ren, X., Sun, J., Liu, X., Zhang, J., Wei, Q., Fan, H., Yu, F., Zhang, X., Two-inch-sized perovskite CH3NH3PbX3 (X= Cl, Br, I) crystals: growth and characterization. *Adv. Mater.*, 27, 5176, 2015.
19. Saidaminov, M.I., Abdelhady, A.L., Murali, B., Alarousu, E., Burlakov, V.M., Peng, W., Dursun, I., Wang, L., He, Y., Maculan, G., Goriely, A., High-quality bulk hybrid perovskite single crystals within minutes by inverse temperature crystallization. *Nat. Commun.*, 6, 1, 2015.
20. Kadro, J.M., Nonomura, K., Gachet, D., Grätzel, M., Hagfeldt, A., Facile route to freestanding CH 3 NH 3 PbI 3 crystals using inverse solubility. *Sci. Rep.*, 5, 11654, 2015.
21. Saidaminov, M.I., Abdelhady, A.L., Maculan, G., Bakr, O.M., Retrograde solubility of formamidinium and methylammonium lead halide perovskites enabling rapid single crystal growth. *Chem. Commun.*, 51, 17658, 2015.
22. Zhumekenov, A.A., Burlakov, V.M., Saidaminov, M.I., Alofi, A., Haque, M.A., Turedi, B., Davaasuren, B., Dursun, I., Cho, N., El-Zohry, A.M., Bastiani, M.D., The role of surface tension in the crystallization of metal halide perovskites. *ACS Energy Lett.*, 2, 1782, 2017.
23. Dirin, D.N., Cherniukh, I., Yakunin, S., Shynkarenko, Y., Kovalenko, M.V., Solution-grown CsPbBr3 perovskite single crystals for photon detection. *Chem. Mater.*, 28, 8470, 2016.
24. Chen, L., Tan, Y.Y., Chen, Z.X., Wang, T., Hu, S., Nan, Z.A., Xie, L.Q., Hui, Y., Huang, J.X., Zhan, C., Wang, S.H., Toward Long-Term Stability: Single-Crystal Alloys of Cesium-Containing Mixed Cation and Mixed Halide Perovskite. *J. Am. Chem. Soc*, 141, 1665, 2019.
25. Kawachi, S., Atsumi, M., Saito, N., Ohashi, N., Murakami, Y., Yamaura, J.I., Cs-Mixed Lead Halides., J., Phys. *Chem. Lett.*, 10, 6967, 2019.
26. Liu, Y., Sun, J., Yang, Z., Yang, D., Ren, X., Xu, H., Yang, Z., Liu, S., 20-mm-Large Single-Crystalline Formamidinium-Perovskite Wafer for Mass Production of Integrated Photodetectors. *Adv. Optical Mater.*, 4, 1829, 2016.
27. Babu, R., Giribabu, L., Singh, S.P., Recent Advances in Halide-Based Perovskite Crystals and Their Optoelectronic Applications. *Cryst. Growth Des.*, 18, 2645, 2018.

28. Baikie, T., Fang, Y., Kadro, J.M., Schreyer, M., Wei, F., Mhaisalkar, S.G., Graetzel, M., White, T.J., Synthesis and crystal chemistry of the hybrid perovskite (CH 3 NH 3) PbI 3 for solid-state sensitised solar cell applications. *J. Mater. Chem. A*, 1, 5628, 2013.
29. Rao, H.S., Chen, B.X., Wang, X.D., Kuang, D.B., Su, C.Y., A micron-scale laminar MAPbBr 3 single crystal for an efficient and stable perovskite solar cell. *Chem. Commun.*, 53, 5163, 2017.
30. Yang, D., Xie, C., Sun, J., Zhu, H., Xu, X., You, P., Lau, S.P., Yan, F., Yu, S.F., Amplified spontaneous emission from organic–inorganic hybrid lead iodide perovskite single crystals under direct multiphoton excitation. *Adv. Opt. Mater.*, 4, 1053, 2016.
31. Han, Q., Bae, S.H., Sun, P., Hsieh, Y.T., Yang, Y., Rim, Y.S., Zhao, H., Chen, Q., Shi, W., Li, G., Yang, Y., Single crystal formamidinium lead iodide (FAPbI3): insight into the structural, optical, and electrical properties. *Adv. Mater.*, 28, 2253, 2016.
32. Saidaminov, M.I., Haque, M.A., Almutlaq, J., Sarmah, S., Miao, X.H., Begum, R., Zhumekenov, A.A., Dursun, I., Cho, N., Murali, B., Mohammed, O.F., Inorganic lead halide perovskite single crystals: phase-selective low-temperature growth, carrier transport properties, and self-powered photodetection. *Adv. Opt. Mater.*, 5, 1600704, 2017.
33. Rakita, Y., Kedem, N., Gupta, S., Sadhanala, A., Kalchenko, V., Kulbak, M., Friend, R.H., Cahen, D., Hodes, G., Low-temperature solution-grown CsPbBr3 single crystals and their characterization. *Cryst. Growth Des.*, 16, 5717, 2016.
34. Dang, Y., Zhou, Y., Liu, X., Ju, D., Xia, S., Xia, H., Tao, X., Formation of hybrid perovskite tin iodide single crystals by top-seeded solution growth. *Angew. Chem. Int. Ed.*, 55, 3447, 2016.
35. Lian, Z., Yan, Q., Lv, Q., Wang, Y., Liu, L., Zhang, L., Pan, S., Li, Q., Wang, L., Sun, J.L., High-performance planar-type photodetector on (100) facet of MAPbI 3 single crystal. *Sci. Rep.*, 5, 16563, 2015.
36. Lian, Z., Yan, Q., Gao, T., Ding, J., Lv, Q., Ning, C., Li, Q., Sun, J.L., Perovskite CH3NH3PbI3 (Cl) single crystals: rapid solution growth, unparalleled crystalline quality, and low trap density toward 108 cm−3. *J. Am. Chem. Soc*, 138, 9409, 2016.
37. Dong, Q., Fang, Y., Shao, Y., Mulligan, P., Qiu, J., Cao, L., Huang, J., Electron-hole diffusion lengths> 175 μm in solution-grown CH3NH3PbI3 single crystals. *Science*, 347, 967, 2015.
38. Dang, Y., Ju, D., Wang, L., Tao, X., Recent progress in the synthesis of hybrid halide perovskite single crystals. *CrystEngComm*, 18, 4476, 2016.
39. Dang, Y., Liu, Y., Sun, Y., Yuan, D., Liu, X., Lu, W., Liu, G., Xia, H., Tao, X., Bulk crystal growth of hybrid perovskite material CH 3 NH 3 PbI 3. *CrystEngComm*, 17, 665, 2015.
40. Jiang, H. and Kloc, C., Single-crystal growth of organic semiconductors. *MRS Bull.*, 38, 28, 2013.

41. Stoumpos, C.C., Malliakas, C.D., Peters, J.A., Liu, Z., Sebastian, M., Im, J., Chasapis, T.C., Wibowo, A.C., Chung, D.Y., Freeman, A.J., Wessels, B.W., Crystal growth of the perovskite semiconductor CsPbBr3: a new material for high-energy radiation detection. *Cryst. Growth Des.*, 13, 2722, 2013.
42. Rodová, M., Brožek, J., Knížek, K., Nitsch, K., Phase transitions in ternary caesium lead bromide. *J. Therm. Anal. Calorim.*, 71, 667, 2003.
43. He, Y., Matei, L., Jung, H.J., McCall, K.M., Chen, M., Stoumpos, C.C., Liu, Z., Peters, J.A., Chung, D.Y., Wessels, B.W., Wasielewski, M.R., High spectral resolution of gamma-rays at room temperature by perovskite CsPbBr3 single crystals. *Nat. Commun.*, 9, 1, 2018.
44. Zhang, P., Zhang, G., Liu, L., Ju, D., Zhang, L., Cheng, K., Tao, X., Anisotropic optoelectronic properties of melt-grown bulk CsPbBr3 single crystal. *J. Phys. Chem. Lett.*, 9, 5040, 2018.
45. Chung, I., Lee, B., He, J., Chang, R.P., Kanatzidis, M.G., All-solid-state dye-sensitized solar cells with high efficiency. *Nature*, 485, 486, 2012.
46. Brissaud, M., Dolin, C., Le Duigou, J., McArdle, B.S., Sherwood, J.N., The purification and growth of large ultra-pure "plastic" single crystals of trimethylacetic acid. *J. Cryst. Growth*, 38, 134, 1977.
47. Braescu, L., Shape of menisci in terrestrial dewetted Bridgman growth. *J. Colloid Interface Sci.*, 319, 309, 2008.
48. Daub, M. and Hillebrecht, H., Synthesis, Single-Crystal Structure and Characterization of (CH3NH3) 2Pb (SCN) 2I2. *Angew. Chem. Int. Ed.*, 54, 11016, 2015.
49. Liao, W.Q., Zhang, Y., Hu, C.L., Mao, J.G., Ye, H.Y., Li, P.F., Huang, S.D., Xiong, R.G., A lead-halide perovskite molecular ferroelectric semiconductor. *Nat. Commun.*, 6, 1, 2015.
50. Zhang, T., Yang, M., Benson, E.E., Li, Z., van de Lagemaat, J., Luther, J.M., Yan, Y., Zhu, K., Zhao, Y., A facile solvothermal growth of single crystal mixed halide perovskite CH 3 NH 3 Pb (Br 1− x Cl x) 3. *Chem. Commun.*, 51, 7820, 2015.
51. Fang, Y., Dong, Q., Shao, Y., Yuan, Y., Huang, J., Highly narrowband perovskite single-crystal photodetectors enabled by surface-charge recombination. *Nat. Photonics*, 9, 679, 2015.
52. Mitzi, D.B., A layered solution crystal growth technique and the crystal structure of (C6H5C2H4NH3) 2PbCl4. *J. Solid State Chem.*, 145, 694, 1999.
53. Ye, T., Fu, W., Wu, J., Yu, Z., Jin, X., Chen, H., Li, H., Single-crystalline lead halide perovskite arrays for solar cells. *J. Mater. Chem. A*, 4, 1214, 2016.
54. Arya, S., Mahajan, P., Gupta, R., Tailor, N.K., Datt, R., Srivastava, R., Satapathi, S., Radhakrishnan, S., Gupta, V., A comprehensive review of synthesis and applications of single crystal perovskite Halides. *Prog. Solid State Chem.* 60, 100286, 2020.

5

Synchrotron-Based Techniques for Analysis of Perovskite Solar Cells

Umar Farooq[1†], Ruby Phul[2†], Mohd Shabbir[3], Rizwan Arif[4] and Akrema[5*]

[1]*Department of Chemistry, Galgotias University, Greater Noida, India*
[2]*Department of Electrical-Electronics Engineering, Abdullah Gul University, Kayseri, Turkey*
[3]*Department of Applied Science and Humanities, GL Bajaj Institute of Technology & Management, Greater Noida, India*
[4]*Department of Chemistry, School of Basic and Applied Sciences, Lingaya's Vidyapeeth, Faridabad, India*
[5]*Department of Chemistry, Faculty of Natural Sciences, Jamia Millia Islamia, New Delhi, India*

Abstract

Perovskite solar cells (PSCs) have been proven as promising material, owing to their remarkable performance attained in the field of photovoltaic (PV) and optoelectronics. These materials possess direct band gaps, high absorption coefficient, low exciton binding energy, and long carrier diffusion lengths. However, the primary electronic structure, size and spatial distribution, and other significant properties of perovskite materials are not fully understood due to a lack of precise and high-spatiotemporal resolution characterization techniques. A synchrotron provides high-brilliance X-ray beams that can easily penetrate deeper into the matter to explore the material's properties at the atomic or molecular level within a short time. Herein, we discuss synchrotron techniques for determining the structure and properties of perovskite materials within the bulk, at the surfaces and the interfaces. This chapter is dedicated to providing the latest findings on developing synchrotron-based techniques tailored for PSC. The method of characterization is also discussed with sample preparation. Finally, we focused on the state-of-the-art strategies for gaining in-depth knowledge of mechanism

**Corresponding author*: akrema007@gmail.com
†*Contributed equally*

Khursheed Ahmad and Waseem Raza (eds.) Perovskite Materials for Energy and Environmental Applications, (123–146) © 2022 Scrivener Publishing LLC

enhancing the photovoltaic performance and providing decisive answers to the outstanding science problems in the perovskite field, pushing forward technological development.

***Keywords*:** Perovskite, solar cells, synchrotron radiation, x-rays, optoelectronics, operando measurement

5.1 Introduction

The demand for energy increases directly with an increase in population. Currently, most of the energy demands are fulfilled by fossil fuels [1]. However, certain limitations (like nonrenewable nature, emission of toxic gases, etc.) are associated with the usage of fossil fuels for the sustainable future of the planet. Therefore, to have a sustainable energy source, different alternative environment, benign pathways of generating energy are being explored [2, 3]. Perovskite solar cells (PSCs) development is considered a vital breakthrough to achieve the planet's energy sustainability among these pathways. However, worldwide, only 2% of energy is produced by using solar cells. To achieve the goals of the Paris Agreement—2015, more solar energy is being utilized to keep global warming below 2° [4]. Therefore, it is crucially essential to globally accelerate the development and installation of more photovoltaic solar cells. For more successful PV cells utilization, the main task is to reduce PV energy's levelized cost.

Because of high efficiency, the ideal bandgap (1.12 eV), low manufacturing cost, silicon wafer-based solar cells are considered the most significant photovoltaic technology [5–7]. Despite the achieved record efficiency of 26.3% for Si wafer-based solar cells, tremendous efforts are being made to achieve the theoretical efficiency of 29%, which remains a significant challenge. A necessary fabrication approach of tandem solar cells (TSCs) has been developed to obtain efficiency beyond this limit. In TSCs, several layers of different cells are being sandwiched in a multijunction configuration. The top layer consists of cells developed from wide bandgap material absorbing in the visible range of the solar spectrum.

In contrast, the bottom layer consists of a cells made of narrow bandgap material absorbing in an infrared region [8]. For the bottom layer low bandgap, low processing/production cost and state-of-the-art efficiency make crystalline Si cells a most favorable choice. In search of a solution-processible and cheap alternative for Si wafer technology, different inorganic and organic materials have been developed.

Among these materials, perovskites act as an essential alternative to the light harvester in solar cells [9–11]. Perovskites have been named after the mineralogist L.A. Perovski. These are significant materials with the general chemical formula ABX_3 (X could be oxygen or any halogen). Simultaneously, "A" and "B" represent cations occupying different sites in the perovskite structure. Cation A is usually large and occupies Cubo octahedral sites and is shared with 12 X anions, whereas, cation B is smaller in size and occupies an octahedral site connected with six X anions. Similarly, ABX_3 is the general formula of inorganic-organic perovskites, where A site is composed of methylammonium iodide (MAI), formamidinium iodide (FAI), Cs, or their mixture, site B is occupied by divalent cations like Pb, Sn, or their mixture and I and/or Br (halides) at X site. Figure 5.1 represents the typical structure of perovskites.

Because of their excellent electrical properties (ferroelectricity and superconductivity), oxide perovskites are the most studied perovskites. However, halide-based perovskites have gained more attention with the development of layered organometallic halide perovskite materials. It was observed that with an increase in dimensionality, the layered organometallic halide perovskite materials show semiconductor to metal transition as well as their bandgap reduces on change in dimensionality from 2D to 3D [12]. This decrease in the bandgap of the 3D halide perovskites is very beneficial for solar cell applications. Inorganic-organic metal halides are considered as auspicious absorbing materials for next-generation solar cells. These materials have attained the certified highest single-junction photoconversion efficiency (PCE) of 25.2% and satisfy the requirements to act as perfect partners for tandem solar cells. These requirements include (a) broad range solar absorption spectrum, (b) tuneable bandgap and band structure, (c) ease of processing, (d) high PCE and compatibility with Si [4, 13, 14]. Because of the possibility of tunning the bandgap, the perovskites can be combined into all perovskite-based tandem solar cells with top and bottom

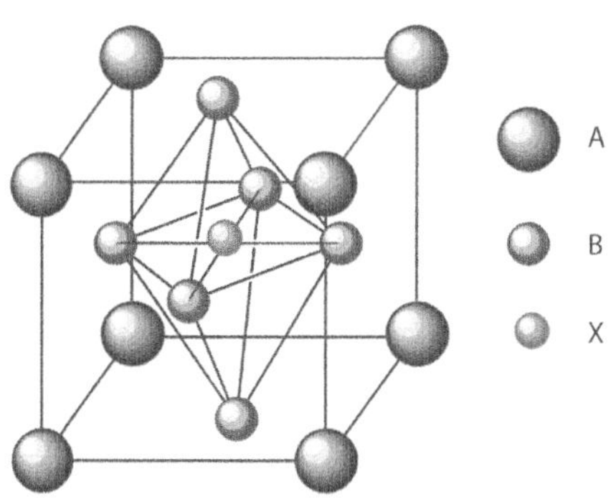

Figure 5.1 Typical lattice structure of a perovskite.

layers of tandem solar cells composed of perovskite absorber. However, apart from several advantages, certain disadvantages are associated with the PSC, particularly in inorganic-organic hybrid solar cells. These disadvantages include the use of lead and less stability under ambient conditions. Also, there is a reduction in their power output under illumination for planer perovskite solar cell due to slow ionic drift, which acts as a bottleneck for their practical application. Therefore, it is imperative for researchers to achieve high efficiency, long-term stability, and fabrication methods to develop large-area PSC. Under accelerated testing conditions, PSCs have retained only a few thousand hours equal to or less than one year of typical operation. For commercial application, the operation limit of PSCs should be equal to 20 years. Therefore, to achieve long-term stability, scalability use of perovskite materials' accurate characterization techniques is needed.

Extensive characterization techniques have been employed to understand the factors responsible for high efficiency, degradation and low scalability of

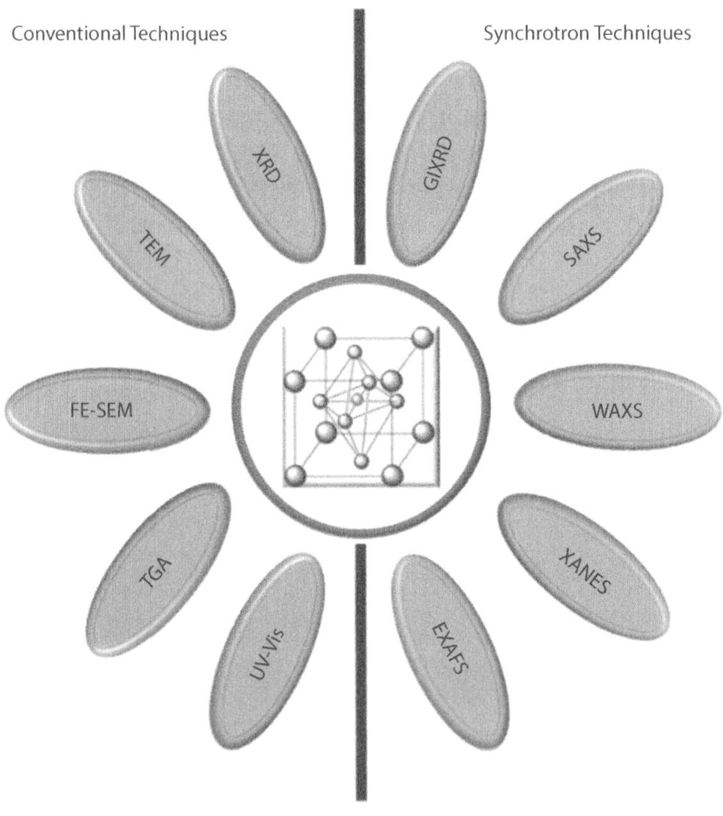

Figure 5.2 Schematic illustration of characterization techniques for PSCs.

perovskite materials. Figure 5.2 summarizes the characterization techniques used to evaluate the chemical composition, morphology, structural, and optoelectronic properties of perovskites. Electronic band structure plays an essential role in controlling the properties of photovoltaic materials. Therefore, it is vital to elucidate the band structure of the perovskite materials used in photovoltaic solar cells. Different characterization techniques like UV-visible spectroscopy, ultraviolet photoelectron spectroscopy, X-ray photoelectron spectroscopy (XPS), kelvin probe forced microscopy, and X-ray fluorescence has been employed to analyze the chemical composition and band structure of perovskite materials [15–19]. For morphological analysis, electron microscopy is being used, while as for crystallographic characterization, X-ray diffraction is used. Besides, atomic force microscopy is also being used to determine the quality of perovskite-based thin films. However, using these characterization, techniques using conventional X-ray sources do not provide in-depth chemical and structural information.

There are many inconsistencies in halide perovskite results, making it difficult to have an actual relationship between composition-microstructure properties and performance. These inconsistencies are attributed to not having decisive, nondestructive characterization techniques to analyze the dynamics and structure of halide perovskites and small-scale devices. However, synchrotron-based characterization techniques have provided new opportunities to thoroughly investigate the microstructure of the perovskite materials, which provides new opportunities for research with a vast avenue for discoveries and advances. Concerning the characterization of halide perovskite materials and their extended family members, improved and matured synchrotron X-ray techniques have been already used to quantify the crystal structures, electronic energy landscape, and chemical composition. Synchrotron-based techniques provide unique opportunities to determine the much more complex crystal structures and symmetries of halide perovskites and their derivatives, such as 2D layered materials. For example, the *in situ* X-ray technique, grazing incidence wide-angle X-ray scattering (GIWAXS) provides an avenue for determining transient processes and metastable intermediates formed in halide perovskites on crystallization, transformation and degradation process [20, 21]. In addition to more profound penetration power and wide energy tunability of X-rays the state-of-art synchrotron techniques have much lasting improved brightness that could be useful in converting many conventional scatterings and spectroscopic techniques into spatial imaging modalities that helps in having multidimensional mapping with a high spatiotemporal resolution for capturing the events of interest occurring in materials and devices under investigation. This chapter aims to provide an overview of different layouts where synchrotron-based X-ray characterization techniques furnish the advantage in acquiring the detailed mechanism of PSCs with their structural analysis.

5.2 Synchrotron Techniques, Their Limitations and Advantages

The synchrotron-based X-ray characterization technology has been proved as a potent tool for PSC research. In a synchrotron facility, charged particles *viz.* electrons are accelerated to the energy ~1.5 to 8 GeV (close to the speed of light) and filled into a storage ring to hold at fixed energy. When these high-energy electrons are accelerated/bent (by magnetic devices) around the ring, electromagnetic radiation is emitted from infrared to X-rays. This highly intense and focused radiation originates the energies from 0.1 keV (λ = 12.4 nm; soft X-rays) to 100 keV (λ= 0.0124 nm; hard X-rays) that is well suited for the investigation at 1 to 100 nm scale [22, 23]. The advantage of using synchrotron-based X-ray techniques is their almost nondestructive nature and option to perform *in/ex situ* or operando experiments. A considerable number of experiments could be done in real-time to demonstrate the structural changes, properties and mechanism of processes [24, 25]. Besides these advantages, the utmost limit of a synchrotron is its availability and access. The synchrotron-based X-ray characterization techniques have been applied to acquire the multidirectional information about the crystallization and degradation mechanism of perovskite materials, the interface regulation behaviour of thin films, and the structural activity of the PSC. The synchrotron radiation X-ray technology has also been used to affirm the multidimensional spatial structure and electronic energy-state structure of materials viz. synchrotron radiation X-ray diffraction/scattering (SR-XRD, SAXS, WAXS), and X-ray absorption spectroscopy (XAS, XANES, EXAFS), and so on.

5.3 Synchrotron Radiation X-Ray Diffraction/Scattering (SR-XRD)

The XRD is the most widely used technique for identifying the phase purity and crystal structure at atomic resolution. The XRD works on the basic principle of Bragg's equation:

$$n\lambda = 2d_{hkl} \sin \theta$$

where λ is the wavelength of the incident X-ray, n is an integer, d_{hkl} is the lattice planes' spacing with the Miller indices *h, k, l*, and θ is the incident angle (the angle between the incident ray and the scattering plane).

The working principle of XRD and SR-XRD is the same, but the difference is in the generation of X-rays. In XRD, the X-rays are generated from the electrons hitting a metal anode such as Co, Cu, Mo with fixed energy for a particular metal. While in SR-XRD, X-rays furnishes a broad selection of energy range and are much more intense. The high intensity of these X-rays provides high-quality diffraction data with higher resolution. However, these high energy X-rays possess a very high penetrating power; therefore, they are suitable for different samples. Also, collimated X-rays from synchrotron can achieve micron to a few hundred-nanometer spatial resolutions, allowing them to map the crystal structure distribution on a surface. Further, SR-XRD is categorized as angle dispersive (ADXRD) and energy-dispersive XRD (EDXRD) [26]. The experiments are carried out in the transmission mode in both these diffraction techniques, requiring high energy X-rays and an extremely thin sample.

For an ADXRD, the source is a monochromatic beam, and the diffraction pattern is acquired from all the solid angles. However, EDXRD uses polychromatic photons as the incident source (SR X-rays), and the detector is placed at a fixed angle. The EDXRD technique gives in-depth information about the change in the crystal structure, strain profiling and phase change of the material/sample under extreme conditions. Energy-dispersive XRD is an ideal technique for unstable materials under XRD as the data is collected at a fast rate and gathers information about the structural changes during *in situ* chemical reactions. Nevertheless, the grazing incidence XRD (GIXRD) technique is used for the thick samples and/or thin films. Unlike conventional XRD, in GIXRD the incident beam and detector are located on the same side of the sample. The conventional XRD of thin films produces week diffraction signals from the films and intense signal from the substrate. To overcome this issue, the diffraction pattern is recorded at a fixed grazing angle of diffraction, which is slightly above the critical angle for total reflection of the material. One can alter the X-ray's penetration depth from 1 nm to 100 nm by changing the incident angle [27].

Yang *et al.* [28] reported the preparation of $CH_3NH_3PbI_{3-x}Cl_x$ perovskite films by deposition technique followed by rapid and gradient thermal annealing. The GIXRD spectra (Figure 5.4) recorded at three different incident angles 1.00, 0.40, and 0.20° (Figures 5.3a, b, and c) with the probe depth of ~60, ~400, and ~3000 nm, respectively revealed that the perovskite films prepared by rapid thermal annealing (r-perovskite film) consist of dual-phase with a primary phase dominant in the interior and secondary phase with a smaller (110) interplanar space (d(110)). The composition of the secondary phase increases with a decrease in depth, so it is

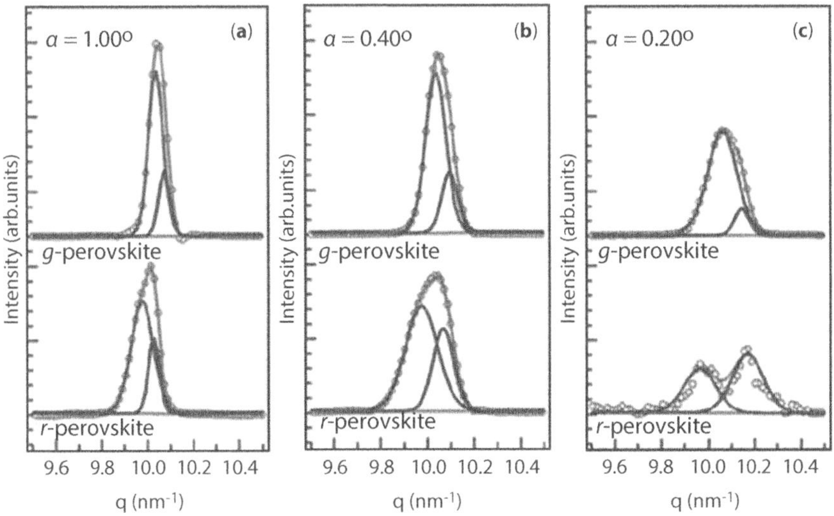

Figure 5.3 GIXRD spectra (open cycle) around the (110) peaks for the r-perovskite and g-perovskite thin films collected at three different X-ray incident angles of (a) 1.00°, (b) 0.40°, and (c) 0.20°. The purple and blue curves are two Gaussian peaks used to fit the experimental data. Reprinted with permission from [28]. Copyright (2017) American Chemical Society.

present majorly on the surface, resulting in the hysteresis in PSC's. Further, the perovskite films prepared by gradient thermal annealing (g-perovskite film) showed a uniformly distributed phase structure in pin-hole free morphology with less uncoordinated Pb and I ions, negligible hysteresis, and enhanced PSC efficiency. The authors concluded that the crystalline phase purity of the perovskite has an important effect on the J-V hysteresis effect of the device.

One case study utilized the synchrotron-based X-ray diffraction technique to study the crystallization and morphology of perovskite film grown on graphene oxide (GO). In this work by Wu *et al.* [29], GO used as a hole conductor and $CH_3NH_3PbI_{3-x}Cl_x$ as an absorber in inverted planar heterojunction (PHJ) perovskite solar cell, attains 12% power conversion efficiency (PCE). Figures 5.4a and b show the 2D GIXRD profiles of mixed halide perovskite films on GO and PEDOT/PSS. The sample shows a series of extremely similar diffraction profiles with spotty patterns, indicating that the highly textured perovskite crystal domains grew on both GO and PEDOT/PSS layer. Further, Figure 5.4c reveals that the radially integrating intensity along the ring at q ~10 nm^{-1} assigned to the (110) plane of perovskite structure showed relatively sharper peaks at the azimuth angle of

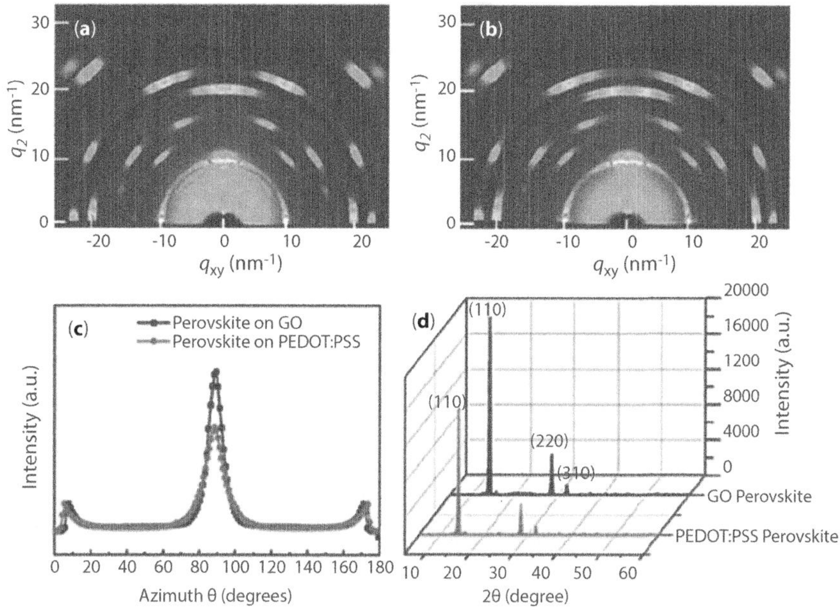

Figure 5.4 2D GIXRD profiles of perovskite films on (a) ITO/GO (~2 nm) and (b) ITO/PEDOT/PSS substrates. (c) Radially integrated intensity plots along the ring at q=10nm^{-1}, assigned to the (110) plane of $CH_3NH_3PbI_{3-x}Cl_x$ perovskite structure. (d) Conventional XRD spectra of perovskite films on quartz/GO (2 nm) and quartz/PEDOT/PSS substrates. Republished with permission of [Royal Society of Chemistry], from [29]; permission conveyed through Copyright Clearance Center, Inc.

90° for perovskite grown on GO, suggesting a higher preferential in-plane orientation than that of PEDOT/PSS. Figure 5.4d displays the conventional diffraction spectra of the $CH_3NH_3PbI_{3-x}Cl_x$ perovskite films grown on PEDOT/PSS and GO substrates which shows strong peaks at 14.1°, 28.5°, and 31.9° corresponds to (110), (220), and (310) planes, respectively, confirms orthorhombic crystal structure of $CH_3NH_3PbI_{3-x}Cl_x$ perovskite.

5.4 *In Situ* XRD

The *in situ* XRD has been employed to investigate the structural changes during a chemical reaction to get information about the mechanism of PSC's to a greater extent [30, 31]. The researchers have also focused on investigating pressure-induced structural changes and altering perovskite materials' physical and chemical properties under high pressure [32–35].

The degradation of $CH_3NH_3PbI_3$ thin films in a humid environment by combining the *in situ* XRD with microscopy and gravimetric analysis was investigated by Zhao *et al.* [31]. They proposed a plausible degradation mechanism depicting the formation of monohydrate species that agrees with Yang's results [30]. The diffraction patterns collected from *in situ* SR-XRD experiment, indicating the degradation of $CH_3NH_3PbI_3$/FTO with/without moisture, is shown in Figure 5.5. Figures 5.5a-c are *in situ* XRD patterns recorded during one test with a time sequence of a → b → c, and the corresponding integrated curves are presented in Figure 5.5e. Figure 5.5a represents the diffraction pattern of as-grown $CH_3NH_3PbI_3$/FTO film in a moisture-free atmosphere, shows that the film is polycrystalline with a preferred orientation along the normal direction. During the experiment, the moisture/humidity around the sample was increased with time. Figure 5.5b shows the gradual decrease in diffraction peak of perovskite due to degradation of $CH_3NH_3PbI_3$ film in the presences of humidity (RH=80 ± 5%), and the reflections corresponding to PbI_2 become more prominent than $CH_3NH_3PbI_3$. The final degradation product was primarily PbI_2 (Figure 5.5c), as there were no peaks from other phases or products. On repeating the *in situ* experiments, a metastable phase was observed during the degradation process (Figure 5.5d and its integral curve is plotted in Figure 5.5f).

Further, Stone *et al.* studied the phase transformation from precursor to perovskite at different annealing temperatures (95 to 105°C) via *in situ* XRD measurements [36]. Figure 5.6 represents the *in situ* XRD and

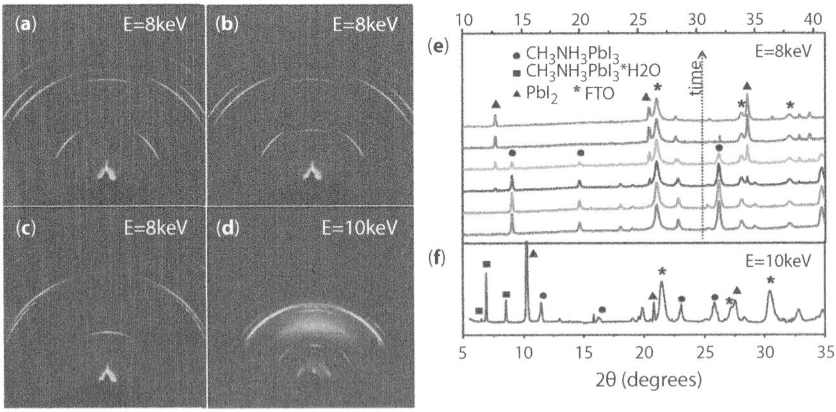

Figure 5.5 *In situ* XRD patterns. (a–d) Typical diffraction patterns of spin-coated $CH_3NH_3PbI_3$/FTO in degradation process and (e–f) the corresponding integral curves obtained by using Fit2D program. Adapted from Ref [31].

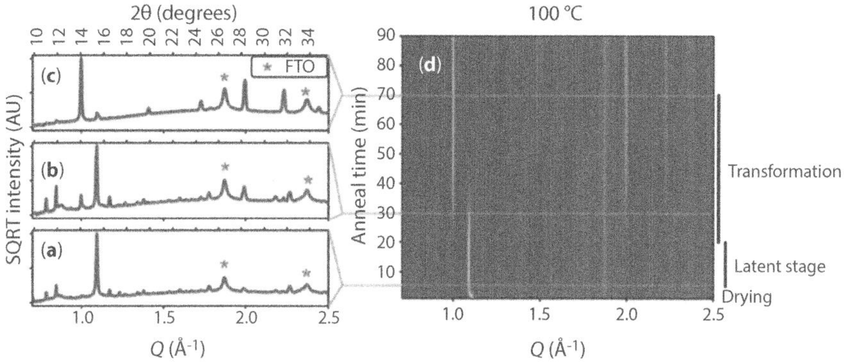

Figure 5.6 *In situ* X-ray diffraction of perovskite annealing at 100 °C. a–c XRD scans while annealing showing square root intensity to enhance small peaks. XRD scans shown are, (a) 5 minutes, the end of drying, (b) 30 minutes, during the transformation, (c) 70 minutes, near the end of the transformation, and (d) Waterfall composite plot: each horizontal slice corresponds to an XRD pattern at a certain time. Adapted from Ref [36].

waterfall composite plot of structural transformation from precursor to perovskite at 100°C measured for 5, 30, and 70 minutes. The first step (first 5 minutes) shows the crystallization of precursor phase (Figure 5.6a) as the film dries, in the second step (5 to 20 minutes), the precursor scattering remains constant with no reflection for perovskite (Figure 5.6b), and the last step (20 to 70 minutes) revealed the loss of precursor with the growth of perovskite phase (Figure 5.6c). The waterfall composite plot (Figure 5.6d) corresponds to each stage of perovskite formation from the precursor with time.

5.5 Small-Angle X-Ray Scattering

The structural characterization of a nanosystem, having typical dimensions from 1 nm to a micron, could be done with small-angle X-ray scattering (SAXS). So, the structure is determined at small angles or large distance. Small-angle X-ray scattering determines the size, shape, surface structure, relative positions of particles, and structural factor analysis. It is a non-destructive technique and provides structural data averaged over macroscopic sample volumes. Small-angle X-ray scattering does not require extensive sample preparation, as the scattering contrast is determined by the inherent electron density differences native to the sample under study.

A highly collimated beam of monochromatic X-rays is transmitted through a sample on the order of 1 mm in thickness. The scattered X-rays are collected on a 2D area detector azimuthally at 360° at a continuous range of scattering angles that deviate from the directly transmitted beam (Figure 5.7). The scattering angle is 2θ, defines a probe length (D).

$$D = 2\pi/q$$

where $q = 4\pi \sin(\theta)/\lambda$ and, λ is the wavelength of X-rays (0.154 nm for Cu X-ray source).

Schlipf *et al.* reported a two-step protocol for preparing $CH_3NH_3PbI_{3-x}Cl_3$ perovskite films where a PbI_2 precursor film was immersed in a mixed solution of CH_3NH_3I and CH_3NH_3Cl. They have analyzed both $CH_3NH_3PbI_{3-x}Cl_3$ and PbI_2 films via grazing-incidence small-angle X-ray scattering (GISAXS) to investigate the in-depth morphology of the films [37]. With the help of scattering results, they established a correlation between the crystal sizes inside the films and observed that the smaller crystals were accumulated within the bulk compared to the surface. Therefore, these results elaborate the crystallization process of perovskite films that is significant for controlled film growth and the high efficiency of PSC.

Recently, Pineda and coworkers have investigated nucleation processes and growth of mixed halide perovskite films via GISAXS and optical microscopic techniques. They have determined the influence of additives like 1,8-diiodooctane (DIO) and hydriodic acid on the *in situ* crystallization dynamics of mixed halide perovskite films [38]. The presence of additives significantly increases the nucleation rate, affects the morphology of the crystals, and promotes heterogeneous nucleation. However, the additives do not alter the crystalline phase or crystal packing of the perovskite film. A growth mechanism was suggested by observing the development of the size of the nuclei, and a model was proposed based on the radius of gyration, power exponent, and effective radius.

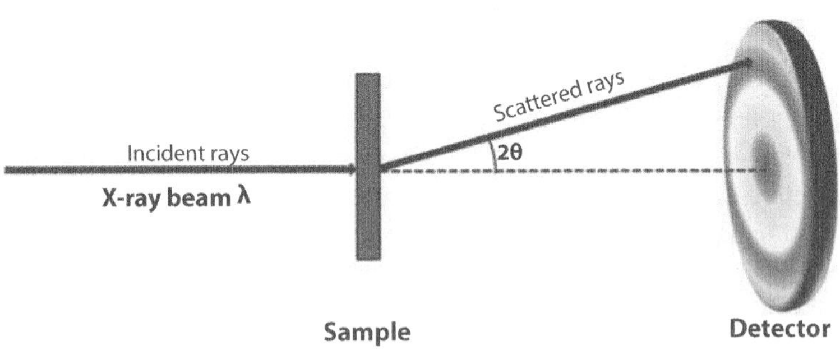

Figure 5.7 Schematic illustration of synchrotron-based X-ray scattering technique.

5.6 Wide-Angle X-Ray Scattering

The crystallographic structure, atomic position, chemical composition and chemical stoichiometric information of PSC's are determined by the wide-angle X-ray scattering (WAXS) technique. The crystallographic orientation plays a vital role in PSC device performance and stability, as it furnishes unique electronic properties. The WAXS measurements could be done by analyzing Bragg's peaks scattered to wide angles. The working principle of WAXS is similar to that of SAXS, with the only difference is that in WAXS, the distance from the sample to the detector is shorter and therefore, the diffraction peaks observes at larger angles.

Barrows and co-worker studied the formation of $CH_3NH_3PbI_{3-x}Cl_x$ perovskite from the precursor during thermal annealing using GIWAXS and GISAXS techniques. They have fabricated several photovoltaic devices by altering the annealing time for perovskite films, and the change in device performance was interpreted on account of X-ray scattering techniques [39]. Further, the synchrotron-based GIWAXS measurements were used to determine the crystal distortion and strain-driven texture formation in black $CsPbI_3$ thin films annealed at 330°C followed by cooling at room temperature [40]. Recently, Alhazmi *et al.* reported the crystallization dynamics of $MAPbI_{3-x}Cl_x$ and $FAPbI_{3-x}Cl_x$ perovskite films during spin-casting using an *in situ* WAXS technique and determined the differences between the crystallization pathways for these perovskite materials, resulting in the difference in the performance of devices fabricated by $MAPbI_{3-x}Cl_x$ and $FAPbI_{3-x}Cl_x$ perovskite films [41].

5.7 Synchrotron Radiation-Based X-Ray Absorption Techniques

The X-ray absorption spectroscopy (XAS) determines the absorption behaviour of the materials as a function of incident photon energy. It is a powerful technique that can characterize all forms of matter (crystalline or amorphous). For the materials with long-range structural order, diffraction-based characterization techniques could obtain reliable structural information. On the contrary, XAS could investigate the structure of disordered solids, liquids, and amorphous materials. One of the main advantages of XAS is its atomic selectivity, through which the local structure of each constituent of a sample could be studied. The XAS could be deduced by measuring the absorption coefficient of the material on interaction with incident X-rays using Beer's Law;

$$I_t = I_0 e^{-\mu l}$$

where I_0 is the intensity of incident photons, I_t is the intensity of the transmitted photons, l is sample thickness, and μ is the absorption coefficient, which depends on (i) type of atoms in the material, (ii) their distributed and bonding environment, and (iii) polarization and energy of the X-rays. Experimentally, it is determined by comparing the intensities of incident light and the ones passing the material with known thickness.

In an X-ray absorption process, a core electron is excited to an empty state on the absorption of the incident photon, and the energy range is the X-ray absorption edge itself. Figure 5.8 depict the different spectral regions of X-ray absorption process. The spectral features within the energy range of 50 to 100 eV, above the edge, are known as X-ray absorption near edge structure (XANES). The XANES structure can also be simulated with the multiple scattering formalism. The spectral features at significantly high energy (up to 500 eV or above) are referred to as extended X-ray absorption fine structure (EXAFS), and the analysis yields for many systems an accurate determination of the local geometry [42, 43]. The XAS could analyze the elemental composition, electronic structure and oxidation state of the samples under observation. One could acquire element-specific

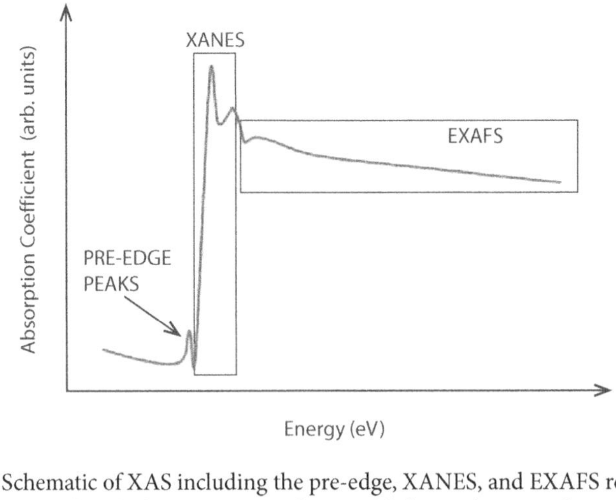

Figure 5.8 Schematic of XAS including the pre-edge, XANES, and EXAFS regions. Reprinted from [46] with the permission of AIP publishing. The range for XANES is 50 to 100 eV, whereas for EXAFS, the range can go up to 500 eV or above.

information from XAS, suggesting that if two or more types of atoms of a particular element are present, the X-ray absorption spectral shape will be a linear combination of all individual sites. This aspect limits the usefulness of XANES and EXAFS in systems with atoms in various sites [44, 45].

5.8 X-Ray Absorption Near Edge Structure

The X-rays have energy lower than 2000 eV, referred to as soft X-rays, suitable for analyzing the elements with low atomic number. So, this range is appropriate for XANES, as the energy separation between adjacent absorption edges are very small. The XANES region infers about: (1) the edge energy position, which is used to identify the oxidation state of the element of interest using reference compounds with known oxidation states; (2) the shape of the XANES spectrum, depending on the local geometric structure around the central absorbing atom; and (3) the pre-edge structures, which are sensitive to the coordination environment of the element of interest [26, 47]. The XANES technique has been employed to characterize the perovskites structures, such as to investigate the alignment of the cationic groups [48, 49], to determine the band structure of formamidinium lead halide [50], and to deduce the degradation mechanism of $MAPbI_3$ [51].

Figure 5.9 shows the XANES spectra at the C K-edge of (i) freshly prepared $MAPbI_3$ films and (ii) $MAPbI_3$ film prepared with a similar procedure but prolonged exposure to X-rays before the XANES measurement. A sharp peak at ~282 eV, followed by a doublet at higher energy, is seen in the XANES spectra of both films. The ratio between the low energy and high energy peak of the two samples were different. The measured spectra of the two films were compared with the simulated spectrum of $MAPbI_3$, which indicates that the measured spectra are in good agreement with the simulated spectra, but the low energy feature was not reproduced in the simulated spectra. This comparative study shows that the low energy peak does not belong to a perfect $MAPbI_3$ structure.

Wilks *et al.* [52] reported that the ionizing energy of incident X-rays might damage the organic molecules and results in new fragments which may show different spectral features. On account of the studies by Wilks *et al.*, Xu *et al.* [51] proposed a degradation pathway for $MAPbI_3$ structure as shown in Figure 5.10. They proposed that if the C-N bond elongates in $CH_3NH_3^+$ species, a new peak will appear at low energy, and the separation between the new peak and the main peak will increase as the C–N bond length increases. Either the elongated C-N bond will relax, or the $CH_3NH_3^+$ species will further dissociate in CH_3^+ and NH_3, where the CH_3^+ rapidly

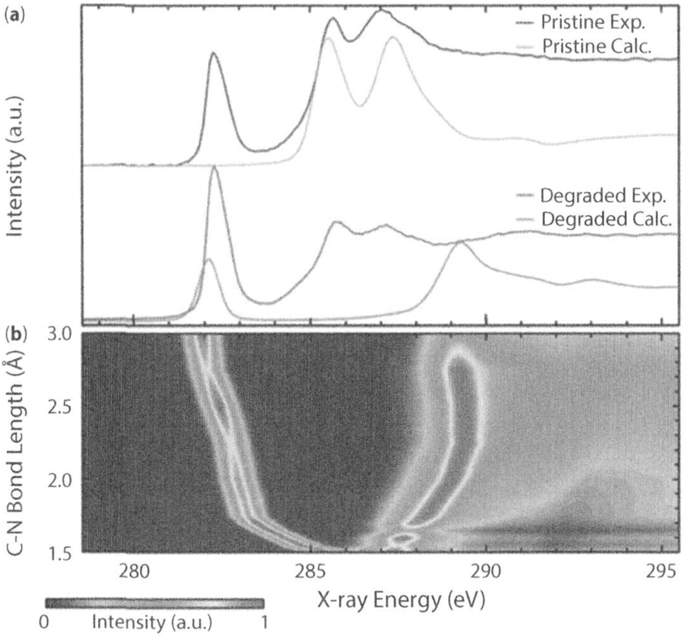

Figure 5.9 Measured and calculated XANES spectra of C K-edge of $MAPbI_3$ film. Reprinted with permission from [51]. Copyright (2016) American Chemical Society.

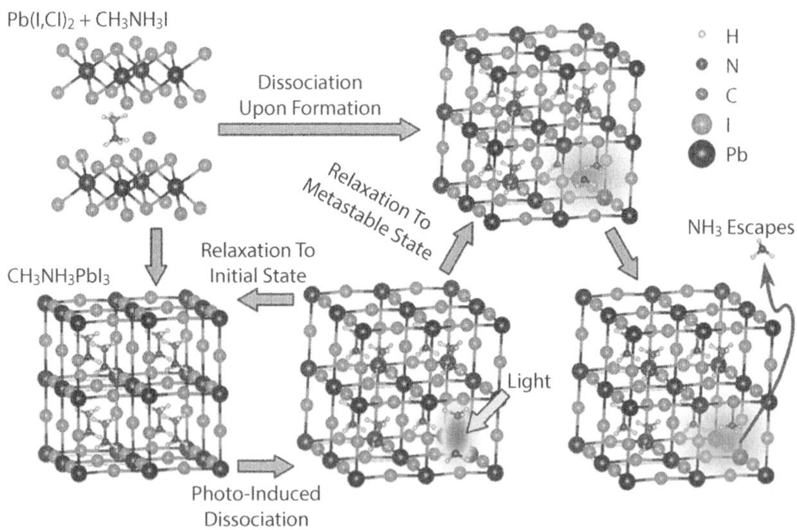

Figure 5.10 Hypothesized process of forming a CH_3I-PbI_2 defect site in $MAPbI_3$ through dissociation, either induced during the perovskite formation or by light (or X-ray) radiation. The NH_3 produced will eventually diffuse out of the lattice. Adapted from Ref [51].

stabilize by bonding with I^- species of the Pb-I lattice and NH_3 migrates out of the perovskite lattice as a neutral molecule.

5.9 Extended X-Ray Absorption Fine Structure

The EXAFS is determined by single scattering by the nearest neighbour atoms of the outgoing photoelectron (mean free path few Å to an nm) produced by the x-ray absorption. Extended X-ray absorption fine structure applies to the structural determination of the materials with low crystallinity or even the amorphous materials, either in the as-synthesized state or after prolonged cycling. The sensitivity of EXAFS is typically below 100 ppm and can be as low as 10 ppm, depending on the element under investigation. In literature, Pb L_3-edge EXAFS has been used to demonstrate the temperature and light-induced degradation of $MAPbI_3$. Further, the degradation products of $MAPbI_3$ have been identified by their characteristic bonding environment [53, 54].

The EXAFS measurements have been used to study the impact of the introduction of different dopants in the pristine lead halide lattice. Yong *et al.* reported that doping of Ni ion enhances violet luminescence of $CsPbX_3$ nanocrystals irrespective of the halide ion. The Pb L_3-edge EXAFS study shows that the enhanced luminescence was due to an improved short-range order of the $CsPbX_3$ lattice [55]. Recently, Bi and co-workers reported the effect of Cu doping on the thermal stability and photoluminescence quantum yield of $CsPbBr_3$ quantum dots (QDs) [56]. They studied the local coordination environment and the local order around the central atoms of QDs and the Pb L_3-edge EXAFS in *k*-space and R-space of Cu-doped $CsPbBr_3$ are shown in Figure 5.11. The periodic amplitude shown by both the samples (Figure 5.11a) indicated that doped sample have similar short-range order that of undoped sample. Therefore, the authors confirmed that Cu^{2+} doping does not affect the short-range order of the perovskite lattice. On the other hand, in Figure 5.11b, $CsPb_{0.93}Cu_{0.07}Br_3$ QDs show a symmetric peak in the *R* space depicting the improvement in the local coordination environment due to the suppression of the octahedral distortion with Cu^{2+} doping.

To further prove the location of Cu^{2+} in the doped $CsPbX_3$ structure, Figure 5.11c confirms a relatively symmetric white line near 8990 eV, which is a typical spectral feature of Cu^{2+} in octahedral coordination environment. Figure 5.11d revealed that on Cu^{2+} doping, the Cu-Br bond length (1.5 Å) is much shorter than Pb-Br bond (3.0 Å) due to lattice contraction, resulting in more vital interaction between Pb and Br, which improves the

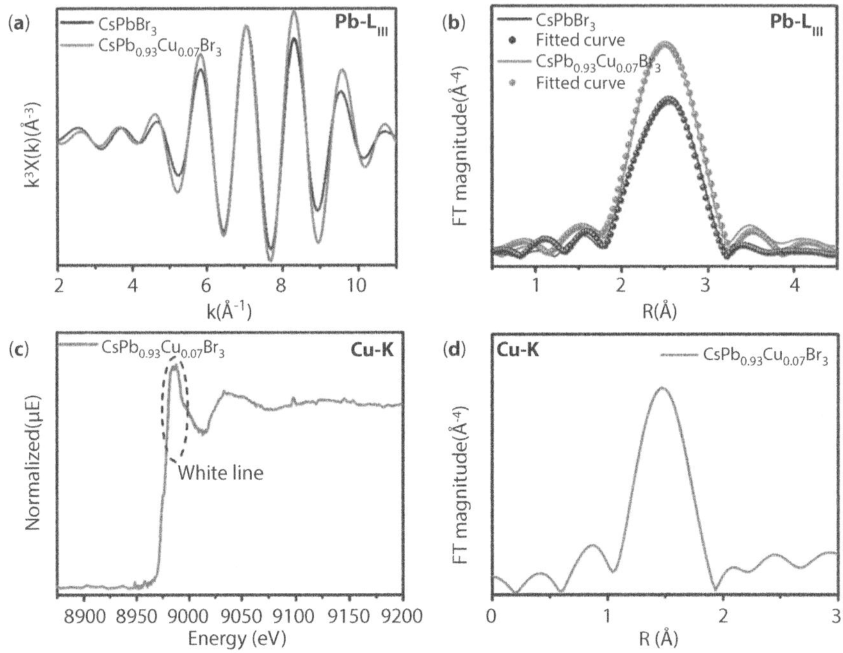

Figure 5.11 EXAFS spectra of $CsPbBr_3$ (black) and $CsPb_{0.93}Cu_{0.07}Br_3$ (red) QDs (a) k_3-weighted Pb L_3-edge EXAFS in k-space and (b) corresponding Fourier transformed R-space EXAFS of $CsPbBr_3$ and $CsPb_{0.93}Cu_{0.07}Br_3$ QDs. (c) Cu K-edge XANES spectrum of $CsPb_{0.93}Cu_{0.07}Br_3$ QDs. (d) k_2-weighted Cu K-edge EXAFS FTs of $CsPb_{0.93}Cu_{0.07}Br_3$ QDs. Adapted from Ref [56].

formation energy of $CsPbBr_3$ quantum dots. The coordination number of Pb is increased, suggesting reducing halide vacancies on doping, enhancing the luminescence intensity of $CsPbBr_3$. EXAFS of the dopant ions was also be measured, which affirms that the *R*-space spectra at the dopant absorption edge show reduced dopant-halide bond length. This has been attributed to the smaller ionic radius of the substitute ions. However, due to the low doping concentration, quantitative EXAFS analysis is limited, presumably due to the low signal to noise ratio at energy well beyond the absorption edge [55, 57].

5.10 Conclusions

In recent times, *in situ* and operando experiments have gained much attraction over the conventional *ex situ* experiments due to some particular advantages. At first, these experiments elude the extensive efforts for making *ex situ* samples and provide increased reliability and precision in the analysis by constantly examining the processes occurring in a single sample. Secondly, the study of time-dependent processes is easily monitored and thus opens a new dimension to study ultrafast processes or nonequilibrium states which may not be easily detected with *ex situ* investigations. Third, there are minimum chances for any contamination and the reaction products can be identified, reliably. The highly reactive species are also can be noticed with *in situ* experiments.

With so many advantages, *in situ* and operando measurements can be sometimes harder to perform due to certain circumstances like the cell must be fully operational under experimental conditions and the apparatus used for experiment must be reliable for data acquisition of high-quality X-ray data. Therefore, different diagnostic tools may be required for different synchrotron techniques. Synchrotron radiation resources are a precious tool for scientific community, therefore, it may not seem necessary to perform *in situ* and operando measurements unless there is a real need for such experiment. Before going to synchrotron experiments, one can get guidance by performing *ex situ* experiments to get critical information regarding synchrotron experiments. We know that X-ray radiation is a weak radiation, but X-rays can induce certain changes in the material under observation. These effects depend on the type of material used, environment, photon energy, beam size and experimental setup. If there is a beam damage, one can minimize it by adapting the sample environment or changing the experimental conditions.

The amount of data generated in synchrotron experiments is very huge and may be difficult to store, handle and process. The mishandling of data may result in the loss of important results, such as structural changes or intermediate processes. The scientific big data mining may be proved next frontier in synchrotron science by collaboration of researchers, synchrotron scientists, computer scientists, and mathematicians. The latest development of next level X-ray facilities at synchrotron sources will provide huge amount of data in a short period of time. However, the proper handling and mining of such a big data may result in some exciting research possibilities for studying fast dynamic behaviour, morphology, chemical complexity. The useful and scientifically important part in the data is often very small and weak in signal. Therefore, properly trained manpower is needed for data mining and extracting the relevant and scientifically important data. At last, the synchrotron radiation-based experiments of perovskite solar cells open a new phase of research for young generation researchers. The *in situ* and operando experiments will probably change the view of studying the experimental processes taking place in the perovskite solar cells, which are not possibly detectable by using conventional methods.

References

1. Ahmad, T., Farooq, U., Phul, R., "Fabrication and photocatalytic applications of perovskite materials with special emphasis on alkali-metal-based niobates and tantalates". *Ind. Eng. Chem. Res.*, 57, 1, 18–41, 2018.
2. Farooq, U., Phul, R., Alshehri, S.M., Ahmed, J., Ahmad, T., "Electrocatalytic and enhanced photocatalytic applications of sodium niobate nanoparticles developed by citrate precursor route". *Sci. Rep.*, 9, 1, 1–17, 2019.
3. Farooq, U., Chaudhary, P., Ingole, P.P., Kalam, A., Ahmad, T., "Development of Cuboidal KNbO3@ α-Fe2O3 Hybrid Nanostructures for Improved Photocatalytic and Photoelectrocatalytic Applications". *ACS Omega*, 5, 32, 20491–20505, 2020.
4. Jošt, M., Kegelmann, L., Korte, L., Albrecht, S., "Monolithic perovskite tandem solar cells: A review of the present status and advanced characterization methods toward 30% efficiency". *Adv. Energy Mater.*, 10, 26, 1904102, 2020.
5. Green, M.A., "Commercial progress and challenges for photovoltaics,". *Nat. Energy*, 1, 15015, 2016.
6. Oh, J., Yuan, H.C., Branz, H.M., "An 18.2%-efficient black-silicon solar cell achieved through control of carrier recombination in nanostructures,". *Nat. Nanotechnol.*, 7, 11, 743–748, 2012.
7. Li, X., Lv, Z., Zhu, H., "Carbon/Silicon heterojunction solar cells: state of the art and prospects,". *Adv. Mater.*, 27, 42, 6549–74, 2015.
8. Ahmad, W., Khan, J., Niu, G., Tang, J., "Inorganic CsPbI3 perovskite-based solar cells: A choice for a tandem device.". *Sol. RRL*, 1, 7, 1700048, 2017.
9. Jena, A.K., Kulkarni, A., Miyasaka, T., "Halide Perovskite Photovoltaics: Background, Status, and Future Prospects". *Chem. Rev.*, 119, 3036–3103, 2019.
10. Green, M.A., Baillie, A.H., Henry, S.J., "The emergence of perovskite solar cells". *Nat. Photonics*, 8, 506–514, 2014.
11. Kim, H.S., Hagfeldt, A., Park, N.G., "Morphological and compositional progress in halide perovskite solar cells". *Chem. Commun.*, 55, 1192–1200, 2019.
12. Mitzi, D.B., Feild, C.A., Harrison, W.T.A., Guloy, A.M., "Conducting tin halides with a layered organic-based perovskite structure". *Nature*, 369, 467–469, 1994.
13. Yu, Z.J., Carpenter, J.V., Holman, Z.C., "Techno-economic viability of silicon-based tandem photovoltaic modules in the United States". *Nat. Energy*, 3, 747–753, 2018.
14. Werner, J., Niesen, B., Ballif, C., "Perovskite/Silicon Tandem Solar Cells: Marriage of Convenience or True Love Story? – An Overview". *Adv. Mater. Interfaces*, 5, 1700731, 2018.

15. You, J., Meng, L., Song, T.B., Guo, T.F., and *et al.*, "Improved air stability of perovskite solar cells via solution-processed metal oxide transport layers". *Nat. Nanotechnol.*, 11, 1, 75–81, 2016.
16. Xiao, Z., Yaun, Y., Shao, Y., Wang, Q., and *et al.*, "Giant switchable photovoltaic effect in organometal trihalide perovskite devices". *Nat. Mater.*, 14, 2, 193–198, 2015.
17. Unger, E.L., Bowring, A.R., Tassone, C.J., Pool, V.L., and *et al.*, "Chloride in Lead Chloride-Derived Organo-Metal Halides for Perovskite-Absorber Solar Cells". *Chem. Mater.*, 26, 24, 7168–7165, 2014.
18. Li, X., Dar, M., II, Yi, C., Luo, J., and *et al.*, "Improved performance and stability of perovskite solar cells by crystal crosslinking with alkylphosphonic acid ω-ammonium chlorides". *Nat. Chem.*, 7, 9, 703–711, 2015.
19. Kim, J.H., Chueh, C.C., Williams, S.T., Jen, A.K., "Room-temperature, solution-processable organic electron extraction layer for high-performance planar heterojunction perovskite solar cells". *Nanoscale*, 7, 17343–17349, 2015.
20. Zhou, Y., Zhou, H., Deng, J., Cha, W., Cai, Z., "Decisive structural and functional characterization of halide perovskites with synchrotron". *Matter*, 2, 2, 360–377, 2020.
21. Kim, M.C., Ham, S.Y., Cheng, D., Wynn, T.A., *et al.*, "Advanced Characterization Techniques for Overcoming Challenges of Perovskite Solar Cell Materials". *Adv. Energy Mater.*, pp1-26, 2001753, 2020.
22. Khodaei, M. and Petaccia, L., "Introductory Chapter: Synchrotron-Based X-Ray Characterization of Nanomaterials", in: *X-ray Characterization of Nanostructured Energy Materials by Synchrotron Radiation*, IntechOpen, London, United Kingdom, 2017.
23. Lin, F., Liu, Y., Yu, X., Cheng, L. *et al.*, "Synchrotron X-ray Analytical Techniques for Studying Materials Electrochemistry in Rechargeable Batteries". *Chem. Rev.*, 117, 13123–13186, 2017.
24. Liu, D. and Kelly, T.L., "Perovskite solar cells with a planar heterojunction structure prepared using room-temperature solution processing techniques". *Nat. Photonics*, 8, 2, 133–138, 2014.
25. Kuai, L., Li, J., Li, Y., Wang, Y., and *et al.*, "Revealing Crystallization Dynamics and the Compositional Control Mechanism of 2D Perovskite Film Growth by *In Situ* Synchrotron-Based GIXRD". *ACS Energy Lett.*, 5, 1, 8–16, 2020.
26. Liu, L. and Dong, Z., "Characterization of Lead Halide Perovskites Using Synchrotron X-ray Techniques", in: *Perovskite Quantum Dots; Springer Series in Materials Science*, vol. 303, Springer, Singapore, 2020.
27. Yang, Y., Yang, L., Feng, S., "Interfacial engineering and film-forming mechanism of perovskite films revealed by synchrotron-based GIXRD at SSRF for high-performance solar cells". *Mater. Today Adv.*, 6, 100068, 2020.
28. Yang, Y., Feng, S., Xu, W., Li, M. *et al.*, "Enhanced Crystalline Phase Purity of CH3NH3PbI3–xClx Film for High-Efficiency Hysteresis-Free Perovskite Solar Cells". *ACS Appl. Mater. Interfaces*, 9, 27, 23141–23151, 2017.

29. Wu, Z., Bai, S., Xiang, J., Yuan, Z. *et al.*, "Efficient planar heterojunction perovskite solar cells employing graphene oxide as hole conductor". *Nanoscale*, 6, 18, 10505–10510, 2014.
30. Yang, J., Siempelkamp, B.D., Liu, D., Kelly, T.L., "Investigation of CH3NH3PbI3 Degradation Rates and Mechanisms in Controlled Humidity Environments Using *In Situ* Techniques". *ACS Nano*, 9, 2, 1955–1963, 2015.
31. Zhao, J., Cai, B., Luo, Z. *et al.*, "Investigation of the Hydrolysis of Perovskite Organometallic Halide CH3NH3PbI3 in Humidity Environment". *Sci. Rep.*, 6, 21976, 2016.
32. Wang, Y., Lu, X., Yang, W., Wen, T. *et al.*, "Pressure-Induced Phase Transformation, Reversible Amorphization, and Anomalous Visible Light Response in Organolead Bromide Perovskite". *J. Am. Chem. Soc*, 137, 34, 11144–11149, 2015.
33. Wang, P., Guan, J., Galeschuk, D.T.K., Yao, Y. *et al.*, "Pressure-Induced Polymorphic, Optical, and Electronic Transitions of Formamidinium Lead Iodide Perovskite". *J. Phys. Chem. Lett.*, 8, 10, 2119–2125, 2017.
34. Liu, G., Kong, L., Gong, J., Yang, W. *et al.*, "Pressure-Induced Bandgap Optimization in Lead-Based Perovskites with Prolonged Carrier Lifetime and Ambient Retainability,". *Adv. Funct. Mater.*, 27, 1604208, 2017.
35. Jaing, S., Fang, Y., Li, R., Xiao, H. *et al.*, "Pressure-Dependent Polymorphism and Band-Gap Tuning of Methylammonium Lead Iodide Perovskite". *Angew. Chem. Int. Ed.*, 55, 6540–6544, 2016.
36. Stone, K.H., Gold-Parker, A., Pool, V.L., Unger, E.L. *et al.*, "Transformation from crystalline precursor to perovskite in PbCl2-derived MAPbI3". *Nat. Commun.*, 9, 3458, 2018.
37. Schlipf, J., Docampo, P., Schaffer, C.J., Körstgens, V. *et al.*, "A Closer Look into Two-Step Perovskite Conversion with X-ray Scattering". *J. Phys. Chem. Lett.*, 6, 7, 1265–1269, 2015.
38. Pineda, E., Alhazmi, N., Ebbens, S.J., Dunbar, A.D.F., "Influence of Additives on the *In Situ* Crystallization Dynamics of Methyl Ammonium Lead Halide Perovskites". *ACS Appl. Energy Mater.*, 4, 2, 1398–1409, 2021.
39. Barrows, A.T., Lilliu, S., Pearson, A.J., Babonneau, D. *et al.*, "Monitoring the Formation of a CH3NH3PbI3–xClx Perovskite during Thermal Annealing Using X-Ray Scattering". *Adv. Funct. Mater.*, 26, 4934–4942, 2016.
40. Steele, J.A., Jin, H., Dovgaliuk, I., Berger, R.F. *et al.*, "Thermal unequilibrium of strained black CsPbI3 thin films". *Science*, 365, 6454, 679–684, 2019.
41. Alhazmi, N., Pineda, E., Rawle, J., Howse, J.R. *et al.*, "Perovskite Crystallization Dynamics during Spin-Casting: An *In Situ* Wide-Angle X-ray Scattering Study". *ACS Appl. Energy Mater.*, 3, 7, 6155–6164, 2020.
42. Iwasawa, Y., "X-ray absorption fine structure for catalysts and surfaces", in: *World Scientific series on synchrotron radiation techniques and applications*, p. 410, World Scientific, River Edge, Singapore, 1996.
43. Hasnain, S.S., *X-ray absorption fine structure*, Ellis Horwood Limited, West Sussex, UK, 1991.

44. Holbrrok, R.D., Galyean, A.A., Gorham, J.M., Herzing, A. *et al.*, "Overview of Nanomaterial Characterization and Metrology.", in: *Characterization of Nanomaterials in Complex Environmental and Biological Media*, pp. 47–87, Elsevier, Amsterdem, Netherlands, 2015.
45. d. Groot, F., "High-Resolution X-ray Emission and X-ray Absorption Spectroscopy". *Chem. Rev.*, 101, 6, 1779–1808, 2001.
46. Er, A.O., Chen, J., Rentzepis, P.M., "Ultrafast time resolved x-ray diffraction, extended x-ray absorption fine structure and x-ray absorption near edge structure". *J. Appl. Phys.*, 112, 031101, 2012.
47. Willmott, P., "Spectroscopic Techniques", in: *An Introduction to Synchrotron Radiation*, pp. 303–366, John Wiley & Sons Ltd, Sussex, UK, 2019.
48. McLeod, J.A., Wu, Z., Shen, P., Sun, B., Liu, L., "Self-Alignment of the Methylammonium Cations in Thin-Film Organometal Perovskites". *J. Phys. Chem. Lett.*, 5, 16, 2863–2867, 2014.
49. Wang, Y., Xia, Z., Liu, L., Xu, W. *et al.*, "The Light-Induced Field-Effect Solar Cell Concept - Perovskite Nanoparticle Coating Introduces Polarization Enhancing Silicon Cell Efficiency". *Adv. Mater.*, 29, 1606370, 2017.
50. Shen, P., Nie, K., Sun, X., Liu, L., McLeod, J.A., "Electronic structure of formamidinium ions in lead triiodide perovskites". *Phys. Status Solidi-Rapid Res. Lett.*, 10, 9, 677-681, 2016.
51. Xu, W., Liu, L., Yang, L., Shen, P. *et al.*, "Dissociation of Methylammonium Cations in Hybrid Organic–Inorganic Perovskite Solar Cells". *Nano Lett.*, 16, 7, 4720–4725, 2016.
52. Wilks, R.G., MacNaughton, J.B., Kraatz, H.B., Regier, T. *et al.*, "Comparative Theoretical and Experimental Study of the Radiation-Induced Decomposition of Glycine". *J. Phys. Chem. A*, 113, 18, 5360–5366, 2009.
53. Abdelmageed, G., Jewell, L., Hellier, K., Seymour, L. *et al.*, "Mechanisms for light induced degradation in MAPbI3 perovskite thin films and solar cells". *Appl. Phys. Lett.*, 109, 233905, 2016.
54. Abdelmageed, G., Mackeen, C., Hellier, K., Jewell, L. *et al.*, "Effect of temperature on light induced degradation in methylammonium lead iodide perovskite thin films and solar cells". *Sol. Energy Mater. Sol. Cells*, 174, 566–571, 2018.
55. Yong, Z.-J., Guo, S.-Q., Ma, J.-P., Zhang, J.-Y. *et al.*, "Doping-Enhanced Short-Range Order of Perovskite Nanocrystals for Near-Unity Violet Luminescence Quantum Yield". *J. Am. Chem. Soc*, 140, 31, 9942–9951, 2018.
56. Bi, C., Wang, S., Li, Q., Kershaw, S.V. *et al.*, "Thermally Stable Copper(II)-Doped Cesium Lead Halide Perovskite Quantum Dots with Strong Blue Emission". *J. Phys. Chem. Lett.*, 10, 5, 943–952, 2019.
57. Ma, J.-P., Chen, Y.-M., Zhang, L.-M., Guo, S.-Q. *et al.*, "Insights into the local structure of dopants, doping efficiency, and luminescence properties of lanthanide-doped CsPbCl3 perovskite nanocrystals". *J. Mater. Chem. C*, 7, 3037–3048, 2019.

6

Recent Progress on Perovskite-Based Solar Cells

Waseem Raza[1]* **and Khursheed Ahmad**[2]

[1]*Department of Chemistry, Indian Institute of Technology Delhi, Hauz Khas, New Delhi, India*
[2]*Discipline of Chemistry, Indian Institute of Technology Indore, Simrol, Khandwa Road, Indore, India*

Abstract
The global energy crisis and environmental pollution are increasingly spotlighted due to the fast development of industries and population growth. Energy is a fundamental requirement of human life and increasing rapidly at the global level, which represents one of the biggest issues in the 21st century. Currently, over 80% of the globe's energy demand is mainly addressed by the consumption of fossil fuels, which stimulate global warming and have a deleterious impact on the environment as well as depleted in the near future. Therefore, it is crucial to develop a new type of alternating energy to fossil fuels that is clean, cheap, renewable, and long-lasting. Solar energy is considered as the most abundant, renewable, and clean energy source, which can satisfy both growing energy demand for the future and the environmental pollution issues. Moreover, solar energy gives 10,000 times more power than the requirement of the entire planet and has become the prime choice as a new alternative energy for the near future. Therefore, conversion of solar energy to usable fuel such as electric power is one of the most promising approach of sustainable energy research. A photovoltaic device or solar cells has been considered as the most promising technologies to harvest solar light and directly convert it to electric power and provides a non-polluting energy source. Among different state-of-the-art developed solar cells, perovskite solar cells are remarkably encouraging and promising hotspots of research due to their incredible energy conversion efficiency, easy fabrication, and low cost.

**Corresponding author*: waseemrazambd@gmail.com

Khursheed Ahmad and Waseem Raza (eds.) Perovskite Materials for Energy and Environmental Applications, (147–166) © 2022 Scrivener Publishing LLC

Keywords: Perovskite, solar cells, photovoltaic, efficiency, band gap, absorption spectrum, hole transport layer, electron transport layer

6.1 Introduction

The continuously increasing energy demand represents one of the biggest challenge faced by mankind, this may be due to the fast-growing population and rapid industrialization. Currently, this demand is mainly fulfilled by fossil fuels, which are limited and will be depleted in the near future as the consumption rate is very high. Moreover, the burning of fossil fuels produces a large amount of harmful gases such as CO_2, SO_x, and NO_x into the atmosphere, which are polluting the environment. The increased concentration of harmful gases in the environment promoting the risk of global warming and acid rain. In addition to this, the direct dumping of different organic and inorganic waste into water systems also polluting the water reservoirs and causing waterborne diseases leading to environmental problems [1]. The ultimate global challenge is to achieve environmental sustainability for our society, and therefore, it is very important to develop the alternative long-lasting, clean and renewable energy source to replace fossil fuels. Among all alternative energy sources, solar energy is considered as one of the most promising alternative energy source for fossil fuels due to its easy availability, cleanness, and accessibility. Moreover, sun is the fundamental and ultimate source of renewable energy for our planet, which can provide endless carbon emission-free energy for mankind. Around 3.85 yottajoules of solar energy reaches the earth every year, which is approximately 10^4 times higher than the world energy consumption annually [2–4]. However, the availability of solar light depends on day time, weather, and seasonal condition. Therefore, it is critical to converting sunlight to usable or storable energy, which can be used on requirement. In this connection, different approaches, such as solar heating, solar thermal electricity, photocatalysis, and photovoltaic, have been involved to harvest the solar energy [5–8]. Among all, photovoltaic (PV) has been regarded as the most efficient and promising approach, which can directly convert solar radiation into electric power using photovoltaic effect. Photovoltaic is a successful case of effective utilization of solar energy to generate electricity using semiconducting materials, which can fulfill the global energy demand and environmental problems. Moreover, PV is the crucial machinery that can provide a carbon-free (decarbonized economy) power supply to achieve a sustainable society. PV system is made up of solar arrays (solar panels), a solar inverter, and other electrical accessories, and solar cells, which are building blocks of the PV system. Therefore, solar cells are the electrical device in the PV system which can directly convert solar radiation to electric power. The first PV solar cell was built by E. Becquerel in 1839 at the age of 19 years. The first solid-state PV solar cell with 1% efficiency was created by C. Fritts in 1883 with the help of a selenium semiconductor having a thin layer of gold [9]. However, the first practical PV solar cell was constructed by C. S. Fuller, G. Pearson, and D. Chapin in 1954 based on silicon solar cells with 4.5 to 6% efficiency [9–12]. Different types of PV solar cells have been developed such as first-generation (1G) solar cells based on single-crystalline silicon, second-generation (2G) solar cells based on cadmium telluride (CdTe), copper indium gallium selenide (CIGS), and amorphous silicon, and the third-generation (3G) solar cells based on organic cells (OSCs), dye-sensitized solar cells (DSSCs), multijunction cells, quantum dot sensitized solar cells (QDSCs) and perovskite solar cells (PSCs) [13–17]. Currently, 1G solar cells based on crystalline silicon are commercial and dominating the PV market, as well as exhibiting remarkable stability with a maximum power conversion efficiency (PCE) of about 25% in the lab, but theoretical limit of the PCE is about 31% [18]. The PCE of 2G solar cells is lower than 1G but they were significantly less expensive than 1G solar cells because of their lower processing cost and materials. 1G solar cells made up of nontoxic crystalline silicon with about 25 years performance warranty and showing fairly high PCE with negligible degradation, therefore leading the PV solar cells market [19]. Although 1G solar cells exhibiting high PCE, but not free from some fundamental drawbacks, such as high manufacturing and installation costs. In addition to this, relatively large global demand for silicon solar cells and device formation of high temperature is very expensive due to the high melting point of silicon limiting its future development [20]. Therefore, it is crucial to explore new PV materials with high PCE and low production costs to replace or complete the state-of-the-art PV technology for the future development of solar cells. To fulfill the gap, scientists have been in quest of developing 3G solar cells with the aim of high PCE with lower cost. Among all 3G solar cells, PSCs have recently been attracted much attention as well as emerged as a hot research area and competing materials for silicon PV due to its comparable PCE, easy fabrication, long charge carrier lifetime, low binding energy, low defect density, and low cost [21, 22]. The term "Perovskite" originated from the mineral name of calcium titanate ($CaTiO_3$), which was discovered by Gustav Rose (German mineralogist) in the Ural Mountains of Russia in 1839 and named after a Russian mineralogist Lev Aleksevich Perovski [23]. Since then, the term perovskite is generally applied to the class of materials having the same type of crystal structure and stoichiometry as the $CaTiO_3$. A perovskite is an organic/inorganic hybrid and possesses a general formula of ABO_3. In this, A and B are cations stand for rare-earth or alkaline-earth metals and transition metals of 3d, 4d, and 5d configurations, and O stand for oxygen which bond to both cation. The ABO_3 structure of perovskite is ideally cubic, where A atom is bigger than B, and form the corner of cubic cells, coordinated by 12-fold cub-octahedral, B atoms are in the center and surrounded by an octahedral of anions with 6-fold coordination and the oxygen atom are placed in the faces centers as shown in Figure 6.1 [24]. The oxide perovskites show good activity in various fields, such as dielectric, piezoelectric, photo-catalysis, ferroelectric, pyroelectric, and so on, but not good for PV applications. Some compounds like $LiNbO_3$, $BiFeO_3$, and $PbTiO_3$ exhibit some PV activity due to the ferroelectric polarization effect, which makes them suitable for PV application [25]. However, halide perovskite, a new class of perovskite differs from oxide perovskite in having halide anion instead of oxide anions with general formula ABX_3 present the excellent semiconducting properties that are desired for PV solar cells. Moreover, the A is a monovalent metallic cation from group I of the periodic table instead of divalent in the case of oxide perovskite, B is divalent cations form transition metals instead of tetravalent and X is the non-metallic anion (halide) as shown in Figure 6.2.

Halide perovskites were discovered in the 1890s, but their comprehensive study, first time was performed by Wells *et al.* in 1893 [27]. Mitzi's group investigate the physical properties of two-dimensional (2D) halide perovskite materials having organic group in 1990s [28]. In 2006, the Miyasaka group used perovskites as the first photovoltaic absorber, which provide only 2.2% PCE [29]. After this, PSCs have been booming development and exhibiting fast improvement in efficiency to 3.8% in 2009, 6.5% in

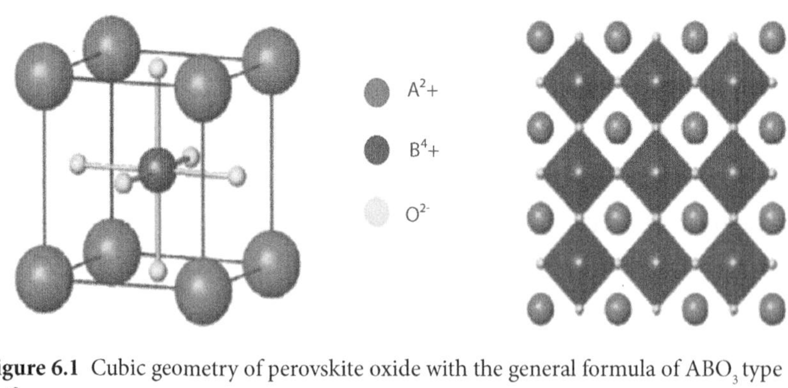

Figure 6.1 Cubic geometry of perovskite oxide with the general formula of ABO_3 type [24].

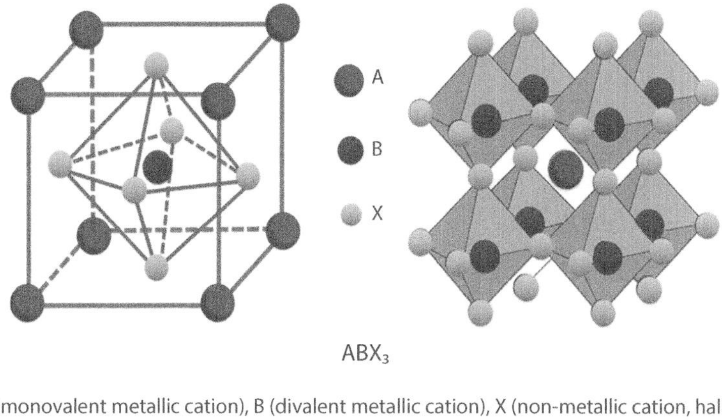

Figure 6.2 Schematic presentation of cubic crystal of halide perovskite with the general formula of ABX_3 type [26].

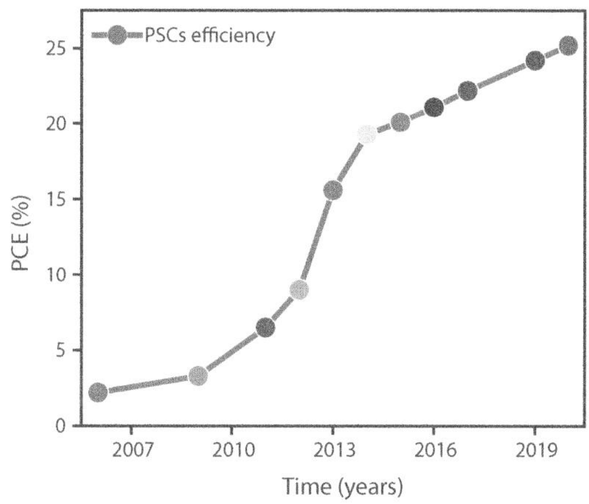

Figure 6.3 PCE for PSCs from 2006 to 2020, data obtained from Best Research-Cell Efficiency Chart [34].

2011, 20.1% in 2015, 22.1 in 2017, 24.2% in 2019, and 25.5 in 2020 [30–34]. Figure 6.3 exhibits the significant improvement in the PCE of PSCs as a function of years. The fast improvement in PSCs performance may be due to their unique properties such as small bandgap, high dielectric constant, lower excitation binding energies, high absorption coefficient, high charge carrier mobility and length, and so on. Therefore, this chapter will provide a state-of-the-art summary and discussion about the recent progress on development and engineering of PSCs materials toward future directions.

6.2 Device Structure and Working Principle of PSCs

The performance of PSCs device greatly depends on the structure and engineering, as well as the choice of materials of the device. Therefore, different architectures have been engineered since solid-state PSCs were discovered to meet the high PCE. Generally, a conventional PSC device is made up of layer-by-layer stacking architecture, which is composed of transparent conducting oxide-coated glass (TCO), a light-absorbing perovskite layer, an electron transport layer (ETL), a hole transport layer (HTL), and a counter electrode (CE). The architecture of most PSCs devices is like a sandwich, in which an active perovskite layer is placed between HTL and ETL layers. By architecture, PSCs can be classified into mesoscopic and planner structure depending on the layers. The mesoscopic/planar PSCs are further categorized into regular n-i-p or inverted p-i-n devices depending on the stacking of different layers, such as ETL/perovskite/HTL or HTL/perovskite/ETL, respectively, as shown in Figure 6.4.

The working principle of a typical PSC is based on built-in potential formed at the p-n junction because of the PV effect. On the illumination of perovskite film, electrons are excited and then dissociate into free electron-hole pairs because of low excitation binding energy. After that photo-generated free electrons and holes move in opposite direction which is built in the current. The electrons are collected by n-type ETL layer and holes are received by p-type HTL layers against the barrier electric field and inhibit the electron-holes recombination in TCO. ETL/HTL must have conduction band minimum (CBM), known as lowest unoccupied molecular orbital (LUMO), and the valence band maximum (VBM) known as highest occupied molecular orbital (HOMO). However, the work function

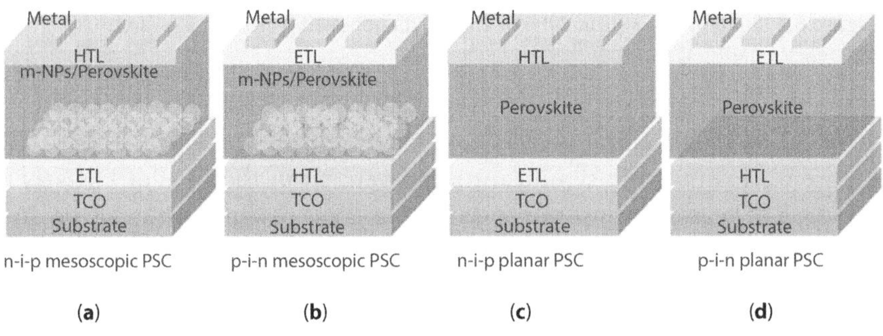

Figure 6.4 Schematic presentation of solid-state device architecture of (a) mesoscopic regular n-i-p, (b) mesoscopic inverted p-i-n, (c) planar regular n-i-p, and (d) planar inverted p-i-n PSCs [35].

of TCO should match well with the CBM and VBM of ETL/HTL to collect electrons and holes at electrodes. Therefore, the intrinsic mechanism of both regular n-i-p and inverted p-i-n architectures are the same but the direction of charge is different. The performance of the device can be affected by the energy level alignment at the interface, which determines the charge transfer/transport as well as selection of semiconductor. The semiconductor with a low bandgap between 1.4 to 1.1 eV can absorb more light and will generate a higher current. Thus, fabricating high-efficiency PSCs is depends on energy level alignment at various interface and selection of low band semiconductors [36].

6.3 Perovskite-Based Solar Cells

Perovskite solar cells have been considered as one of the most promising materials for the next-generation solar cells with high efficiency due to their low-cost, flexible substrate, and low-temperature processing. In 2009, Kojima *et al.* report on the potential use of the first organic-inorganic halide perovskites material ($MAPbI_3$, where MA = $CH_3NH_3^+$) as a light sensitizer in a DSSCs for solar energy conversion using volatile liquid electrolyte [30]. However, PCE and stability of reported material was very low due to corrosion effect of liquid electrolyte. To improve the PCE of PSCs, a solid-state hole transporting material (HTM, Spiro-OMeTAD) was used instead of liquid electrolyte leading to enhancement in PCE (9.7%) as shown in Figure 6.5 [37]. The introduction of HTM in place of liquid electrolyte gives a historical change as it not only increase the PCE but also improved the long-term stability [37]. The bandgap and VBM of

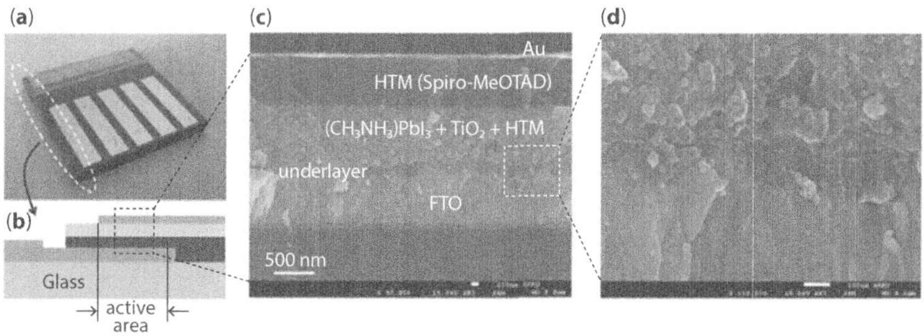

Figure 6.5 Schematic presentation of (a) solid-state device, (b) cross-section structure, (c) cross-section scanning electron microscopic (SEM) image of device and (d) FTO-active layer (under) interface [37].

the synthesized perovskite film were investigated using diffuse reflectance and ultraviolet photoelectron spectroscopic (UPS) analysis.

The bandgap of pristine TiO_2 and $MAPbI_3$ deposit TiO_2 film was found to be 3.2 and 1.5 eV as shown in Figure 6.6 (a & b). The valance band and conduction band energy (E_{VB} & C_{VB}) of the $MAPbI_3$ deposit TiO_2 film were calculated with the help of UPS spectrum as shown in Figure 6.6 (c) and found to be −5.43 and −3.93 eV, respectively. Figure 6.6 (d) exhibits the bandgap edge positions and charge separation for TiO_2 and $MAPbI_3$ deposit TiO_2 film. Giordano *et al.* reported the Li-doped mesoporous titania (ETL) for PSCs, which gives faster electronic transportation with negligible hysteresis loss (less than 0.3%) [38]. Generally, $MAPbI_3$ is used in PSCs but some scientists also used $FAPbI_3$ (FA = formamidinium) PSCs as the optoelectronic properties are closely related to $MAPbI_3$ [39]. However, the bandgap of $FAPbI_3$ is lower than $MAPbI_3$, which allows more absorption from the solar spectrum but the fabrication cost of stable film of $FAPbI_3$ is difficult than $MAPbI_3$ film. The regular n-i-p configuration of

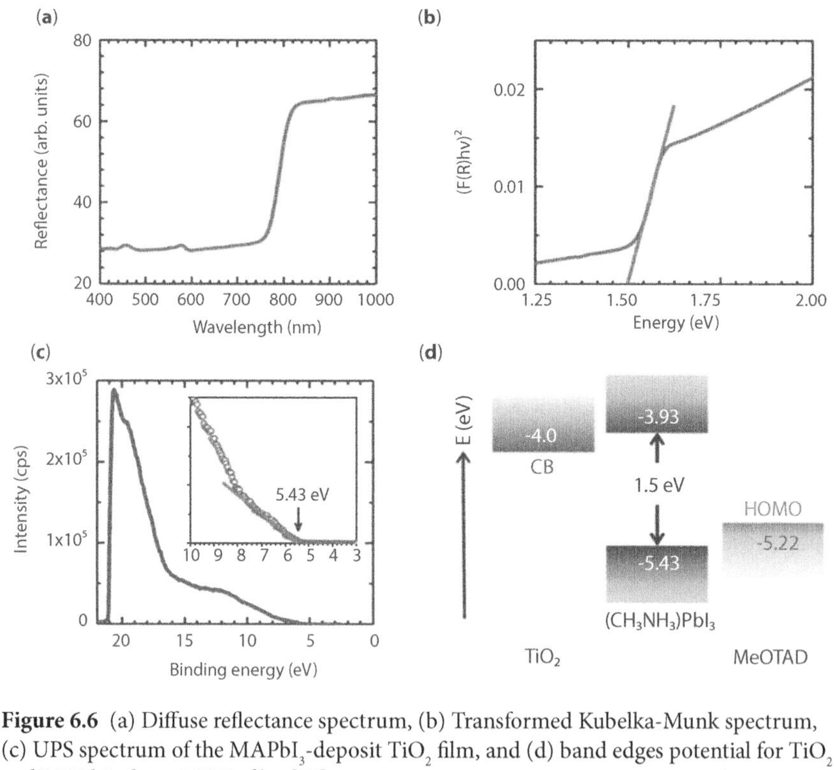

Figure 6.6 (a) Diffuse reflectance spectrum, (b) Transformed Kubelka-Munk spectrum, (c) UPS spectrum of the $MAPbI_3$-deposit TiO_2 film, and (d) band edges potential for TiO_2 and $MAPbI_3$-deposit TiO_2 film [37].

$FAPbI_3$ exhibits the highest PCE than inverted p-i-n configuration because regular PSC shows negligible hysteresis for I-V as compared with the inverted one. Li *et al.* synthesized the $FA_{0.81}MA_{0.15}PbI_{2.51}Br_{0.45}$ as an absorber film in PSCs with uniform morphology using a flash vacuum processing method [40]. Bi *et al.* synthesized the perovskite films using poly (methyl methacrylate) (PMMA) as a new approach for getting higher PCE [41].

Yang *et al.* introduced defect engineering to synthesize the perovskite film using multiple cations and mixed halide anions for getting higher PCE [42]. The most common light-absorbing material in PSCs is $MAPbI_3$, where X can be Cl, Br, or I, and unit cell parameters are different for all halide anions, such as 5.68 for Cl, 5.92 for Br, and 6.72 for I. Moreover, the phase transition with a decrease in temperature is also observed from orthorhombic ($T < 160$ K), tetragonal ($162.2 < T < 327.4$ K), and tetragonal ($T > 327.4$ K) due to the large size of $MAPbI_3$ which leads to distortion in structure [43]. The bandgap of $MAPbI_3$ is a direct bandgap and found to be 1.5 eV, whereas the bandgap of $MAPbBr_3$ is found to be 2.3 eV which is wider than $MAPbI_3$. However, the bandgap of $FAPbI_3$ is found to be 1.48 eV which is lower than $MAPbI_3$, but the stability is much lower than $MAPbI_3$. However, stability (moisture as well as thermal) is an important part of device manufacture using perovskite film. The organic-inorganic perovskites are sensitive to thermal stressing more than inorganic ones. The most common light-absorbing material in PSCs is $MAPbI_3$, which exhibits less thermal stability due to presence of organic volatile MA cation [44]. To improve the thermal stability of perovskite-based solar cells, scientists start the synthesis of MA-free perovskite materials for solar cells. The stability of inorganic PSCs was found to be higher than organic-inorganic PSCs such as $CsPbI_3$ and $CsPbI_2Br$ exhibit thermal stability up to 400°C [45–47]. Li *et al.* in 2018 reported the synthesis of highly stable (400°C) all inorganic $CsPbI_2Br$ perovskite absorbing layer with long-term thermal stability [46]. Figure 6.7 (a_1, a_2) shows the XRD spectra of the as-prepared $CsPbI_2Br$ film, which form δ-perovskite (non-perovskite) if the iodine content is higher at low temperature. Upon increasing the heating with elevated temperature, the color of the film also changed from transparent (Figure 6.7 (b) a_1) to light brown (Figure 6.7 (b) a_2). The color of the film is found to change from light brown to dark brown upon heating at higher temperature (160°C for 10 min) as shown in Figure 6.7 (b) a_3. The XRD peaks also exhibits the formation of the final desire film of $CsPbI_2Br$ perovskite in the cubic phase as shown in Figure 6.7 (a) a_3. Figure 6.7 (c) shows the top view SEM image of prepared $CsPbI_2Br$ perovskite after a two-step temperature-control method indicating uniform morphology. They also fabricated the solar cell device using all-inorganic PSCs and the structure is shown in Figure 6.8 (a)

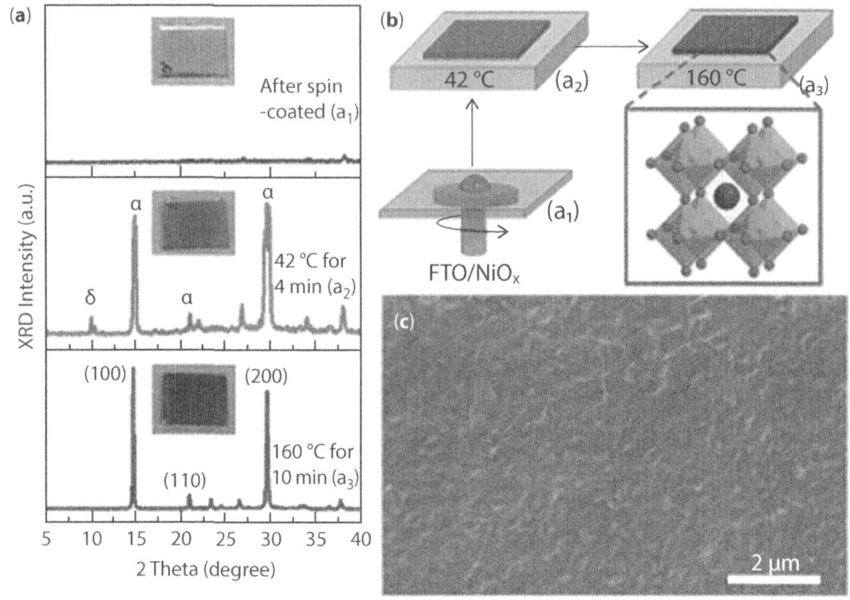

Figure 6.7 (a) XRD pattern of prepared $CsPbI_2Br$ perovskite film at different temperatures, (b) change in color of $CsPbI_2Br$ perovskite film of two-step temperature-control process, and (c) SEM image of prepared $CsPbI_2Br$ perovskite film [46].

which is made up of FTO/NiO_X/$CsPbI_2Br$/ETL/Ag where ETL consists of ZnO, $C_{60,}$ and ZnO@C_{60} bilayer respectively to check the efficiency of the synthesized film. It could be seen from Figure 6.8 (b) that ETL displays a uniform layer by layer morphology. Figure 6.8 (c) exhibits the band edges position and Gibbs free energy of prepared $CsPbI_2Br$ film, ZnO, $C_{60,}$ and ZnO@C_{60} respectively. The CBM and VBM were analyzed using UPS and CBM for $CsPbI_2Br$ film was found to be 4.16 eV. The CBM for ZnO and C_{60} were found to be 4.2 and 4.5 eV, respectively. However, the Gibbs free energy was calculated using equation 6.1 and found to be 0.04, and 0.34 eV for ZnO ETL and C_{60} ETL, respectively.

$$\Delta G = (\text{perovskite})E_{CBM} - (\text{ETL})E_{CBM} \tag{6.1}$$

They also showed that the electron extraction efficiency of C_{60} ETL is higher than ZnO ETL, resulting reduction in interface recombination. Moreover, the electron extraction of ZnO@C_{60} bilayer is found to be improved as compare to single ZnO and C_{60} layer, it may be due to multiple injection pathways and reduction in trap-state density. Moreover, the promising electron extraction ability of the ZnO@C_{60} bilayer may also be

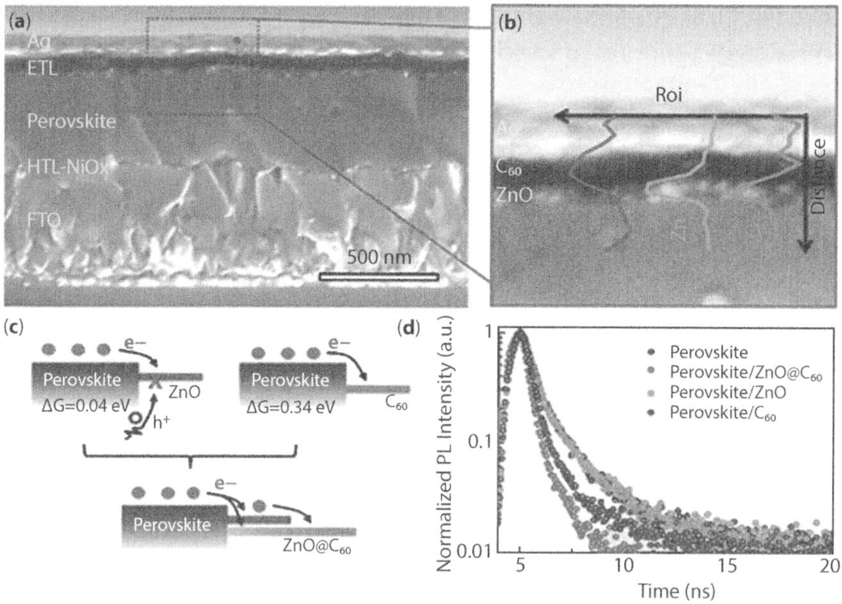

Figure 6.8 (a) Cross-sectional SEM image of the $CsPbI_2Br$ perovskite film with HTL and ETL on FTO, (b) corresponding enlarge SEM image obtained from red area showing presence of different element, (c) band edges potential for $CsPbI_2Br$ based ZnO, C_{60}, and ZnO@C_{60} respectively, and (d) time-resolved photoluminescence spectra of the synthesized $CsPbI_2Br$ perovskite film with different ETLs [46].

due to good gradient energy band matching between perovskite layers and ZnO@C_{60} bilayer. Therefore, extraction of photogenerated electrons is easier from the perovskite layer to the metal electrode. Figure 6.8 (d) exhibits the time-resolved photoluminescence spectra of the synthesized $CsPbI_2Br$ perovskite film with different ETLs. It could be seen from Figure 6.8 (d) that photoluminescence quenching of perovskite/C_{60} and perovskite/ZnO@C_{60} is faster than that of perovskite/ZnO, and results are in line with the previous assumption as perovskite/ZnO@C_{60} exhibits higher electron extraction and lower recombination rate.

Moisture also influences the stability of the PSCs device, it was observed that unpacked cells degraded within a few hundred hours after the exposure to humid air (50%) [48, 49]. Water in humid air can easily penetrate the perovskite structure and form a strong bond with organic cation in the case of organic-inorganic perovskite crystal ($MAPbI_3$) leading to weakening the bond between organic and PbI_6 and perovskite structure [49]. Moreover, the formation of volatile hydroiodic acid (HI) takes place because of the protonation of iodide, resulting formation of yellow-colored lead iodide (PbI_2) as decomposed product. The moisture stability of perovskite films can also be enhanced by tuning the lattice parameter and bond environment as well as changing the composition [50]. It has been reported that substitution of halide anion is a promising way to improve the moisture stability [51]. To improve the stability of organic-inorganic $MAPbI_3$ perovskite film, Noh *et al.* change the composition of halide anion (I) with mixed halide anions ($I\text{-}_XBr_X$) [52]. They reported that change of I anion with a small fraction of Br anion (20% to 29%) exhibits better performance and enhanced dark storage compared with cells with less Br anion. The shrinkage of lattice parameter of $MAPb(I\text{-}_XBr_X)_3$ is observed after substitution of iodide with bromide leading to strengthens of organic-lead halide bonding which in turn increases the moisture stability [52]. They also report that tuning in the band gap of $MAPb(I\text{-}_XBr_X)_3$ hybrid has been achieved by substitution of I with Br ions. It is reported that $MAPbI_3$ has a distorted three dimensional (3D) tetragonal structure, whereas $MAPbBr_3$ has cubic perovskite structure at room temperature as shown in Figure 6.9 (a and b). The difference in the structure of above two perovskite seems to

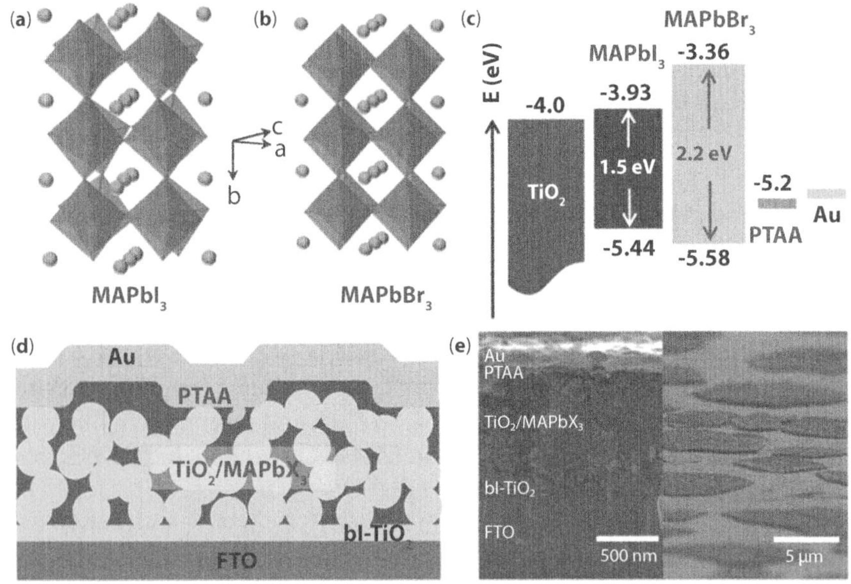

Figure 6.9 (a) Crystal structure of $MAPbI_3$ in distorted tetragonal phase, (b) Cubic crystal of $MAPbBr_3$ at room temperature, where Red polyhedron is $PbX_{6/2}$ [X = I and Br] and green spheres are MA, (c) The CBM and VBM of $MAPbI_3$, $MAPbBr_3$, and TiO_2, (d) Schematic presentation of device configuration containing hybrid heterojunction of $MAPbX_3$ layer, and (e) Cross-sectional and titled view of FESEM image of heterojunction solar cell device [52].

originate due to difference in the ionic radius of Br and I anion (1.96 and 2.2 A^0) respectively. It could be seen from Figure 6.9 (c) that these two perovskite have suitable band edge positions for the formation of heterojunction solar cell. Figure 6.9 (d) exhibits the structure of device formed with the help of hybrid heterojunction of $MAPbX_3$. Upon irradiation of light on heterojunction solar cell, then generation of electron-hole pairs takes place form the hybrid $MAPbX_3$ perovskite film. The photogenerated electrons were transferred to TiO_2 and holes were capture by polytriarylamine (PTAA) or counter electrode. Figure 6.9 (e) exhibits the cross-sectional and titled SEM image of 3D bilayer with a pillared $MAPbX_3$ overlayer. This hybrid material have enough potential to harvest more light from solar spectrum and provide high-efficiency solar cells at a relatively low cost. The solar cell device made up of porous layer of metal oxide (TiO_2), which was covered by sunlight-absorbing hybrid $MAPbX_3$ and a hole transporting materials (PTTA) and counter electrode (Au) as shown in Figure 6.9 (e). Therefore, hybrid solar cells have been attracting a great deals of scientists attention in the same field.

Scientists used another approach to improve the moisture stability of PSCs, where they replaced the halide anion with thiocyanate anion (SCN). The modified PSCs with SCN ion shows higher stability as compare to $MAPbI_3$ film, this may be due to a stronger interaction between Pb^{2+} and SCN^- than Pb^{2+} and I^- anion [53, 54]. Jiang *et al.* reported the synthesis of $MAPb(SCN)_2I$ perovskite film where two iodide ions are replaced by pseudohalide thiocyanate ions (SCN–) [54]. They also tested the stability of prepared PSCs and found to be higher than conventional $MAPbI_3$ film when exposed to humid (95%) air. The stability against humidity can also be increased by synthesizing more stable perovskite structures using multiple cations (alloying) such as MA^+ and FA^+.

In addition to the issue of stability for PSCs, lead toxicity is also a big hurdle for industrialization due to its various health effect therefore leaving a negative impact on users leading to avoiding its market acceptance of the final product. World Health Organization (WHO) has announced that Pb is one of the ten most toxic elements for the environment, as well as human health. Therefore, European Union strictly and clearly prohibit the use of hazardous substances in electrical and electronic devices to avoid its exposure to the environment and people [55]. To avoid the issue of Pb toxicity in PSCs solar cells, scientists start the manufacture of less toxic materials by replacing the Pb with another low toxic metal. According to ionic size and Goldschmidt tolerance factor, which predicts stability and formability of perovskite structures, different cations can replace the Pb from perovskites. There are a wide range of elements available in the periodic table which can be considered for alternative of Pb-free PSCs, such as group 14 elements like Sn^{2+}, Ge^{2+}, alkaline earth metals such as Be^{2+}, Mg^{2+}, Ca^{2+}, Sr^{2+}, and Ba^{2+}, transition metals such as V^{2+}, Mn^{2+}, Fe^{2+}, Co^{2+}, Ni^{2+}, Pd^{2+}, Cu^{2+}, Zn^{2+}, Cd^{2+}, and Hg^{2+}, lanthanides like Eu^{2+}, Tm^{2+}, and Yb^{2+}, and p-block elements like Ga^{2+}, and In^{2+}, However, keep in mind the stability and efficiency of PSCs, Sn^{2+}, Ge^{2+}, Mg^{2+}, Mn^{2+}, Ni^{2+}, and Co^{2+} looks promising alternative of Pb^{2+} [56]. To reduce the toxicity of Pb, Ogomi *et al.* synthesized the Sn and Pb mixed ($MASn_{0.5}Pb_{0.}5I_3$) PSCs as the light absorber in 2014 first time and they got efficiency up to 4.18% [57]. The reason for choosing Sn for replacing Pb in PSCs may be because of its similar ionic radii (1.35 A^0) as $Pb2^+$ (1.49 A^0). Therefore, Noel *et al.* report the synthesis of the first lead-free perovskite $MASnI_3$ film with a direct bandgap of about 1.3 eV, which is close to 1.5 eV for $MASnI_3$ film and reporting 6.4% PCE [58]. Figure 6.10 (a) exhibits the XRD spectrum of $MASnI_3$ and the structure is almost identical to the most widely used perovskite film of $MAPbI_3$ and corresponding to the tetragonal conformation of the perovskite structure. Figure 6.10 (b) shows

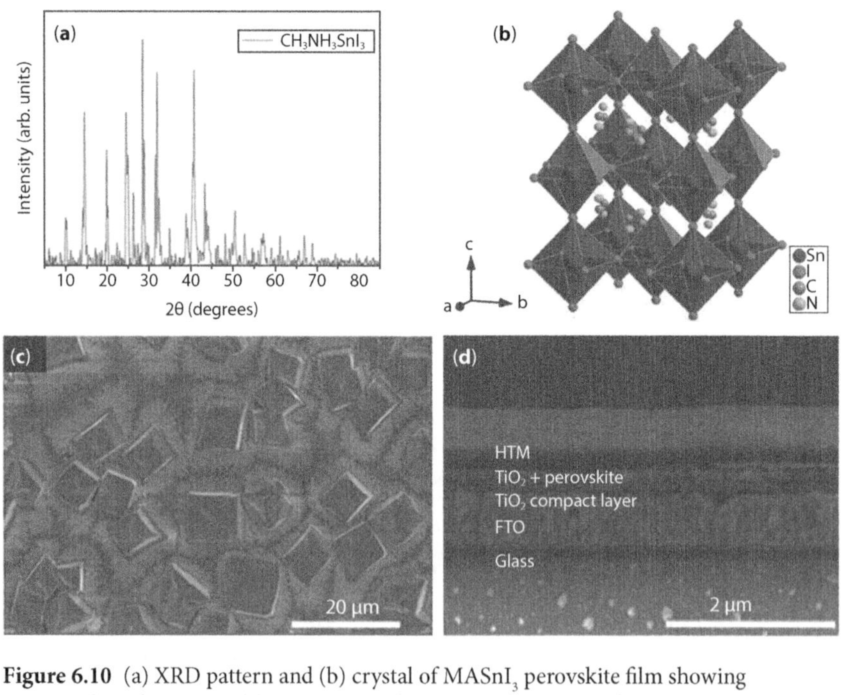

Figure 6.10 (a) XRD pattern and (b) crystal of $MASnI_3$ perovskite film showing tetragonal conformation, (c) SEM image of $MASnI_3$ (top-view) and (d) Cross-sectional view of complete solar cell device with $MASnI_3$ as an absorbing layer [58].

the corresponding crystal structure of $MASnI_3$ presenting the tetragonal conformation of the structure. They also analyze the surface morphology of the synthesized $MASnI_3$ film using SEM and showing randomly oriented crystals as presented in Figure 6.10 (c). The perovskite solar cell device containing $MASnI_3$ film as a light absorber with ETL and HTL is shown in Figure 6.10 (d).

Different perovskite film was synthesized by various scientists to further increase the PCE (14.35%), but the PCE comparable to lead halide-based PSCs did not achieve [59].

6.4 Conclusion

This chapter summarizes the recent progress in the development of perovskite-based solar cells and discussed in term of composition, structure, morphology, device architecture, stability, and interfacial engineering. The remarkable properties, such as small exciton binding energy, tunable bandgaps, large absorption coefficient, ambipolar charge transport properties, and long carrier diffusion, of perovskite materials, making them a promising light-absorbing material for solar cells. According to the recent, Best Research-Cell Efficiency PV Chart, the efficiency of single-junction PSCs has already achieved a 25.2% in a small period of time. To further improve the efficiency of PSCs, additives have the great possibility to promote the improvement in the crystal growth, perovskite film morphology, and electronic quality of film to address long term stability against the moisture, temperature, and toxicity issues. Various research focuses in this field are interface engineering, controlling morphologies of crystallization, changing cations and anions, and architecting the sandwiched device are carried out to improve the stability. However, the most efficient perovskite material are still based on toxic Pb cation. For commercialization, environmental, and human health safety PSCs should be based on non-toxic/Pb-free materials. However, replacing toxic Pb with Sn may be a promising way to fabricate less toxic PSCs but unfortunately, Sn suffers from easy oxidation. Therefore, it is necessary to study the other alternative metal ions.

References

1. Wang, W., Tadé, M.O., Shao, Z., Research progress of perovskite materials in photocatalysis- and photovoltaics-related energy conversion and environmental treatment. *Chem. Soc Rev.*, 44, 15, 5371–5408, 2015.

2. Kumar, A., Kumar, A., Krishnan, V., Perovskite Oxide Based Materials for Energy and Environment-Oriented Photocatalysis. *ACS Catal.*, 10, 17, 10253–10315, 2020.
3. Naseri, A., Samadi, M., Pourjavadi, A., Moshfegh, A.Z., Ramakrishna, S., Graphitic carbon nitride (g-C3N4)-based photocatalysts for solar hydrogen generation: Recent advances and future development directions. *J. Mater. Chem. A*, 5, 45, 23406–23433, 2017.
4. Zhang, G., Liu, G., Wang, L., Irvine, J.T.S., Inorganic perovskite photocatalysts for solar energy utilization. *Chem. Soc Rev.*, 45, 21, 5951–5984, 2016.
5. Sadhishkumar, S. and Balusamy, T., Performance improvement in solar water heating systems - A review. *Renew. Sustain. Energy Rev.*, 37, 191–198, 2014.
6. Chen, G., Seo, J., Yang, C., Prasad, P.N., Nanochemistry and nanomaterials for photovoltaics. *Chem. Soc Rev.*, 42, 21, 8304–8338, 2013.
7. Kitano, M. and Hara, M., Heterogeneous photocatalytic cleavage of water. *J. Mater. Chem.*, 20, 4, 627–641, 2010.
8. Siva Reddy, V., Kaushik, S.C., Ranjan, K.R., Tyagi, S.K., State-of-the-art of solar thermal power plants - A review. *Renew. Sustain. Energy Rev.*, 27, 258–273, 2013.
9. Wang, W., Tadé, M.O., Shao, Z., Research progress of perovskite materials in photocatalysis- and photovoltaics-related energy conversion and environmental treatment. *Chem. Soc Rev.*, 44, 15, 5371–5408, 2015.
10. Blandford, R. and Watkins, M., April 25, 1954: Bell Labs demonstrates the first practical silicon solar cell. *Am. Phys. Soc News*, 18, 42, 4–5, 2009.
11. Tsokos, K.A., *Physics for the IB Diploma Full Colour*, Cambridge University Press, 2010.
12. Chapin, D.M., Fuller, C.S., Pearson, G.L.,) A new silicon p-n junction photocell for converting solar radiation into electrical power. *J. Appl. Phys.*, 25, 5, 676–677, 1954.
13. Acik, M. and Darling, S.B., Graphene in perovskite solar cells: Device design, characterization and implementation. *J. Mater. Chem. A*, 4, 17, 6185–6235, 2016.
14. Ren, Y., Sun, D., Cao, Y., Tsao, H.N., Yuan, Y., Zakeeruddin, S.M., Wang, P., Grätzel, M., A Stable Blue Photosensitizer for Color Palette of Dye-Sensitized Solar Cells Reaching 12.6% Efficiency. *J. Am. Chem. Soc*, 140, 7, 2405–2408, 2018.
15. Xie, C., Classen, A., Späth, A., Tang, X., Min, J., Meyer, M., Zhang, C., Li, N., Osvet, A., Fink, R.H., Brabec, C.J., Overcoming Microstructural Limitations in Water Processed Organic Solar Cells by Engineering Customized Nanoparticulate Inks. *Adv. Energy Mater.*, 8, 13, 1702857, 2018.
16. Baek, S.W., Lee, S.H., Song, J.H., Kim, C., Ha, Y.S., Shin, H., Kim, H., Jeong, S., Lee, J.Y., A hydro/oxo-phobic top hole-selective layer for efficient and stable colloidal quantum dot solar cells. *Energy Environ. Sci.*, 11, 8, 2078–2084, 2018.

17. Jeon, N.J., Na, H., Jung, E.H., Yang, T.Y., Lee, Y.G., Kim, G., Shin, H.W., Il Seok, S., Lee, J., Seo, J., A fluorene-terminated hole-transporting material for highly efficient and stable perovskite solar cells. *Nat. Energy*, 3, 8, 682–689, 2018.
18. Shockley, W. and Queisser, H.J., Detailed balance limit of efficiency of p-n junction solar cells. *J. Appl. Phys.*, 32, 3, 510–519, 1961.
19. Comello, S., Reichelstein, S., Sahoo, A., The road ahead for solar PV power. *Renew. Sustain. Energy Rev.*, 92, 744–756, 2018.
20. Mohamad Noh, M.F., Teh, C.H., Daik, R., Lim, E.L., Yap, C.C., Ibrahim, M.A., Ahmad Ludin, N., Mohd, Y., Bin, A.R., Jang, J., Mat Teridi, M.A., The architecture of the electron transport layer for a perovskite solar cell. *J. Mater. Chem. C*, 6, 4, 682–712, 2018.
21. Wang, B., Iocozzia, J., Zhang, M., Ye, M., Yan, S., Jin, H., Wang, S., Zou, Z., Lin, Z., The charge carrier dynamics, efficiency and stability of two-dimensional material-based perovskite solar cells. *Chem. Soc Rev.*, 48, 18, 4854–4891, 2019.
22. Chen, W., Li, X., Li, Y., Li, Y., A review: Crystal growth for high-performance all-inorganic perovskite solar cells. *Energy Environ. Sci.*, 13, 7, 1971–1996, 2020.
23. Tanaka, H. and Misono, M., Advances in designing perovskite catalysts. *Curr. Opin. Solid State Mater. Sci.*, 5, 5, 381–387, 2001.
24. Assirey, E.A.R., Perovskite synthesis, properties and their related biochemical and industrial application. *Saudi Pharm. J.*, 27, 6, 817–829, 2019.
25. Yuan, Y., Xiao, Z., Yang, B., Huang, J., Arising applications of ferroelectric materials in photovoltaic devices. *J. Mater. Chem. A*, 2, 17, 6027–6041, 2014.
26. Yi, Z., Ladi, N.H., Shai, X., Li, H., Shen, Y., Wang, M., Will organic-inorganic hybrid halide lead perovskites be eliminated from optoelectronic applications? *Nanoscale Adv.*, 1, 4, 1276–1289, 2019.
27. Wells, H.L., Über die Cäsium- und Kalium-Bleihalogenide. *Z. für Anorg. Chemie*, 3, 1, 195–210, 1893.
28. Mitzi, D.B., Feild, C.A., Harrison, W.T.A., Guloy, A.M., Conducting tin halides with a layered organic-based perovskite structure. *Nature*, 369, 6480, 467–469, 1994.
29. Kojima, A., Teshima, K., Miyasaka, T., Shirai, Y., Novel Photoelectrochemical Cell with Mesoscopic Electrodes Sensitized by Lead-Halide Compounds (2). *ECS Meet. Abstr.*, MA2006-02, 7, 397, 2006.
30. Kojima, A., Teshima, K., Shirai, Y., Miyasaka, T., Organometal halide perovskites as visible-light sensitizers for photovoltaic cells. *J. Am. Chem. Soc*, 131, 17, 6050–6051, 2009.
31. Im, J.H., Lee, C.R., Lee, J.W., Park, S.W., Park, N.G., 6.5% efficient perovskite quantum-dot-sensitized solar cell. *Nanoscale*, 3, 10, 4088–4093, 2011.
32. Yang, W.S., Noh, J.H., Jeon, N.J., Kim, Y.C., Ryu, S., Seo, J., Seok, S., Il High-performance photovoltaic perovskite layers fabricated through intramolecular exchange, in: *Science (80-.)*, vol. 348, pp. 1234–1237, 2015.

33. Yang, W.S., Park, B.W., Jung, E.H., Jeon, N.J., Kim, Y.C., Lee, D.U., Shin, S.S., Seo, J., Kim, E.K., Noh, J.H., Seok, S., Il (2017) Iodide management in formamidinium-lead-halide-based perovskite layers for efficient solar cells, in: *Science (80-.)*, vol. 356, pp. 1376–1379.
34. NREL National Renewable Energy Laboratory, in: *Best research-cell efficiency char*, 2021, https://www.nrel.gov/pv/cellefficiency. html.
35. Li, H. and Zhang, W., Perovskite Tandem Solar Cells: From Fundamentals to Commercial Deployment. *Chem. Rev.*, 120, 18, 9835–9950, 2020.
36. Tai, Q., Tang, K.C., Yan, F., Recent progress of inorganic perovskite solar cells. *Energy Environ. Sci.*, 12, 8, 2375–2405, 2019.
37. Kim, H.S., Lee, C.R., Im, J.H., Lee, K.B., Moehl, T., Marchioro, A., Moon, S.J., Humphry-Baker, R., Yum, J.H., Moser, J.E., Grätzel, M., Park, N.G., Lead iodide perovskite sensitized all-solid-state submicron thin film mesoscopic solar cell with efficiency exceeding 9%. *Sci. Rep.*, 2, 1, 1–7, 2012.
38. Giordano, F., Abate, A., Correa Baena, J.P., Saliba, M., Matsui, T., Im, S.H., Zakeeruddin, S.M., Nazeeruddin, M.K., Hagfeldt, A., Graetzel, M., Enhanced electronic properties in mesoporous TiO2 via lithium doping for high-efficiency perovskite solar cells. *Nat. Commun.*, 7, 1, 1–6, 2016.
39. Yang, W.S., Noh, J.H., Jeon, N.J., Kim, Y.C., Ryu, S., Seo, J., Seok, S., Il High-performance photovoltaic perovskite layers fabricated through intramolecular exchange, in: *Science (80-.)*, vol. 348, pp. 1234–1237, 2015.
40. Li, X., Bi, D., Yi, C., Décoppet, J.D., Luo, J., Zakeeruddin, S.M., Hagfeldt, A., Grätzel, M., A vacuum flash-assisted solution process for high-efficiency large-area perovskite solar cells, in: *Science (80-.)*, vol. 353, pp. 58–62, 2016.
41. Bi, D., Yi, C., Luo, J., Décoppet, J.D., Zhang, F., Zakeeruddin, S.M., Li, X., Hagfeldt, A., Grätzel, M., Polymer-templated nucleation and crystal growth of perovskite films for solar cells with efficiency greater than 21%. *Nat. Energy*, 1, 10, 16142, 2016.
42. Yang, W.S., Park, B.W., Jung, E.H., Jeon, N.J., Kim, Y.C., Lee, D.U., Shin, S.S., Seo, J., Kim, E.K., Noh, J.H., Seok, S., Il Iodide management in formamidinium-lead-halide-based perovskite layers for efficient solar cells, in: *Science (80-.)*, vol. 356, pp. 1376–1379, 2017.
43. Li, C., Soh, K.C.K., Wu, P., Formability of ABO3 perovskites. *J. Alloys Compd.*, 372, 1–2, 40–48, 2004.
44. Turren-Cruz, S.H., Hagfeldt, A., Saliba, M., Methylammonium-free, high-performance, and stable perovskite solar cells on a planar architecture, in: *Science (80-.)*, vol. 362, pp. 449–453, 2018.
45. Ahmad, W., Khan, J., Niu, G., Tang, J., Inorganic CsPbI3 Perovskite-Based Solar Cells: A Choice for a Tandem Device. *Sol. RRL*, 1, 7, 1–9, 2017.
46. Liu, C., Li, W., Zhang, C., Ma, Y., Fan, J., Mai, Y., All-Inorganic CsPbI2Br Perovskite Solar Cells with High Efficiency Exceeding 13%. *J. Am. Chem. Soc*, 140, 11, 3825–3828, 2018.
47. Sutton, R.J., Eperon, G.E., Miranda, L., Parrott, E.S., Kamino, B.A., Patel, J.B., Hörantner, M.T., Johnston, M.B., Haghighirad, A.A., Moore, D.T., Snaith,

H.J., Bandgap-Tunable Cesium Lead Halide Perovskites with High Thermal Stability for Efficient Solar Cells. *Adv. Energy Mater.*, 6, 8, 1–6, 2016.
48. Yang, J., Siempelkamp, B.D., Liu, D., Kelly, T.L., Investigation of CH3NH3PbI3degradation rates and mechanisms in controlled humidity environments using *in situ* techniques. *ACS Nano*, 9, 2, 1955–1963, 2015.
49. Christians, J.A., Miranda Herrera, P.A., Kamat, P.V., Transformation of the excited state and photovoltaic efficiency of CH3NH3PbI3 perovskite upon controlled exposure to humidified air. *J. Am. Chem. Soc*, 137, 4, 1530–1538, 2015.
50. Noh, J.H., Im, S.H., Heo, J.H., Mandal, T.N., Seok, S., Il Chemical management for colorful, efficient, and stable inorganic-organic hybrid nanostructured solar cells. *Nano Lett.*, 13, 4, 1764–1769, 2013.
51. Wang, Z., Shi, Z., Li, T., Chen, Y., Huang, W., Stability of Perovskite Solar Cells: A Prospective on the Substitution of the A Cation and X Anion. *Angew. Chemie - Int. Ed.*, 56, 5, 1190–1212, 2017.
52. Noh, J.H., Im, S.H., Heo, J.H., Mandal, T.N., Seok, S., Il Chemical management for colorful, efficient, and stable inorganic-organic hybrid nanostructured solar cells. *Nano Lett.*, 13, 4, 1764–1769, 2013.
53. Tai, Q., You, P., Sang, H., Liu, Z., Hu, C., Chan, H.L.W., Yan, F., Efficient and stable perovskite solar cells prepared in ambient air irrespective of the humidity. *Nat. Commun.*, 7, 1, 1–8, 2016.
54. Jiang, Q., Rebollar, D., Gong, J., Piacentino, E.L., Zheng, C., Xu, T., Pseudohalide-Induced Moisture Tolerance in Perovskite CH3NH3 Pb(SCN) 2 I Thin Films. *Angew. Chemie*, 127, 26, 7727–7730, 2015.
55. Prüss-Ustün, A., Wolf, J., C. Corvalán, R.B., Neira, M., Preventing disease through healthy environments. *World Heal. Organ.*, 12, 1–106, 2016.
56. Hoefler, S.F., Trimmel, G., Rath, T., Progress on lead-free metal halide perovskites for photovoltaic applications: a review. *Monatshefte fur Chemie*, 148, 5, 795–826, 2017.
57. Ogomi, Y., Morita, A., Tsukamoto, S., Saitho, T., Fujikawa, N., Shen, Q., Toyoda, T., Yoshino, K., Pandey, S.S., Ma, T., Hayase, S., CH3NH3SnxPb(1-x) I3 perovskite solar cells covering up to 1060 nm. *J. Phys. Chem. Lett.*, 5, 6, 1004–1011, 2014.
58. Noel, N.K., Stranks, S.D., Abate, A., Wehrenfennig, C., Guarnera, S., Haghighirad, A.A., Sadhanala, A., Eperon, G.E., Pathak, S.K., Johnston, M.B., Petrozza, A., Herz, L.M., Snaith, H.J., Lead-free organic-inorganic tin halide perovskites for photovoltaic applications. *Energy Environ. Sci.*, 7, 9, 3061–3068, 2014.
59. Yang, S., Fu, W., Zhang, Z., Chen, H., Li, C.Z., Recent advances in perovskite solar cells: Efficiency, stability and lead-free perovskite. *J. Mater. Chem. A*, 5, 23, 11462–11482, 2017.

7

$BiFeO_3$-Based Materials For Augmented Photoactivity

Rashmi Acharya*, **Lopamudra Acharya and Kulamani Parida**†

Center for Nano Science and Nano Technology, ITER, Siksha 'O' Anusandhan Deemed to be University, Bhubaneswar, Odisha, India

Abstract

Semiconductor-mediated photocatalysis has gained immense attention in the past few decades owing to its potential application in environment and energy fields. Several oxide semiconductors, like TiO_2, $SrTiO_3$, ZnO, SnO_2, and so on, have been applied in various photocatalytic applications. Owing to the wide band-gap energy, these semiconductor photocatalysts absorb only ultraviolet radiation, which consists of about 5% of sunlight and hence, their practical applications are greatly limited. Therefore, visible-light-responsive photocatalysts have garnered tremendous importance. Perovskite bismuth ferrite (BFO) is considered as a third-generation visible light active photocatalyst because of its exceptional band gap energy (2.0–2.7 eV) corresponding to visible region, chemical stability, ferromagnetic, and ferroelectric features. Nevertheless, its photocatalytic activity is far away from expectation because of low specific surface area, rapid rate of charge carriers' recombination, inadequate conduction band edge position with respect to oxygen (or H_2) reduction level. Thus, various strategies are adopted to improve the photoactivity of BFO. Heterostructure construction through elemental doping and compositing with other semiconductor has been found beneficial.

The present chapter consists of four sections. First section introduces the significance of photocatalysis, basic principle of visible light responsive photocatalytic reactions, various applications of semiconductor-based photocatalysis, and importance of BFO as a photocatalyst. The structure, physicochemical properties, and photocatalytic activity of pristine BFO were discussed in the next section.

**Corresponding author*: drrashmiacharya75@gmail.com; rashmiacharya@soa.ac.in

†*Corresponding author*: kulamaniparida@soa.ac.in

Khursheed Ahmad and Waseem Raza (eds.) Perovskite Materials for Energy and Environmental Applications, (167–216) © 2022 Scrivener Publishing LLC

In the third section, we have described strategies, particularly elemental doping and heterojunction construction applied in improving photocatalytic performance of BFO. Augmented visible light-induced photoactivity was highlighted in terms of extended light absorption range, reduction in recombination rate, and modulation of band structures. Fourth section concludes with merits, limitations, and perspectives of modification strategies for BFO.

***Keywords*:** BFO, elemental doping, heterojunction, visible light responsive, photoactivity

7.1 Introduction

Rapid consumption of fossil fuels poses growing challenges in environmental pollutions and energy demands in the past few decades. Global energy supply is projected to be twice of that of currently required by 2050 owing to fast population growth and vast industrialization [1, 2]. Recently, the global energy requirement is usually met by fossil fuels like coal, natural gas, and petroleum products which are being depleted rapidly. Fossil fuel consumption will undoubtedly lead to detrimental emissions which create several environmental issues. Further, supply of clean and safe drinking water has made the effort to remove toxic heavy metal ions and organic pollutants from natural aqueous ecosystems [3, 4]. Several technologies such as reverse osmosis, ultrafiltration, adsorption, biological degradation, absorption with or without chemical reaction, ion exchange, and photocatalysis have already been used for the treatment of water [5–9]. The adsorption process involves the disadvantage of only transfer of the pollutants to another phase without their destruction [10]. An ideal technology should be applied to all kinds of pollutants in such a way that it should not be energy consuming and should not emit environmentally detrimental by-products [11]. Moreover, the increasing CO_2 level in the atmosphere owing to excess fossil fuels consumption leads to global warming and energy crisis. In recent years, attentions are made to reduce CO_2 emission through conversion of solar energy into chemical energy [12]. In this context, low cost and effective technologies are required to solve energy and environmental problems [13, 14]. Therefore, material scientists and engineers across the globe have pursued research to overcome these obstacles for effective energy conversion and environmental protection [15].

Semiconductor-mediated photocatalysis has been considered as an economically viable technology among different renewable energy ventures as it uses clean and inexhaustible solar energy [16–20]. Fundamental mechanism of this technology basically involves four key stages which were presented in Scheme 7.1. Stage 1 is associated with light harvesting, which is strongly dependent on structure and surface morphology of photocatalysts. Harvestation of light can mainly be improved through the construction of the hierarchical macroporous or mesoporous architectures as these structures utilise light efficiently through their multiple reflections and scattering effects [21]. Stage 2 involves excitation of charges, which is dependent on unique electronic structures of the photocatalyst. In this process, an electron can jump from the valence band (VB) to the conduction band (CB) of a semiconductor by absorbing suitable amount of radiant energy. As a result, a hole (h^+) is created in the VB.

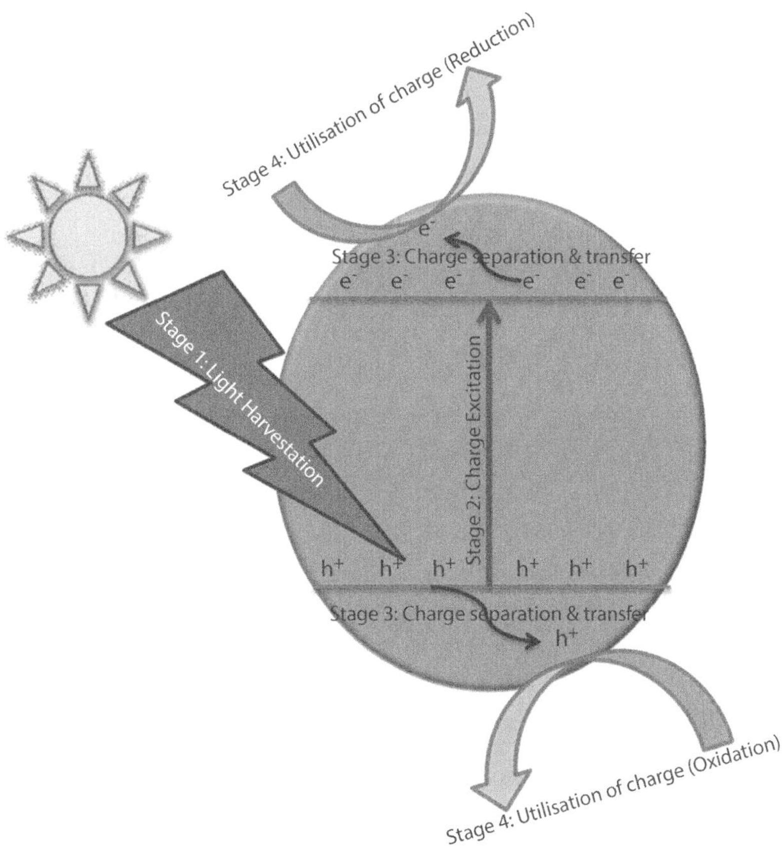

Scheme 7.1 Fundamental mechanism of semiconductor-mediated photocatalysis.

Third step (stage 3) is involved in charge separation and transfer. The hole-electron pairs recombination on the surface and in the bulk of a semiconductor might cause in efficient charge separation and their transport to surface/interface active sites. It is usually considered as a decisive factor to determine the quantum efficiency. By minimizing the charge carriers' diffusion length or building electric fields at the interface could cause the reduction in the recombination rates which significantly boost the photoactivity of the semiconductor [22]. In stage 4, there occurs utilization of charges through numerous surface active electro-catalytic reactions. The energetic excitons emigrate to the semiconductor surface without recombination and are captured by the co-catalysts or surface active sites. The photoexcited electrons undergo reduction reaction at the CB while the holes are responsible for oxidation reaction at the VB with the surface adsorbed reactants. However, these reactions occur only if the reduction potential is placed below the CB minimum potential and oxidation potential should be less positive than VB maximum potential. Further, these redox reactions taking place at the semiconductor surface depend on the large onset over potential and sluggish kinetics. Incorporation of appropriate co-catalysts (electro-catalysts) can supress these two restrictive factors [23]. In addition, the co-catalysts (electro-catalysts) can boost the photostability and ability for charge separation of a semiconductor [24]. Overall, the photocatalytic quantum efficiency (η_c) for generation of H_2 is usually determined by the cumulative effect of light harvestation efficiency (η_{abs}), charge separation efficiency (η_{cs}), charge utilization efficiency (η_{cu}), charge migration, and transfer efficiency (η_{cmt}). This relationship is presented in Equation 7.1 [22]:

$$\eta_c = \eta_{abs} \times \eta_{cs} \times \eta_{cmt} \times \eta_{cu} \tag{7.1}$$

Photocatalytic technology has attracted substantial interdisciplinary attention because of its diverse potential applications in energy fields and environmental issues. Future energy and environmental crisis can conveniently be addressed by direct conversion of solar energy to chemical energy and energy fuels [25–30]. Along with sun light, photocatalysis needs a robust semiconductor to carry out various catalytic reactions like splitting of water to generate sustainable H_2 energy [31–37], conversion of CO_2 to hydrocarbon fuels [38–41], bacteria disinfection [42, 43], pollutants degradation [44, 45] and selective organic transformations [46, 47] and nitrogen (N_2) fixation [48]. These photocatalytic reactions are briefly described below.

7.1.1 Photocatalytic Water Splitting

Hydrogen (H_2) is regarded as a renewable, sustainable, green, and clean energy source to meet global energy crisis and to overcome emerging CO_2 emissions [49]. The conventional processes like partial oxidation, coal gasification, steam reforming, and so on, used for generation of H_2 energy from natural fuels are restricted due to stringent environmental regulations and high cost [50]. H_2 generation via photocatalytic splitting of water is considered as a green technique because of use of inexhaustible solar energy and abundantly available water resources. In photocatalytic water splitting, water is oxidized by h_{vb}^+ present at VB to produce O_2 as per Equation 7.2 only if the band edge VB potential is more positive than the oxidation potential for O_2 evolution ($E^o_{O2/H2O}$ = 1.23 V vs NHE).

$$2H_2O + 4h_{VB}^+ \rightarrow O_2 + 4H^+ \qquad (7.2)$$

The H^+ ions so formed are reduced by e_{cb}^- at CB to produce H_2 gas as shown in Equation 7.3 only when the potential at CB is more anodic than H_2 redox potential ($E^o_{H+/H2}$ = 0 V vs NHE).

$$2H^+ + 2e_{CB}^- \rightarrow H_2 \qquad (7.3)$$

7.1.2 Photocatalytic Conversion of CO_2

Environmental issues like climate changes and global warming are now persisting mainly due to the rising concentration of atmospheric greenhouse gases particularly CO_2. The existing technologies used to minimise CO_2 level are expensive and energy consuming [51, 52]. Recently, semiconductor-mediated visible light responsive photocatalytic CO_2 reduction has been emerged as a sustainable and viable approach for minimizing concentration of atmospheric CO_2 as well as simultaneous production of energy fuels like CH_4 [53]. Photocatalytic CO_2 conversion of involves CO_2 and H_2O molecules adsorption on the semiconductor surface. These adsorbed species are then converted to CH_4 and O_2 under required light energy irradiation as per equation 7.4.

$$CO_2 + 2H_2O \xrightarrow[\text{Semiconductor}]{hv} CH_4 + 2O_2 \qquad (7.4)$$

The mechanism of photocatalytic reduction of CO_2 is associated with the release of e_{cb}^- at CB and h_{vb}^+ at VB under necessary light energy irradiation. Then CO_2 is reduced by e_{cb}^- to form CH_4 only when the CB minimum potential is placed above CO_2/CH_4 redox couple (– 0.24 V vs. NHE) [54]. Similarly, more positive VB maximum potential than the oxidation potential of H_2O/H^+ couple (+ 0.82 V vs. NHE) allows water oxidation by h_{vb}^+ [55].

7.1.3 Photocatalytic Fixation of Nitrogen

The semiconductor-based photocatalytic conversion of nitrogen (N_2) to NH_3 is gaining tremendous attention principally due to application of prolific ingredients like water and air as well as clean, inexpensive and abundantly available solar light as driving force. In photocatalytic fixation of N_2, irradiation of required radiant energy on the semiconductor surface promotes electron to the CB (e_{cb}^-) with creation of holes (h_{vb}^+) at the VB. Then water is oxidized by the so formed h_{vb}^+ to produce O_2 along with the abstraction of protons (H^+ ions) after satisfying required thermodynamic conditions. That is the maximum VB potential should be more positive than the O_2 evolution potential. Adsorbed N_2 molecules are then reduced by e_{cb}^- on the surface of the semiconductor with the help of abstracted protons, only when the CB potential lies above N_2/NH_3 redox couple [48]. The overall photocatalytic fixation of N_2 is expressed in the equation shown below.

$$N_2 + 3H_2O \xrightarrow{hv} 2NH_3 + 1.5O_2 \tag{7.5}$$

7.1.4 Selective Organic Transformation for the Synthesis of Fine Chemicals

Selective organic transformations through photocatalysis are now gaining importance in comparison to the conventional processes for fine chemicals synthesis mainly due to two reasons. Firstly, it limits the use of oxidizing agents, reducing agents, and heavy metal catalysts, which are the chemical reagents detrimental to the environment. Secondly, this process avoids the energy consuming conditions like high temperature and high pressure [56–59]. During semiconductor-based photocatalysis, the molecular O_2 combine with e_{cb}^- as mentioned in the Equation 6 to form strong oxidizing superoxide anionic species ($O_2^{\bullet -}$). Similarly, hydroxyl radicals ($^{\bullet}OH$) with strongly oxidizing property are formed by the reaction of either adsorbed water molecules or surface hydroxyl groups (OH) with h_{vb}^+ as per Equation 7.7.

$$e_{cb}^- + O_2 \rightarrow O_2^{\bullet -} \tag{7.6}$$

$$h_{vb}^{+} + {}^{-}OH \text{ (or } H_2O) \rightarrow {}^{\bullet}OH \quad (7.7)$$

These reactive species react under suitable reaction conditions with given organic compounds to carry out selective organic transformations. For instance, h_{vb}^{+} and $O_2^{\bullet -}$ are the active species which carry out selective oxidation of alcohol to aldehyde on CdS/TiO_2 surface under irradiation of visible light [60].

7.1.5 Photodegradation of Pollutants

The photocatalytic degradation of pollutants was first carried out in 1977 by Frank and Bard to reduce CN^- in water [61, 62]. Thereafter, a substantial research was pursued to decompose on lethal organic compounds and reduce toxic ions of heavy metals [Cr (VI)] through semiconductor photocatalysis upon suitable light energy illumination. The organic pollutants photodegradation proceeds through the production of $O_2^{\bullet -}$ and $^{\bullet}OH$ as reactive species were mentioned in Equation 7.6 and 7.7. Further, another active species $^{\bullet}OOH$ can be obtained by the reaction of $O_2^{\bullet -}$ species with H^+ ions as shown in Equation 7.8.

$$O_2^{\bullet -} + H^+ \rightarrow {}^{\bullet}OOH \quad (7.8)$$

The organic pollutants undergo decomposition reaction by these reactive species so formed ($^{\bullet}OH$, $O_2^{\bullet -}$ and $^{\bullet}OOH$) along with photogenerated electrons and hole pairs to form harmless products like H_2O and CO_2 (Equation 7.9).

$$^{\bullet}OOH + \text{dyes} \rightarrow CO_2 + H_2O \quad (7.9)$$

The photoreduction of toxic, mutagenic, and carcinogenic aqueous Cr (VI) to less toxic Cr (III) over semiconductor photocatalyst has now been imperative to restrict chromium contamination among downstream users and aquatic lives. The photocatalytic process involves oxidation of water by the holes at VB to liberate H^+ ions followed by reduction of Cr (VI) species by electrons at CB in presence of protons. These redox reactions are presented in equation 7.10 and 7.11 respectively [63].

$$Cr_2O_7^{2-} + 14H^+ + 6\,e_{cb}^{-} \rightarrow 2Cr^{3+} + 7H_2O \quad (7.10)$$
$$E^{o}_{Cr(VI)/Cr(III)} = 1.33\text{ V}$$

$$2H_2O + 4\,h_{vb}^{+} \rightarrow O_2 + 4H^{+} \qquad (7.11)$$
$$E^{o}_{O2/H2O} = 1.23\ V$$

In 1972, Fujishima and Honda had carried out landmark event to split water over TiO_2 electrodes using ultraviolet (UV) radiation [64]. Then, TiO_2-mediated photocatalytic decomposition of organic pollutants in aqueous suspensions was investigated by Carey and his co-investigators in 1976 [65]. After 3 years of this invention, semiconductor powders, including TiO_2, ZnO, CdS, and so on, were used in aqueous medium by Inoue and co-workers for photocatalytic conversion of CO_2 to various organic compounds [66]. Since then, significant development in designing highly efficient semiconductor photocatalysts has been made. Several photocatalysts potentially active in UV region, such as TiO_2, ZnO, $SrTiO_3$, and so on and in visible region, like CdS, $BiVO_4$, g-C_3N_4, and so on, have extensively been employed to harness solar energy for different redox reactions [67–71]. Development of highly efficient semiconductor photocatalyst system to overcome the upcoming energy shortages and emerging environmental pollution has now become a hot research field. In recent years, efforts are given to develop visible-light-active photocatalysts for successful utilization of the solar spectrum, which consists of about 43% of visible light. In search of efficient visible light active semiconductor photocatalysts, perovskite bismuth ferrite (BFO) has stimulated ripples of excitement among the research communities as one of the next generation photocatalysts because of its narrow band gap (2.0–2.7 eV), exceptionally high chemical stability, ferroelectric, and ferromagnetic properties. Nevertheless, its photocatalytic activity is far away from expectation because of low specific surface area, rapid charge carriers recombination, inadequate conduction band edge position with respect to the oxygen (or H_2) reduction potential [72, 73]. Thus, various strategies are adopted to improve the photoactivity of BFO. Heterostructure construction through elemental doping and compositing with other semiconductor has been found beneficial. In this chapter, we have briefly discussed the structure, physicochemical properties, and photocatalytic activity of pristine BFO. Various modification strategies, such as elemental doping and heterojunction construction in particular adopted to optimise the photoactivity of BFO were described. Augmented visible light-induced photoactivity was highlighted in terms of extended light absorption range, reduction in recombination rate, and modulation of band structures. Various photocatalytic applications of modified BFO were then depicted along with discussion of plausible mechanism. Fourth section concludes with merits, limitations, and perspectives of modification strategies for BFO.

7.2 Structure, Physicochemical, and Photocatalytic Activity of $BiFeO_3$

BFO possesses perovskite structure. Perovskites are represented by the general formula ABX_3 [74]. In this case, A and B are cations with A larger in size than that of B whereas X is either a halide or an oxide. The crystals properties are controlled by the cations. Perovskite belongs to the space group of Pm3m, and it has cubic symmetry. B cations are 6-fold co-ordinated and A cations are 12 fold cubo octohedral co-ordinated and are surrounded by an octahedron of X-anions. The structure of ABO_3 perovskites is presented in Figure 7.1. ABO_3 type of perovskites exhibits lattice distortions due to which crystal phases are changed with a sequence orthogonal, tetragonal, monoclinic, and triclinic phases. Different degree of orientation in structure has profound impact on electronic and optical properties of perovskite semiconductors. ABO_3 materials are regarded as suitable photocatalysts in comparison to the other photosensitive semiconductors because of their facile design, adequate band gap energy along with required photophysical properties exhibited by the A and B cations [75]. ABO_3 perovskite photocatalysts have been classified into four different groups, namely (i) Titanates: $CaTiO_3$, $FeTiO_3$ (ii) Tantalates: $KTaO_3$, (iii) Ferrites: $LaFeO_3$ (iv) others like $LaCaO_3$, $LaNiO_3$ [76, 77]. At room temperature, the multiferroic BFO exhibits a rhombohedral distorted perovskite structure with space group R3c. It exhibits Neel temperature at 370°C and curie temperature at 830°C. This material can easily be prepared at room temperature. However, it sometimes appears along with impure phases like $Bi_2Fe_4O_9$ and $Bi_{25}FeO_{39}$ [78, 79]. The magnetoelectric coupling property of BFO enables this multiferroic material to be used in piezoelectric devices, spintronics, nonvolatile memory, and sensors [80–82]. BFO is a widely studied photocatalyst due to its remarkable chemical stability, appropriate band gap energy to utilise visible light of solar radiation and

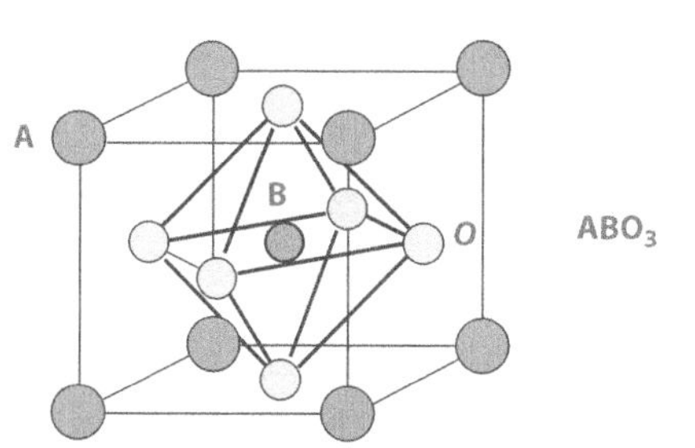

Figure 7.1 Structure of perovskite with formula ABO_3.

efficient separation of excitons resulted from its room temperature multiferroic behaviour [83–85].

Photocatalytic performance of a semiconductor is influenced by its crystalline nature, surface morphology, and particle size. Different synthesis methods, such as hydrothermal [86], sol-gel [87], solid-state reaction [88, 89], and so on, have been employed to achieve desired shape, size, and morphology of BFO particles. BFO samples hydrothermally prepared by Gao *et al.* decomposed 72% of MO under solar light irradiation [90]. Pure phase BFO prepared by Wang *et al.* through co-precipitation route degraded only 26% of MO [91]. Li and his co-investigators synthesized nanosized BFO (NS-BFO) by a simple hydrothermal technique. The specific surface area for NS-BFO was found to be10.60 m^2/g, which is increased by fourfold from that of micro-sized BFO (MS-BFO). Barrett-Joyner-Halenda (BJH) average pore size for NS-BFO was also found to be much larger (23.8 nm) as compared to that obtained for MS-BFO (17.2 nm). The synthesized nanoparticles exhibited enhanced photocatalytic efficiency under solar irradiation for degradation of methyl orange (MO) due to appreciably large photocurrent value, rapid charge transfer, and exceptionally high specific surface area [92]. Solid state reaction route was used by Pattnaik *et al.* to fabricate BFO nanoparticles. It was observed that by rising the synthesis temperature from 350 to 500 °C, the photoactivity was enhanced and further rise of temperature led to decrease the photocatalytic activity. The sample prepared at 500°C decolorized 77% a 10 ppm Congo red (CR) solution under solar irradiation. UV-Vis diffused reflectance spectra

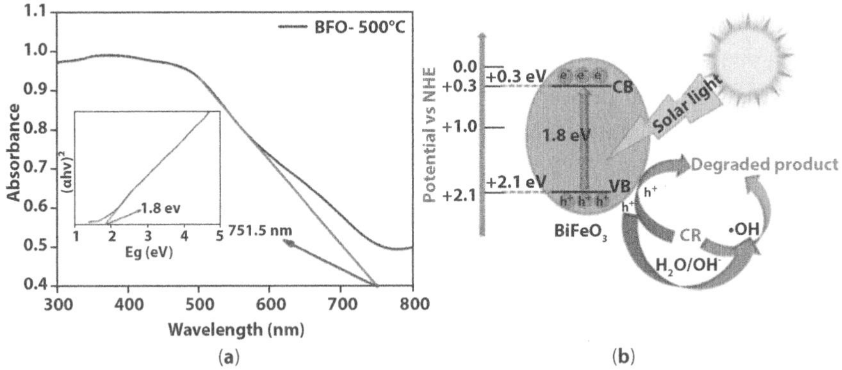

Figure 7.2 (a) UV-Vis DRS spectra of BFO nanoparticles synthesized at 500°C showing absorption maximum at 751.5 nm and the direct band gap energy of 1.7 eV (shown in the inset), (b) The plausible mechanism for visible light responsive BFO-mediated photocatalytic CR degradation (reproduced from ref. no. [93]).

of this sample showed an absorbance at 751.5 nm corresponding to visible region and band gap energy of 1.7 eV (Figure 7.2 (a)). The augmented photocatalytic performance of the sample might be attributed to inhibited excitons recombination, lowest band gap energy, and smallest crystallite size (30 nm). The plausible mechanism for visible light responsive BFO-mediated photocatalytic CR degradation was presented in Figure 7.2(b) [93]. Although BFO harnesses solar radiation (visible light) to display photoactivity, its performance needs to be improved.

7.3 Elemental Doping in BFO

Nowadays, BFO, a visible light active semiconductor material with narrow band gap of ~ 2.2 to 3 eV is regarded as one of the promising photocatalyst because of its nontoxicity, good chemical stability, ferroelectric, and ferromagnetic properties [93–95]. However, the photocatalytic performance of BFO is hindered by the fast recombination rate of photo-induced charge carriers and low conduction band (CB) position [96, 97]. To overcome these obstacles, introduction of transition metals and rare earth metal into BFO lattice has been reported as one of the successful approach [98, 99]. This is because dopants could act as e^-/h^+ trapping sites that promote the electron-hole separation efficiency. Also the doping causes insertion of an energy level below the CB of BFO [100, 101]. Considering the literature reports, it has been concluded that the rare earth elements are substituted on Bi^{3+} site whereas transition metal ions are doped in Fe^{3+} site in accordance with their ionic radius [96, 102, 103].

Elemental doping resulted in some structural and optical modifications in BFO. These structural and optical changes enable the visible light active BFO to exhibit augmented photoactivity. Several characterization techniques, such as XRD, UV-Vis DRS spectroscopy, photoluminescence, and so on, have been used to identify such modifications in BFO. In this section, we have discussed some of these characterization techniques to explain the modifications that are caused by elemental doping, which enhance the photoactive performance of BFO.

7.3.1 PXRD Studies

The PXRD pattern for pristine BFO shows main diffraction peaks at $2\theta=22.37°$ 31.79°, 32.01°, 39.45°, and 45.55° corresponding to (012), (104), (110), (202), and (024) plane, respectively, according to JCPDS card no. 71-2494 [104]. BFO exhibited a rhombohedral structure with R3C space

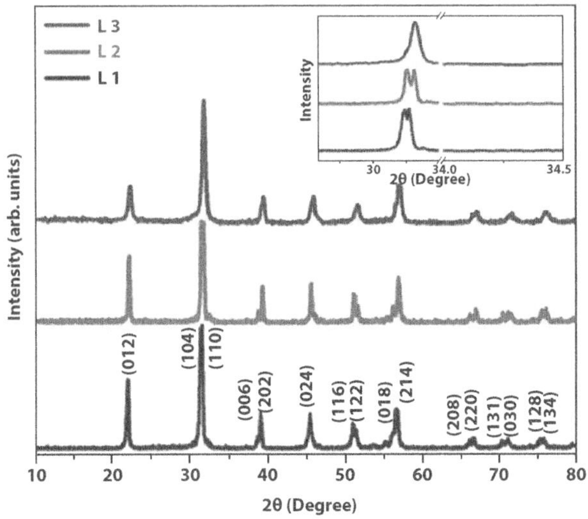

Figure 7.3 PXRD pattern of La doped BFO (reproduced from ref. no. [105]).

group [105]. With the increase of elemental doping content the twin peaks corresponding to (104)/(110) and (006)/(202) reflections are observed to be completely collapsed and also slightly shifted to higher 2θ value. These factors indicated the structural distortion of BFO from rhombohedral to orthorhombic [94, 106, 107].

According to Kumar *et al.* in case of La doped BFO, the twin peaks appear at 31.5° and 33°, 38.5°, and 40° completely collapsed into a singlet peak with increasing the La concentration from L 1 to L 3 as can be seen from the above Figure 7.3. This type of structural changes is attributed to the successful insertion of rare earth metal La into lattice of host metal Bi^{3+} [105]. Similarly, as reported by Ponjar *et al.* merging of twin peaks (006)/(202) and (104)/(110) with red shifting of 2θ value to higher angle indicated the phase change of BFO from rhombohedral to orthorhombic phase [106].

7.3.2 Morphological Studies

Morphology of pure and doped BFO is studied through Field Emission Scanning Electron Microscope (FESEM) and High Resolution Scanning Electron Microscope (HRTEM). Doping of rare earth metals and transition metal ions affects the structural properties of BFO. With the increasing concentration of dopants on the BFO lattice, reduction in particle sizes of BFO was observed due to the difference in ionic radii between the base

Figure 7.4 FESEM images of (a) pure BFO, (b–d) Sm (10%), and Co (5%, 10%, 15%) codoped BFO, respectively (reproduced from ref. no. [108]).

material and dopants. Due to the decrease in particle sizes, the agglomeration of particles increases upon introduction of metal ions to BFO [96].

Rhaman *et al.* have fabricated Sm and Co co-doped BFO taking Sm (10%) and Co (5%, 10%, 15%) and found that the particle size of pure BFO (Figure 7.4 (a)) remarkably decreases from 166 nm to 42 nm with increasing the amounts of dopants as shown in Figure 7.4 (b–d). The particle size of BFO is dependent on the O_2 vacancies created due to the presence of bismuth ion. Thus, the introduction of Sm and Co onto BFO lattice suppressed the formation of O_2 vacancies leading to the reduction of average particle size of doped BFO [108].

7.3.3 XPS Studies

Chemical states and elemental compositions of neat and doped BFO is analyzed through X-ray photoelectron spectroscopy (XPS) analysis. As reported by Zhang *et al.* two peaks at 158.5 and 163.8 eV appeared for Bi 4f spectrum due to Bi $4f_{7/2}$ and Bi $4f_{5/2}$, respectively suggesting the presence

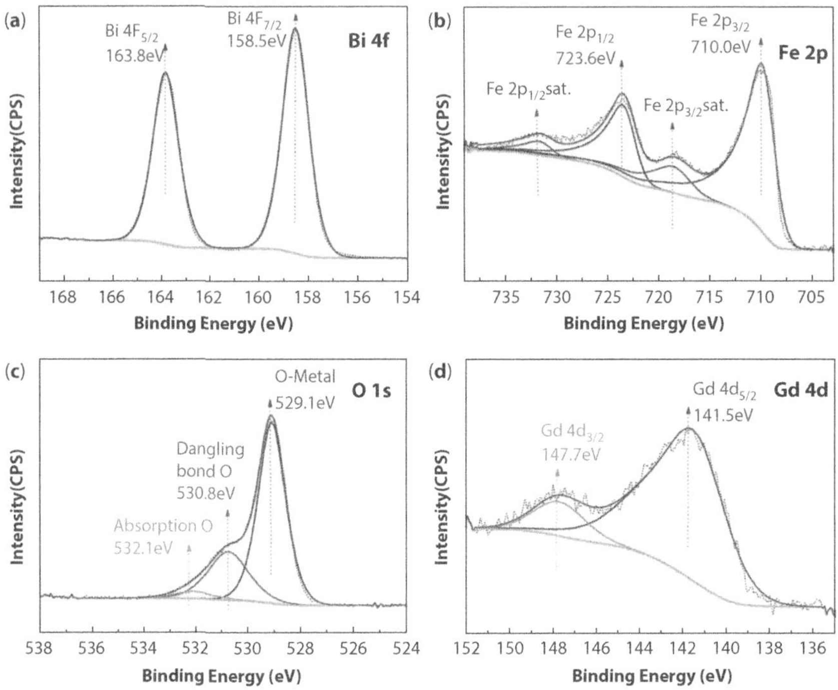

Figure 7.5 XPS spectra of (a) Bi 4f, (b) Fe 2p, (c) O 1s, and (d) Gd 4d of Gd doped BFO (reproduced from ref. no. [109]).

of bismuth in $BiFeO_3$ in Bi^{3+} oxidation state (Figure 7.5 (a)). As observed in Figure 7.5 (b), Fe 2p XPS spectrum is deconvoluted in to two peaks having binding energies around 710.0 and 723.6 eV corresponding to Fe $2p_{3/2}$ and Fe $2p_{1/2}$ in that order. Also the appearance of a satellite peak 8 eV above to the Fe $2p_{3/2}$ (718.2 eV) indicates that Fe is existed in Fe^{3+} state. Three different peaks at 529.1 eV, 530.8 eV, and 532.1 eV obtained for O 1s spectrum could be ascribed to oxygen-metal bond, the dangling bond, and surface adsorbed O_2 bond in that order (Figure 7.5 (c)). After doping of Gd into BFO, XPS spectrum for Gd 4d obtained at 141.5 and 147.7 eV corresponding to Gd 4d 5/2 and Gd 4d 3/2, respectively. This result indicates the presence of Gd in Gd^{+3} valence states as displayed in Figure 7.5 (d) [109].

7.3.4 Optical Property Studies

Doping of metal ions significantly affects the optical properties of BFO by improving the light response and decreasing the band gap energy. The light absorption capacity of BFO shows a red shift to higher wavelength value (visible range) upon metal ion doping [105]. Decrease in the band gap energy of doped BFO has been observed in comparison to pristine BFO. One of the reasons for the reduction of band gap energy after doping is the formation of an impurity level during cation doping either near the valence band or the conduction band of BFO. Another reason for the changes in the band gap energy is the oxygen defects created by the doping of metal ions to A or B site of BFO to maintain the electroneutrality [103, 106]. Doping of elements would create an extra energy level either below the conduction band edge or above the valance band edge of pristine BFO. If the energy level appear below the conduction band edge, that will trap the photo-excited electrons. While the holes will be trapped by the energy level created above the valence band. In this way, the dopants will perform as trapping agents for charge carriers thereby inhibiting the rate of recombination. Thus, the photogenerated electrons and holes will react with adsorbed O_2 and OH^- to generate superoxide ($^{\bullet}O_2^-$) and hydroxyl ($^{\bullet}OH$) radicals to facilitate the photocatalytic activity of doped BFO [105, 109]. The decreased band gap energy and broad visible light absorption ability of doped BFO are favourable for the photocatalytic performance.

As displayed in Figure 7.6 (a), pure BFO showed light absorption from UV to visible light range of about 563 nm. The light absorption intensity of BFO increases with increasing the concentration of Gd from 1 wt% to 3 wt% in BFO. According to Zhang *et al.* also a gradual decrease in the band gap energy was observed as the content of Gd in BFO increases. The calculated band gap energies of pristine BFO, Gd1%-BFO, Gd3%-BFO, and Gd5%-BFO was found to be 2.2, 2.18, 2.16, and 2.1, respectively. With improved light absorption ability and reduced band gap energy doped BFO

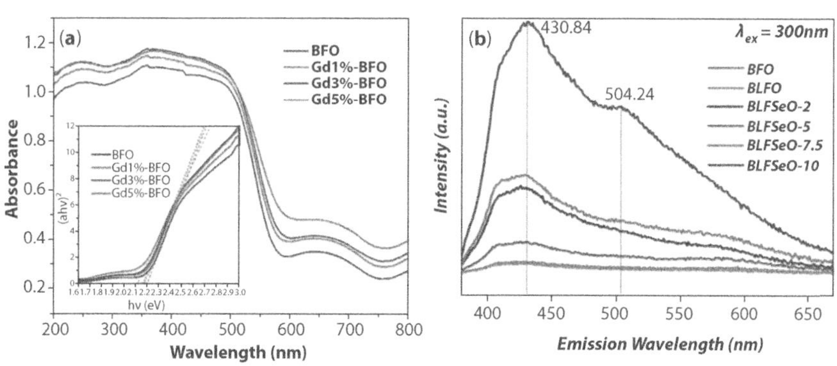

Figure 7.6 (a) UV-Vis DRS spectra of Gd doped BFO with inset showing band gaps of pure and doped BFO (reproducible from ref. no. [109]). (b) Photoluminescence spectra of pure and La^{3+} and Se^{4+} co-doped BFO (reproduced from ref. no. [103]).

is more favorable for better photocatalytic activity in comparison to pure BFO [109]. From the photoluminescence spectra (Figure 7.6 (b)), it can be clearly seen that the La^{3+} and Se^{4+} co-doped BFO (BLFSeOs) have lower PL peak intensity as compared to neat BFO. This indicates that the dopants captured the photo-induced electrons leading to the reduced recombination of e^-/h+ pairs to some optimum value. As the concentration of dopants exceeds the optimum value the doping could cause surface defects which will act as recombination centre for charge carriers [103].

7.3.5 Effect of Doping on Photocatalytic Activity of BFO

Doping of metal ions to BFO lattice has a great impact on the photoactivity of BFO. From literature survey, it has been cleared that the doped BFO exhibited better photocatalytic performance than that of pristine BFO.

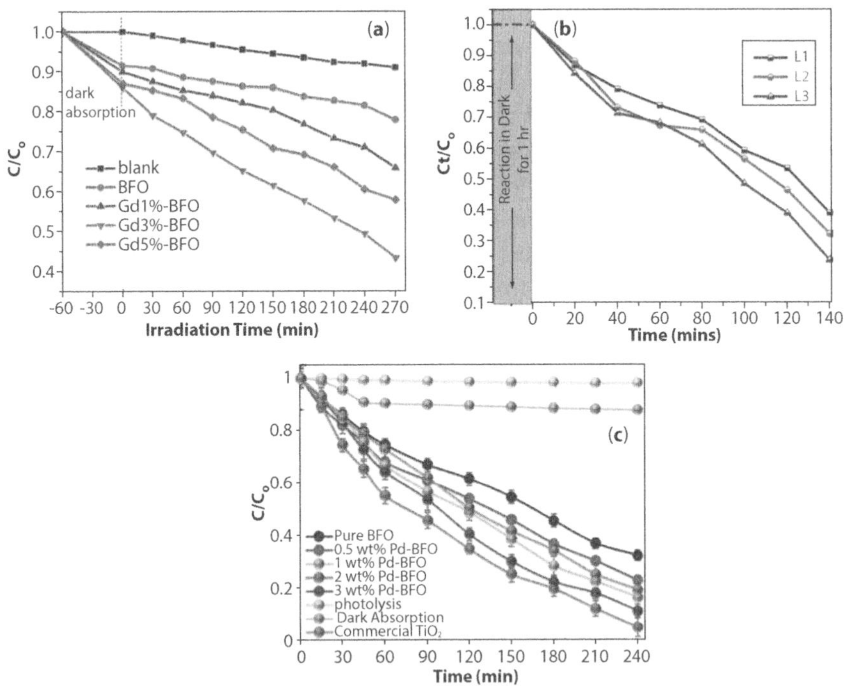

Figure 7.7 (a) RhB degradation efficiency of pure and Gd doped BFO (reproduced from ref. no. [109]). (b) Photocatalytic degradation of MB pristine BFO and La doped BFO (reproduced from ref. no. [105]). (c) Photodegradation of MG by pristine and Pd doped BFO (reproduced from ref. no. [101]).

For example, Zhang *et al.* fabricated a series of Gd doped BFO nanomaterials through sol-gel method and studied the photocatalytic performance by the decomposition of RhB in presence of visible light. Gd-doped BFO exhibited 2.55 times higher photocatalytic RhB degradation efficiency than pure BFO which is due to the enhanced light absorption ability and reduced rate of excitons recombination (Figure 7.7 (a)) [109]. Kumar *et al.* synthesized La-doped BFO using sol-gel synthesis method. The photocatalytic performance of the doped BFO was carried out by the decolourization of methylene blue under visible light. An enhancement in the photocatalytic degradation efficiency of La-doped BFO was observed than that of pure BFO (Figure 7.7 (b)) [105]. Moreover, Jafferi *et al.* investigated the photoactivity of pure and Pd-BFO through MG decomposition in presence of visible light. They found that 2 wt% Pd-BFO displayed maximum efficiency of 95.7% towards MG photodegradation while pure BFO can only degrade 72.3% in 240 min (Figure 7.7 (c)). The augmented photoactivity of doped BFO in comparison to pure material might be due to the improved ability for light absorption as well as minimized rate of recombination. Another important factor is the proper concentrations of dopant ions that could trap photo-induced electrons [101].

7.4 BFO Semiconductor Heterojunction Construction

The coupling between two semiconductors of desired band structure at the nanometric scale leads to form an interface between them which is called heterojunction [110]. Photocatalytic activity of BFO could be also escalated through the construction of semiconductor heterojunctions, which resulted in inducing the band bending and internal electrical field formation. This significantly boosted the spatial charge separation. When an interface is formed between any two semiconductors, the energy difference between the conduction and valence bands of these two phases allows the irreversible spatial charge separation till this energy difference is sufficiently large [111]. The structure of interface plays a vital role in interfacial transfer process efficiency in addition to the importance of formation of these interfaces in the efficient separation of charge carriers [112]. In general, semiconductor heterojunctions can be classified into three categories such as Type I, Type II, and Type III (Scheme II). Among these, heterojunctions of Type II can accelerate spatial separation of charge carriers. As a result, the excitons recombination is retarded, and their lifetime becomes prolonged. Thus, construction of Type II alignment should be

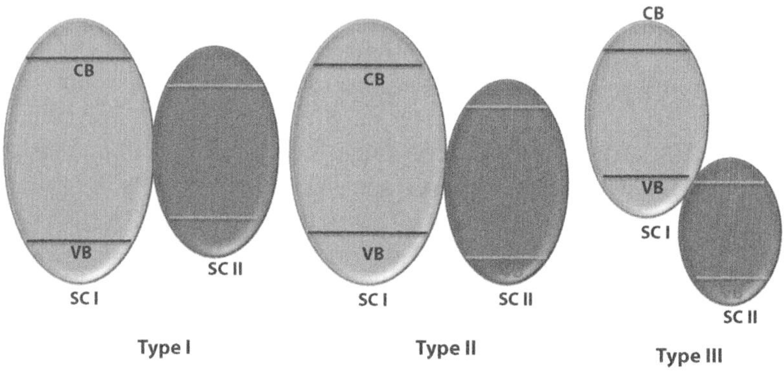

Scheme 7.2 Type I, type II, and type III semiconductor heterojunctions.

highly desired for photocatalysis [113]. BFO has been coupled with various semiconductors to improve its performance. It has been combined with wide band gap semiconductors, like TiO_2, CeO_2, and so on, as well as narrow band gap semiconductors, namely Bi_2WO_6, $BiVO_4$, and so on. Herein, we have separately discussed the importance of these heterojunctions.

7.4.1 Heterojunction Construction With Wide Band Gap Semiconductors

Wide band gap semiconductors have extensively been used in various photocatalytic applications because of their adequate band edge potentials to carryout numerous redox reactions required in photocatalysis. However, their photocatalytic applications are largely restricted mainly because of wide band gap which uses a narrow portion (4–5% UV) of inexhaustible solar light. On the other hand, visible light active BFO exhibits poor photocatalytic performance due to small lifespan of the photo-induced charge carriers. Combination of BFO with wide band gap semiconductors will not only enable the heterojunction photocatalyst to harness solar light but also make it thermodynamically and kinetically strengthened to carry out photocatalytic reactions efficiently [114]. Among the wide band gap semiconductors, TiO_2 has been investigated exhaustibly because of its low cost, high abundance, nontoxicity, strong oxidizing power, excellent photochemical stability, chemical inertness, and environment friendly nature [115–118]. Moreover, splendid hydrophilic properties of photoexcited surfaces of TiO_2 are revealed from tremendous self-cleaning and anti-fogging abilities [119]. Instead of these advantages, the photocatalytic practical applications of TiO_2 system is highly restricted mainly due to very fast recombination rate [120] and wide band gap energy which allows it to absorb only in UV region that contributes 5% of solar light [121, 122]. Several modification strategies have been applied to improve its photocatalytic activity. Among these, coupling of TiO_2 with narrow band gap semiconductor was found to be advantageous. The combination of BFO with TiO_2 for constructing heterojunction nanocomposites extends light absorption range as well as prolong the lifetime of the charge carriers to promote their separation and hence photoactivity is improved. Zhang *et al.* [123] have used pulsed laser deposition technique to fabricate BFO with TiO_2 and investigated its photochemical performance. They observed that TiO_2 can absorb the visible light of wavelength 460 nm when supported on BFO substrate. Yang *et al.* [124] adopted an electrospinning technique to synthesize TiO_2 and BFO nanofibers of one dimensional nanostructure. The synthesized nanocomposites displayed outstanding performance toward photodecomposition of methylene blue under irradiation of visible light. This exceptional activity might be due to comprehensive charge separation across the heterojunction formed between TiO_2 and BFO. In addition, the nanofiber composites could easily be recycled without appreciable reduction in photoactivity due to large length to diameter ratio of the nanofibers. Zhu *et al.* [125] have demonstrated that BFO/TiO_2-NTs covered Ti electrode prepared by ultrasonic immersion route, exhibited significant photoconversion efficacy in presence of visible light than that exhibited by the unmodified TiO_2-NTs/Ti electrodes. The visible light-induced enhanced photoelectrocatalytic performance towards degradation of RhB is principally due to the long lifespan of the photogenerated charge carriers via the applied bias and expanded spectral response stimulated by BFO. Thus, the visible -light responsive photocatalytic activity of TiO_2 is promoted by coupling it with a certain amount of BFO. Moreover, the catalytic activity of the semiconductor is limited if the catalyst nanoparticles possess no pores as the reaction undergoes only on the external surface of the semiconductor. This difficulty might be overcome by using porous nanomaterials which would adsorb the reactants on the surface as well as inside the pores causing the augmentation in the contact area. As a result, the catalytic performance is increased [126–128]. Humayun *et al.* have successfully synthesized porous TiO_2/$BiFeO_3$ (T/P-BFO) nanocrystalline composites and compared their photoactivity with porous BFO. T/P-BFO nanocomposite containing 9 mole percentage TiO_2 exhibited longer life period and excellent charge separation efficiency with respect to the P-BFO. This made the fabricated photocatalyst to perform excellently towards H_2 production through water splitting and decomposition of acetaldehyde and phenol under illumination of visible light. The existence of porous BFO in the composite was evident from HRTEM and N_2 adsorption–desorption isotherms. The presence of fringes corresponding to (110) planes of inter-planar width 0.35 nm and (012) planes with 0.28 nm inter-planar width of the HRTEM image for T/P-BFO composite revealed nanocrystalline phase of TiO_2 and P-BFO, which indicated the existence of a strong interfacial interaction between P-BFO and TiO_2. The N_2 adsorption–desorption isotherms of T/P-BFO nanocomposite confirmed its porous nature. The nanocomposite's specific surface area was found to be 26.31 $m^2 \cdot g^{-1}$, whereas that for P-BFO was found to be only 20.28 $m^2 \cdot g^{-1}$. Using Barrett-Joyner Halenda (BJH) method, the pore size distributions of the nanocomposite and P-BFO was determined from Figure 7.8 (a). Both the samples possess disordered porous structure as confirmed from long and wide hysteresis loops. After coupling with TiO_2, the P-BFO pore size was approximately reduced from 26.7 to 15.47 nm. These differences in the pore size and surface area between the composite and P-BFO imply high photocatalytic performance of the former [129]. Enhanced photoactivity also depends on efficient separation of charges and their transfer, which can be investigated with help of the surface photovoltage (SPV) studies. SPV comprises of two responses, namely SS-SPS and TS-SPV [130]. SPS response of nanocrystalline semiconductors is mainly derived from the separation of charges via diffusion process. The concentration of oxygen plays a vital role for this. The SPS responses were measured for the P-BFO in different atmosphere as well as in N_2 atmosphere. It was enriched remarkably in air and O_2 environment. This suggests that O_2 presence is essential for the appearance of SPS response. SPS of T/P-BFO nanocomposite displays good response in N_2 atmosphere. Such SPS response is primarily due to the significant

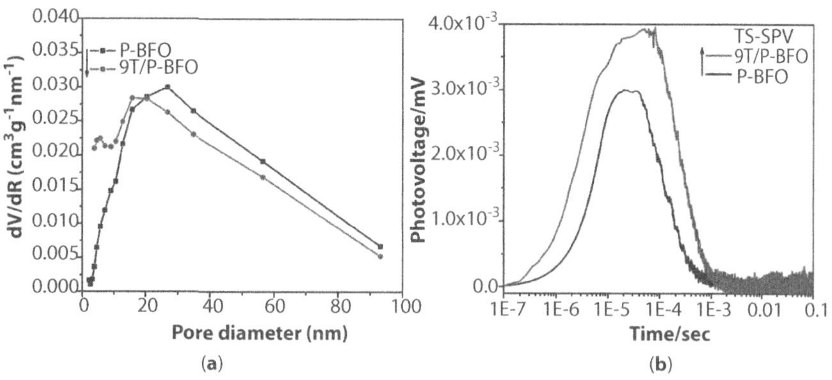

Figure 7.8 (a) Pore diameters (b) TS-SPV responses of T/P-BFO and P-BFO (reproduced from ref. no. [129]).

development in isolation of charge carriers resulting from remarkable spatial transfer of high-energy electrons from P-BFO to TiO_2 under visible-light irradiation [114]. The TS-SPV responses were determined under a laser pulse by irradiating at 532 nm. Figure 7.8 (b) displayed positive responses for all the samples indicating that the positive holes are generated under 532 nm laser irradiation. These were then accumulated on the surface of testing electrode. The sluggish response is obtained primarily due to the diffusion of charges. In air, it is found positive because of the capture of the photo-induced electron by adsorbed O_2. This caused the holes to get diffused at the surface of electrode preferentially. This implies that the nanocomposite shows prolonged charge carrier lifetime with high efficiency in separation with respect to P-BFO [129]. Higher charge separation efficiency of T/P-BFO nanocomposite was also observed from PL spectroscopy. Lower PL signal signifies lower rate of recombination of charges and hence, pronounced affinity for separation. Strong PL signals (blue emission) observed between 350 to 525 nm for T/P-BFO and P-BFO might be due to the transition of electrons from the CB of BFO to its VB [131, 132]. Lowest PL signal detected for T/P-BFO indicates its higher charge separation efficiency. Overall, T/P-BFO heterojunction is illuminated with visible-light of wave length ≤500 nm to generate excitons and the high energy electrons could be transferred to the CB of TiO_2 due to the built in electric field at the heterostructure interface. These considerably hinder the recombination of charge carriers and hence extend its lifetime as well as improve the separation efficiency. The T/P-BFO nanocomposites exhibit poor recombination rate, which enables it to be significantly applied in photocatalysis for pollutants degradation and solar fuel generation.

CeO_2 is another wide band gap semiconducting rare earth metal oxide most frequently used in photocatalysis. Oxygen defect rich CeO_2 has attracted great attention as it exhibits high photocatalytic activity. The presence of significant content of Ce^{3+} in it acts as trapping sites and prolongs the lifespan of the excitons, which in turn is responsible for enhanced photoactivity [133–137]. However, exceptionally high electron-hole pairs recombination rate, slow charge transfer kinetics, large e^-/h^+ binding energy and UV radiation responsive activity restricts the photocatalytic performance of pristineCeO_2 [135]. Therefore, recent researches are now focusing on the fabrication of nanostructured CeO_2 of low-dimensions preferably 1D and 2D. Such low dimensional nanostructured CeO_2 exposes a definite plane which isolates the charge carriers orthogonally through preferential electrons percolation to the back contact with simultaneous accumulation of the holes at the interface of semiconductor/electrolyte. In addition, 1D nanostructures of rod shape channelize the photoexcited electrons efficiently with respect nanostructures of other low-dimensions. As a result, there combination of charge carriers is effectively inhibited, which facilitates pronounced photocatalytic activity [138]. Therefore, it is of great importance to couple 1D CeO_2 with a visible light responsive semiconducting material of suitable band structure to obtain a visible light sensitive photocatalytic system along with effective separation of excitons so as to get enhanced photoactivity. Combination of 1D CeO_2 with 2D BFO will provide proper band alignment for successful isolation and transport of excitons across the heterojunction interface. This will result in remarkable visible light-induced photoactivity in comparison with their single counterparts. Mansingh *et al.* fabricated 1D–2D nitrogen doped carbon modified CeO_2-BFO p-n heteroarchitecture and investigated its photocatalytic performance towards photocatalytic reduction of nitrogen for ammonia synthesis, photoreduction of Cr (VI) followed by phenol photooxidation under application of visible light [139]. It was mentioned that visible light responsive photoactivity is usually linked to optical band gap as well as photon absorption ability which are measured by UV–vis DRS spectrophotometer. The UV–vis DRS spectra for BFO, CeO_2, $MCeO_2$, and $MCeO_2$–BFO (Figure 7.9 (a)) revealed that the absorption band edge position for CeO_2 shifts from 420 to 550 nm for carbons decorated nitrogen-doped on CeO_2 nanorods ($MCeO_2$). In addition, the absorption maximum for $MCeO_2$–BFO shows a red shift as compared to that of CeO_2 and $MCeO_2$, which might be attributed to the incorporation of a very less amount of Bi or Fe in CeO_2 nanorod [140]. This suggested that the composite can extend its visible light absorbance to longer wave length region and hence harvest solar light efficiently to produce large numbers of charge carriers (e^-/h^+) [141]. Further, it was observed from Figure 7.9 (a) that the optical absorbance for $MCeO_2$–BFO lies between BFO and $MCeO_2$ indicating existence of strong binding between BFO and $MCeO_2$. Urbach energy for $MCeO_2$–$BiFeO_3$ was found to be 0.75 eV, which was quiet higher than that obtained for $MCeO_2$ and BFO (0.46 and 0.69 eV, respectively). This high Urbach energy value for $MCeO_2$–$BiFeO_3$ indicated the presence of appreciable distortion or defect (i.e., oxygen vacancy and large amount of Ce^{3+} ions) in the system. This was further evidenced from Raman and PL studies. Raman spectra for $MCeO_2$, BFO and $MCeO_2$–BFO were shown in Figure 7.9 (b). Highly intense bands noticed in low frequency range were assigned to A1 (A1–1, A1–2, A1–3) vibrations of neat BFO. This suggested that bismuth is present at different sites of the perovskite structure. The peaks centred around high frequency range (<240 cm^{-1}) might be attributed to BFO sheet having E phonons. These features correspond to the pristine BFO with rhombohedral structure [142]. The polarization band of sharp tower-like

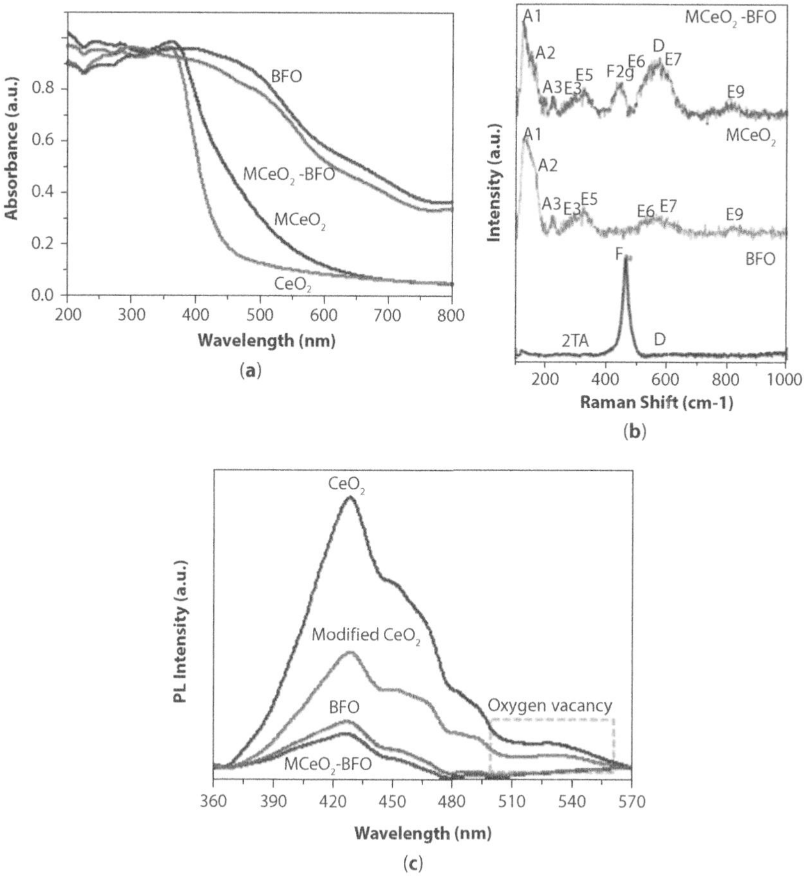

Figure 7.9 (a) UV–vis DRS of BFO, CeO_2, $MCeO_2$, $MCeO_2$–BFO (b) Raman spectra of BFO, $MCeO_2$, $MCeO_2$–BFO (c) Photoluminescence spectra for BFO, CeO_2, $MCeO_2$, $MCeO_2$–BFO composite (reproduced from ref. no. [139]).

shape centred at 464 cm^{-1} in the Raman spectra of $MCeO_2$ indicates the F_{2g} symmetric vibration. This confirmed the presence of CeO_2 with cubic fluorite structure. In addition to it, the appearance of two scattering bands of low-intensity around 297 cm^{-1} (2TA) and 595 cm^{-1} indicated the presence of lattice displaced oxygen and oxygen vacancy (OD) respectively [136]. The characteristic Raman bands for $MCeO_2$–BFO match well with that of BFO and $MCeO_2$ only except slight variation in peak intensity and position. This suggested that formation of a robust heterojunction between BFO and $MCeO_2$ in the composite. Additionally, the magnified intensity of the OD band of $MCeO_2$–BFO heterojunction indicated large number of oxygen vacancies and high of Ce^{3+}content. PL peaks observed (Figure 7.9 (c)) around 420–440 nm depicted the presence of defect in between O 2p and Ce 4f states. The peak around 525 nm is an indicative of presence of oxygen defect [143–145]. The presence of such defects is largely responsible for exciton separation states which resulted in enhanced photocatalytic performance. Further, the low PL peak intensity obtained binary heterostructure indicated a pronounced separation of charge carriers as compared to that of BFO and $MCeO_2$ because of significant transfer of charges through the heterojunction. Owing to visible light activity and pronounced separation and transfer of excitons, $MCeO_2$–BFO heterojunction exhibited remarkable photoactivity by reducing Cr (VI) up to 98.2%, degrading phenol 85% and producing ammonia of 117.77 $\mu mol \cdot h^{-1} \cdot g^{-1}$ under visible light irradiation [139].

$BaTiO_3$ is a perovskite oxide with ferroelectric property. It possesses a wide band gap energy of 3.30 eV and hence photocatalytically active in UV region [146]. To harvest solar light, it is essential to couple with a semiconductor of narrow band gap energy. $BiFeO_3$ is narrow band gap semiconductor (Eg = 2.29 eV) and display visible light responsive photoactivity [147]. $CuBi_2O_4$ is a p-type semiconductor with spinel structure. Being a narrow band gap semiconductor with band gap 1.76 eV, it can utilize a considerable portion of sun light ($\lambda > 705$ nm) [148]. The difference in conduction band minimum potential of BFO (+0.35 eV) and valence band edge potential of $CuBi_2O_4$ (+1.13 eV) is small. Similarly, the difference between CBM of $BaTiO_3$ (+0.08 eV) and the VB of $CuBi_2O_4$ (+1.13 eV) is also small. Considering band edge potentials, a symmetric double Z-scheme system consisting of $BaTiO_3$, BFO and $CuBi_2O_4$ may be constructed [149]. In $BiFeO_3/CuBi_2O_4/BaTiO_3$ double Z-scheme photocatalyst, $CuBi_2O_4$ is used as the PS I since it has negative CB (−0.63 eV) while both BFO (+2.64 eV) and $BaTiO_3$ (+3.38 eV) can be used as the PS II as they have positive VB. As a result, the composite photocatalyst can effectively utilize UV, visible as well as NIR region of solar spectrum. Dionysiou group have compared the photocatalytic activity of pristine $BaTiO_3$, BFO, and $CuBi_2O_4$ with that of $BiFeO_3/CuBi_2O_4/BaTiO_3$ nanocomposite. The hybrid photocatalyst displayed superior performance in decomposing an antibiotic norfloxacin (NFX) in comparison with other three photocatalysts. The pseudo first order dynamics was applied to determine rates of photodegradation process. The apparent degradation reaction rate constants for $BaTiO_3$, $CuBi_2O_4$, $BiFeO_3$, and $BiFeO_3/CuBi_2O_4/BaTiO_3$ are found to be 0.0007, 0.0014, 0.0072 and 0.0113 min^{-1} respectively. This suggested that $BiFeO_3/CuBi_2O_4/BaTiO_3$ exhibited superior photocatalytic performance in removing NFX. It was observed that 1.0 g/L nanocomposite degraded 93.5 %

of 2.5 mg/L NFX within 1 hour under illumination of solar light. The nanocomposite showed intense absorption in the range of 200 to 800 nm whereas neat $BaTiO_3$, $BiFeO_3$, and $CuBi_2O_4$ had absorption edges at 376, 564, and 705 nm, respectively. It indicated that the nanocomposite absorbs maximum portion of sun light. The absorption edge of $BiFeO_3$/$CuBi_2O_4$/$BaTiO_3$ is located in between that of $BaTiO_3$ and $CuBi_2O_4$, which is resulted from the close interaction of the three nanoparticles. The light response range was extended up to UV, visible, and even near-infrared (NIR) region owing to the formation of symmetric double Z scheme $BiFeO_3$/$CuBi_2O_4$/$BaTiO_3$. As a result, this composite photocatalyst is enabled to utilise complete solar spectrum [150]. Moreover, the band gap energy for $BiFeO_3$, $CuBi_2O_4$, $BaTiO_3$, and $BiFeO_3$/$CuBi_2O_4$/$BaTiO_3$ was calculated according to equation 7.12 [151] and was estimated to be 2.29, 1.76, 3.30, and 2.20 eV, respectively.

$$E\ (eV)=1240/\lambda \tag{7.12}$$

Where E is the band gap energy and λ is the absorption wavelength.

Furthermore, the formation of symmetric double Z-scheme system is also responsible for the large VB oxidation surface which promotes effective e--h+ separation that improves the oxidation-reduction ability of the composite photocatalyst. In the photocatalytic system consisting of single Z-scheme $CuBi_2O_4$/$BiFeO_3$, the photo-induced electrons from the CB of BFO (PS II) can migrate to the VB (+1.13 eV) of $CuBi_2O_4$ (PS I), where these are combined with the photogenerated holes. Similarly, for the $CuBi_2O_4$/$BaTiO_3$ single Z-scheme system, the photo-induced electrons can migrate from the CB (+0.08 eV) of $BaTiO_3$ (PS II) to the VB of $CuBi_2O_4$ (PS I) and get discharged by the holes. This led to the availability of more number of e^- at the CB (−0.63 eV) of $CuBi_2O_4$ and more number of h^+ at the VB of $BiFeO_3$ (+2.64 eV) and $BaTiO_3$ (+3.38 eV) owing to their pronounced separation. Thus, the ternary composite with symmetric double Z-scheme exhibits strong redox ability than its single counterparts. The photodegradation mechanism was illustrated in Figure 7.10, which shows that dominant role is played by •OH and h^+ while $•O_2^-$ plays trivial role in the photo degradation of NFX [150].

$SrTiO_3$ is an electronic ceramic material with strong resistance power for photocorrosion [152]. It has a wide band-gap of +3.40 eV with more negative CB potential of −1.26 eV and hence, is commonly used as a photocatalyst for reduction of some inorganic pollutants [153]. Its photocatalytic applications are limited due to poor utilization of solar spectrum,

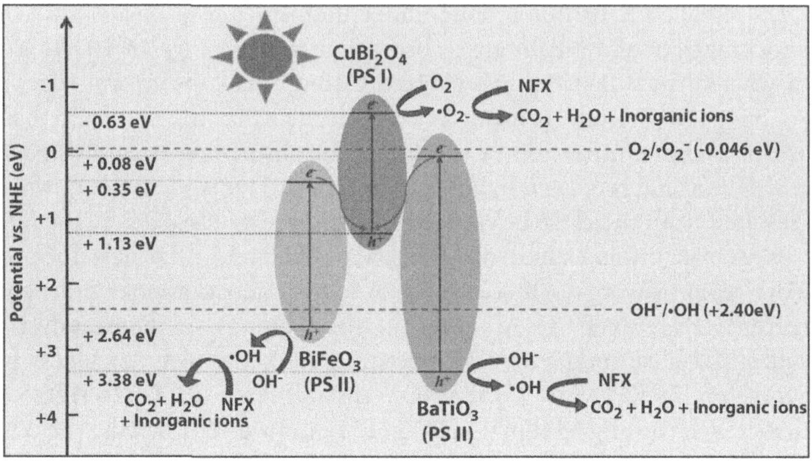

Figure 7.10 Mechanism for photodegradation of NFX over symmetric double Z-scheme $BiFeO_3/CuBi_2O_4/BaTiO_3$ composite (reproduced from ref. no. [150]).

which comprises only 5% of UV radiation. To widen the light response range and improve solar-light utilization, it is often combined with narrow band gap semiconductor [154]. On the other hand, BFO is a highly stable narrow band-gap semiconductor with band gap energy of +2.10 eV and more positive VB potential +2.61 V [155, 156]. Therefore, it can use visible light and possess strong oxidation ability [157]. Moreover, the potential at VB of $SrTiO_3$ is relatively closer to that at CB of BFO due to which construction of Z-scheme photocatalytic system becomes convenient. Further, ZnS is a wide band gap (Eg =+3.43 eV) semiconductor with more positive VB potential of +2.48 V. It provides strong oxidation surface for degradation of pollutants [158]. Qu *et al.* fabricated a novel coated $SrTiO_3$/(BiFeO3@ZnS heterojunction with Z-scheme architecture. The combination of Z-scheme and H-type photocatalytic system effectively inhibits the excitons recombination by accelerating their successful transfer and separation. Moreover, the present composite photocatalytic system can provide a robust oxidation surface and allow the availability of abundant photo-derived electrons at the CB of $SrTiO_3$ as well as holes at VB of ZnS to carry out redox reactions efficiently. As a result, the novel architecture shows an excellent solar light driven photoactivity towards degradation of 2,4-dichlorophenol along with simultaneous reduction of Cr (VI).It was observed that maximum 91.32% degradation of 2,4-dichlorophenol and 97.87% Cr (VI) reduction was achieved under optimized conditions [159].

7.4.2 Heterojunction Construction With Narrow Band Gap Semiconductors

Photocatalysts having absorption range in the visible region have attracted extensive attention in recent years [160–165]. Among various visible light driven photocatalysts, BFO is being exhaustively used in different photocatalytic reactions due to its low cost, high chemical stability, suitable band gap, and excellent photoelectrical properties. However, its commercial visible light active photocatalytic application is still restricted due to poor quantum yield resulted from the high rate of recombination of photogenerated charge carriers. Therefore, it has been modified by coupling with other suitable visible light responsive semiconductors to form heterojunctions through which the photo-induced excitons can easily be separated and transferred to inhibit their recombination. Several narrow band gap semiconductors, like Bi_2WO_4, $BiVO_4$, WO_3, and so on, have been combined with BFO to form heterojunctions for improved photoactivity. Construction of g-C_3N_4/$BiFeO_3$ [166], CuO/$BiFeO_3$ [167], Bi_2MoO_6/$BiFeO_3$ [168], and so on, heterojunctions have proved to be advantageous in augmenting photocatalytic performance of BFO.

Bi_2WO_6 belongs to Aurivillius family and consists of perovskite-like $[WO_4]^{2-}$ layers sandwiched between bismuth oxide $[Bi_2O_2]^{2+}$ layers. This structure facilitates the comprehensive separation of photogenerated charge carriers and formation of internal electric fields between the slabs. These factors are accountable for enhanced photocatalytic activity of Bi_2WO_6. Moreover, unique layered structure along with preferable band composition, visible light absorptivity, and appreciable stability enables Bi_2WO_6 as a robust photocatalyst for degradation of organic pollutants and O_2 evolution under illumination of visible light [169–172]. Owing to low lattice mismatch and matched band structure, it is often combined with BFO to construct heterojunction [173]. After formation of $BiFeO_3$/Bi_2WO_6 heterojunction, the electrons at the CB of BFO could be migrated to the CB of Bi_2WO_6 whereas the photogenerated holes at the VB of Bi_2WO_6 were transferred to the BFO VB. This augments the efficiency of charge separation and hence, photocatalytic activity is improved. Surface morphologies and magnification in interfacial contacts play vital role for further boosting the photoactivity. Tao *et al.* adopted solvothermal technique to synthesize a novel architectured BFO/Bi_2WO_6 nanofiber heterojunction (DH-NFs) through the growth of Bi_2WO_6 nanosheets (NSs) on electrospun BFO nanofibers (NFs). UV-Vis. DRS studies revealed that the absorption maximum for BFO and Bi_2WO_6 were close to 600 and 395 nm, respectively, whereas that for the heterojunction lies in between them. The DH-NFs exhibited approximately 2 times higher charge-separation efficiency in comparison to BFO NFs and Bi_2WO_6 NSs. The rate of oxygen evolution reaction for the composite was determined to be 18.3 and 31.9 times higher than that obtained for BFO NFs and Bi_2WO_6 NSs, respectively. Similarly, DH-NFs exhibited RhB degradation 36.7 and 8.7 times higher than those of BFO NFs and Bi_2WO_6 NSs, respectively. The remarkable photocatalytic activity exhibited by the novel heterojunction might be attributed to (i) effective separation followed by axial transport of electron-hole pairs through the 1D heterojunctions, (ii) facilitation of the spatially separated redox reaction sites and charge carriers by the discrete structure, (iii) the formation of high surface area which might provide enough active sites to carry out photocatalytic reaction [174].

Bismuth oxide (Bi_2O_3) is a direct band gap semiconductor which exhibits visible-light responsive photoactivity. Its superior photocatalytic performance is primarily ascribed to availability of abundant active sites resulted from enormous surface area which is achieved by the crystals with Bi–O layers or special polyhedron structures of Bi–O. In addition, the tough hybridization of 6s and 6p orbitals of Bi with 2p orbitals of O favours the movement of photoproduced holes [175]. Thus, the crystalline structure and morphology of Bi_2O_3 are responsible for its photocatalytic activity [176] Bi_2O_3 exists in different polymorphs among which α-, β-, and γ-forms show high photocatalytic performance. γ-Bi_2O_3 phase exhibits superior photoactivity in comparison to other two forms owing to its unique valence band structure and presence of large number of active sites derived from disordered truncated octahedral BiO5 units [177–179]. Margha *et al.* reported that the γ-Bi_2O_3/$BiFeO_3$ heterojunction exhibited excellent photoactivity in the degradation of RY160 dye under solar light irradiation [180].

$BiVO_4$ (BVO) possesses monoclinic structures and is recognized as a potential visible light responsive photocatalyst having band gap of 2.3 eV. [181]. Owing to the rapid excitons recombination, it exhibits poor performance towards photocatalytic water splitting and pollutants mineralization. The BVO photoanodes show about 1.0 mA cm^{-2} at 1.23 V (vs. RHE) under AM 1.5 G illumination as maximum water oxidation photocurrent density during water splitting [182, 183]. The most effective strategy to minimise the recombination rate is to couple the BVO with another semiconductor of desired band potential to construct a heterojunction [184]. A p-type semiconductor contains large extent of holes while a n-type semiconductor has excess electrons. Combination of these two types of materials resulted in the formation of p–n junction at the interface as pace charge region. As a result, an inner electric field is built at the equilibrium of the electron and hole. This built in electric field acts as a potential barrier which prevents the recombination of excitons due to which photoactivity is boosted. The combination of p-type BFO with n-type BVO resulted in the formation of the p-n junction which shows an improved photoactivity. Fan *et al.* synthesized BFO/BVO heterojunction by a hydrothermal method. The composite photocatalyst with 50% molar ratio of BFO exhibited degradation efficiency about 29.2 times higher than that of neat BFO and about 7.3 times higher than of pristine BVO towards RhB under visible light illumination [185]. Lee and his co-authors have successfully fabricated BFO/BVO p–n heterojunction by a facile ultrasonic/hydrothermal route. About 84% and 95% of tetracycline (TC) was photodegraded in presence of visible light at pH 6.7 and 9.5, respectively, by the heterojunction nanostructure whereas neat BFO decomposed only 31% and 22% at respective pH values. Additionally, the BFO/BVO nanocomposite showed appreciable photocurrent density of 0.36 mA cm^{-2} and 0.23 mA cm^{-2} under irradiation of UV-visible light and visible light, respectively, at 1.0 V vs. Ag/AgCl. These values are 3.0 and 3.28 times higher than those of BVO under identical conditions [186].

Metal–organic frameworks (MOFs) are a new class of crystalline materials which contains organic ligands and metal ions. These have special features such as tunable pore size, high specific surface area, designable framework architecture, good chemical, and thermal stability. These properties enable these to be widely used in the areas of catalysis, photocatalysis, separation, biomedicine, gas storage, luminescence, and chemical sensors [187–189]. Moreover, MOFs are used as potential photocatalysts owing to their desired porosity, superior adsorption affinity, and visible light absorption ability [190–192]. Bargozideh *et al.* have used a one-pot solvothermal method to design UiO-66/$BiFeO_3$ p-n heterojunction photocatalyst by combining a zirconium-based MOF (UiO–66) with BFO. They observed that the prepared nanocomposite with 30 wt% UiO-66 exhibit excellent photocatalytic behaviour by degrading 88.7% and 89.3% of MO and RhB, respectively, in 1 hour under visible-light irradiation. This magnificent photoactivity might be attributed to the efficient separation of photogenerated carriers as evident from the EIS studies (Figure 7.11 (i)). The pronounced charge separation in the composite was resulted from the presence of UiO-66, which successfully captures electrons through Zr clusters of MOF [193].

Bismuth sulfide (Bi_2S_3) is a typical semiconductor photocatalyst with band gap energy ranges from 1.3 to 1.7 eV, which implies deep and wide absorption range in the visible region [194]. This indicates that heterojunction formation between Bi_2S_3 and BFO can effectively widen the absorption intensity and spectral response range in the visible region. Additionally, the band structure of BFO matches well with that of Bi_2S_3. This can predominantly facilitate the separation and transfer of excitons. According to Bhoi *et al.* [195] Bi_2S_3/BFO heterojunction prepared by the combination of BFO nanoplates and Bi_2S_3 nanorods through a two-step hydrothermal method, exhibited improved photocatalytic behaviour towards mineralization of carbendazim pesticide. About 96% of carbendazim was removed in 2 hours under visible light illumination. Ma *et al.* developed a BFO/Bi_2S_3 direct Z- scheme heterojunction by electrospinning followed by *in-situ* anion exchange technique. The obtained Bi_2S_3/BFO heterojunction nanofibers decomposed tetracycline hydrochloride about 5 and 6 times more than that of pure BFO and Bi_2S_3, respectively, in presence of visible light. The exceptionally high photoactivity of Bi_2S_3/$BiFeO_3$ composite might be due to extension in visible light absorption range, comprehensive separation, and transfer of photoinduced charges through direct Z-scheme mechanism [196].

WO_3 is a stable, nontoxic, and visible light active semiconductor with band gap energy of 2.82 eV [197, 198]. Moreover, its high VBM potential of 3.22 V enables the photogenerated holes to act as strong oxidizing agents. In contrast, its low CBM position of 0.4 eV makes the photogenerated electrons with weak reduction ability and hence, recombination of these electrons with the holes is facilitated [199, 200]. Since, CBM position of WO_3 and that of VBM for g-C_3N_4 are located at 0.4 V and 1.4 V, respectively, combination of these two might result into a photocatalyst of Z-scheme configuration [199]. Huang *et al.* [200] reported that WO_3/g-C_3N_4 Z-scheme heterojunction displayed 2.9 fold high in photocatalytic degradation of dye as compared to neat g-C_3N_4. According to Xu *et al.* [201], the decoration of the WO_3 nanorods with few layers of g-C_3N_4 enhanced the efficiency of dye degradation. Therefore, combination of BFO, WO_3 and g-C_3N_4 could result into a Z-scheme photocatalyst, which can display significant photoactivity. Luo group have designed $BiFeO_3$-g-C_3N_4-WO_3 (BFO-CN-WO) Z-scheme heterojunction photocatalyst by a wet chemical method to investigate production of H_2 through water splitting and degradation of 2,4-dichlorophenol (2,4-DCP) under illumination of visible light. Upon interaction of visible light, the three components of the composite photocatalyst generate excitons. Since the CBM positions of BFO and WO are close to the VBM position of CN, there occurs recombination between the photoinduced electrons situated at the CB of BFO and WO and VB holes of CN. As a result, electrons at the CB of CN would undergo WRR to produce H_2 while the holes at the VB of WO and BFO would participate in water/OH- oxidation to release strongly reactive $OH^·$

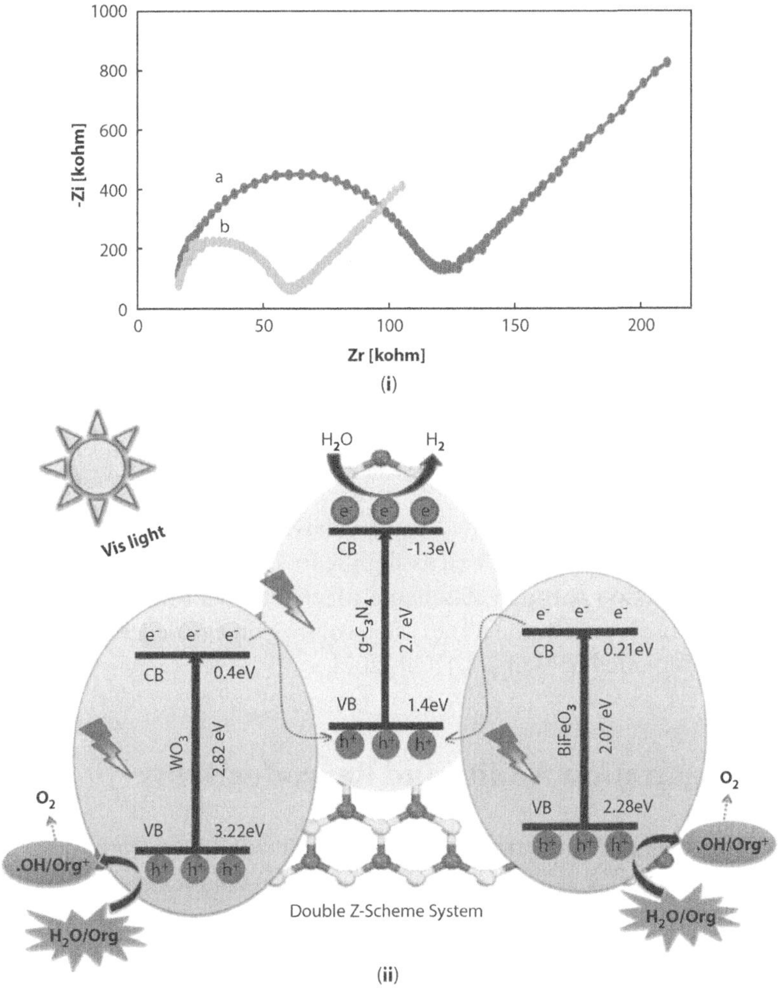

Figure 7.11 (i) Electrochemical impedance spectra of (a) $BiFeO_3$ and (b) UiO-66/$BiFeO_3$ (reproduced from ref. no. [193], (ii) plausible mechanism for photocatalytic H_2 generation and pollutant degradation under visible light illumination by $BiFeO_3$-g-C_3N_4-WO_3 composite through construction of double Z-scheme (reproduced from ref. no. [202]).

species, which decompose pollutants to harmless product as shown in the equations 7.13–7.18.

$$BFO/CN/WO \xrightarrow{hv} e^-_{CB}\ (BFO) + h^+_{VB}(BFO) / e^-_{CB}\ (CN) + h^+_{VB}(CN) / e^-_{CB}\ (WO) + h^+_{VB}(WO) \tag{7.13}$$

$$e_{CB}^{-}(BFO)+h_{VB}^{+}(BFO)/e_{CB}^{-}(CN)+h_{VB}^{+}(CN)/e_{CB}^{-}(WO)$$
$$+h_{VB}^{+}(WO)\rightarrow h_{VB}^{+}(BFO)/e_{CB}^{-}(CN)/h_{VB}^{+}(WO) \quad (7.14)$$

$$e_{CB}^{-}(CN)+2\mathrm{H}^{+}\rightarrow \mathrm{H}_2 \quad (7.15)$$

$$h_{VB}^{+}(BFO)+\mathrm{OH}^{-}\rightarrow{}^{\bullet}\mathrm{OH} \quad (7.16)$$

$$h_{VB}^{+}(WO)+\mathrm{H_2O}/\mathrm{OH}^{-}\rightarrow{}^{\bullet}\mathrm{OH} \quad (7.17)$$

$${}^{\bullet}\mathrm{OH} + 2,4\text{-dichlorophenol} \rightarrow \text{(intermediates)} \rightarrow \text{degraded products} \quad (7.18)$$

In other words, the obtained photocatalyst exhibits strong reduction abilities for H_2 generation as well as significant oxidation efficiency for pollutant degradation through Z-scheme mechanism shown in Figure 7.11 (ii). It was observed that the ternary heterojunction evolved 90 μmol h^{-1} g^{-1} of H_2 and degraded 63% of 2,4-DCP [202].

7.5 Separation Ability and Reproducibility

The BFO-based photocatalysts exhibit significant separation ability and reproducibility. The magnetic properties of BFO enable these visible light responsive photocatalysts to be separated easily and quickly from the treated solution by applying an external magnetic field. Bargozideh *et al.* calculated the saturation magnetizations values 36 and 5 emu g^{-1} for $BiFeO_3$ and $BiFeO_3/MoS_2$, respectively by using vibrating sample magnetometry (VSM) technique. The low value in case of the latter might be due to the presence MoS_2. Though the composite possesses lower saturation magnetization value than the neat BFO, it can easily be separated with the help of an external magnet [203]. Tao *et al.* reported that $BiFeO_3/Bi_2WO_6$ DHNFs could be separated from the solution after 20 minutes by sedimentation whereas the nanocomposite took only 2 minutes to be separated through magnetic separation. This rapid separation of the composite photocatalyst might be attributed to the ferromagnetic behaviour exhibited by BFO as revealed from the narrow S-shaped magnetic hysteresis loops presented in Figure 7.12 (a). Further, it was observed that the photodegradation of RhB and photocatalytic

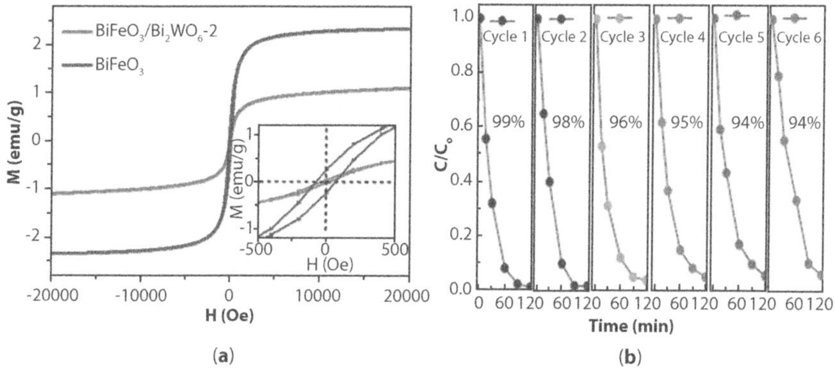

Figure 7.12 (a) Narrow S-shaped magnetic hysteresis loops of BFO and $BiFeO_3/Bi_2WO_6$ DHNFs showing their ferromagnetic behaviour (reproduced from ref. no. [174]), (b) photocatalytic decomposition of RhB under visible light illumination by $Ag_2O/BiFeO_3$-Ag without loss in activity and change in crystal structure for six cycles (reproduced from ref. no. [204]).

oxygen evolution properties exhibited by $BiFeO_3/Bi_2WO_6$ DHNFs was almost after five cycles [174]. Camacho-Escobar *et al.* have mentioned that $Ag_2O/BiFeO_3$-Ag decomposed RhB under visible light illumination without loss in activity and change in crystal structure for six cycles as shown in Figure 7.12 (b) [204]. These evidences proved that BFO-based photocatalytic systems have exceptionally high recyclability ability and facile separation ability.

7.6 Conclusion and Perspectives

Visible light-induced photoactivity of BFO is exponentially improved through elemental doping and heterojunction construction with semiconductors. Doping of metal ions into the BFO crystal lattice created oxygen vacancy. Additionally, doping resulted in the formation of an energy level either above the valence band or below the conduction band of BFO, thereby reducing the band gap energy, which extends light absorption range to longer wave length region of visible spectrum. Further, the photo-excited electrons are trapped by the energy level formed under the conduction band minimum and the holes will be trapped by the energy level created above the valence band. This inhibited the recombination rate of excitons and hence, photocatalytic performance is remarkably augmented. Doping of Gd, La, Pd, and so on, with proper weight percentage are found beneficial to escalate the photocatalytic performance of BFO.

It was observed that 3 wt. % Gd doped BFO exhibited about 2.55 times higher photoactivity in comparison to pristine BFO. Combination of wide band gap semiconductors with BFO resulted in effective separation and transfer of photoinduced charge carriers through the interfacial area of the heterojunction. In addition, narrow band gap energy of BFO assists it to harness energy from visible region while its wide band gap counterpart provides strong redox ability due to adequate position of CB and VB level. As a result, the so formed heterojunction displayed augmented visible light assisted photocatalytic performance. For instance, $SrTiO_3$/($BiFeO_3$@ZnS) heterojunction with Z-scheme architecture decomposed maximum of 91.32% 2,4-dichlorophenol and reduced 97.87% Cr(VI) under solar light irradiation. Coupling of BFO with narrow band gap semiconductor has not only experienced a bathochromic shift in visible light absorption range but also inhibited the recombination rate by substantial transfer of excitons through the heterojunction. This enhanced the photocatalytic ability of BFO. BFO/BVO heterojunction containing 50% molar ratio of BFO degraded RhB by 29.2 times higher than that of neat BFO and about 7.3 times higher than that of pristine BVO under visible light illumination. The BFO-based photocatalytic system exhibited good separation ability due to its magnetic property and high stability up to six consecutive cycles without decrease in photocatalytic efficiency.

It is now the challenge before the scientific community to shift the present photocatalytic system from lab scale research to practical applications. Efforts must be given in future to extend the light absorption range to longer wave length region to harness complete solar spectrum. Future direction of research needs a combination of doped BFO with a wide band gap semiconductor along with a narrow band gap NIR active counterpart for the development of an efficient BFO-based photocatalytic Z-scheme architecture with comprehensive separation charge carriers, complete absorption of solar spectrum and strong redox ability.

7.7 Acknowledgement

Authors extend their gratitude to the authorities of S'O'A (Deemed to be university), Bhubaneswar, Odisha, India for providing necessary facilities to carry out this work.

References

1. Gong, Y., Wang, J., Wei, Z., Zhang, P., Li, H., Wang, Y., Combination of Carbon Nitride and Carbon Nanotubes: Synergistic Catalysts for Energy Conversion. *ChemSusChem*, 7, 2303–2309, 2014.
2. Ong, W.J., Tan, L.L., Ng, Y.H., Yong, S.T., Chai, S.P., Graphitic Carbon Nitride (gC3N4)Based Photocatalysts for Artificial Photosynthesis and Environmental Remediation: Are We a Step Closer To Achieving Sustainability? *Chem. Rev.*, 116, 7159–7329, 2016.
3. Huang, C.H., Chang, K.P., Ou, H.D., Chiang, Y.C., Chang, E.E., Wang, C.F., Characterization and application of Ti-containing mesoporous silica for dye removal with synergistic effect of coupled adsorption and photocatalytic oxidation. *J. Hazard. Mater.*, 186, 1174, 2011.
4. Acharya, R. and Parida, K., A review on adsorptive remediation of Cr (VI) by magnetic iron oxides and their modified forms. *Biointerface Res. Appl. Chem.*, 10, 5266 - 5272, 2020.
5. Acharya, R., Martha, S., Parida, K.M., Remediation of Cr (VI) using clay minerals, biomasses and industrial wastes as adsorbents, in: *Advanced Materials for Wastewater Treatment*, S. ul-Islam, (Ed.), pp. 129–170, John Wiley Scrivener, USA, 2017.
6. Acharya, R., Naik, B., Parida, K.M., Adsorption of Cr (VI) and textile dyes on to mesoporous silica, titanate nanotubes and layer double hydroxides, in: *Nanomaterials in the Wet Processing of Textiles*, pp. 219–260, John Wiley Scrivener, USA, 2018.
7. Acharya, R., Lenka, A., Parida, K., Magnetite modified amino group based polymer nanocomposites towards efficient adsorptive detoxification of aqueous Cr (VI): A review. *J. Mol. Liq.*, 337, 116487, 2021.
8. Mansingh, S., Acharya, R., Martha, S., Parida, K.M., Pyrochlore $Ce_2Zr_2O_7$ decorated over rGO: a photocatalyst that proves to be efficient towards the reduction of 4-nitrophenol and degradation of ciprofloxacin under visible light. *Phys. Chem. Chem. Phys.*, 20, 9872–9885, 2018.
9. Mishra, S. and Acharya, R., Photocatalytic applications of graphene based semiconductor composites: A review. *Mater. Today Proc.*, 35, 164–169, 2021.
10. Kannan, N. and Sundaram, M.M., Kinetics and mechanism of removal of methylene blue by adsorption on various carbons—a comparative study. *Dyes Pigm.*, 51, 25–40, 2001.
11. Jiang, L., Huang, Y., Liu, T., Enhanced visible-light photocatalytic performance of electrospun carbon-doped TiO_2/halloysite nanotube hybrid nanofibers. *J. Colloid Interface Sci.*, 439, 62–68, 2015.
12. Jiang, Z., Sun, H., Wang, T., Wang, B., Wei, W., Li, H., Yuan, S., An, T., Zhao, H., Yu, J., Wong, P.K., Nature-based catalyst for visible-light-driven photocatalytic CO_2 reduction. *Energy Environ. Sci.*, 11, 2382–2389, 2018.
13. Sofianou, M.V., Tassi, M., Psycharis, V., Boukos, N., Thanos, S., Vaimakis, T., Yu, J., Trapalis, C., Solvothermal synthesis and photocatalytic performance

of Mn^{4+}-doped anatase nanoplates with exposed {0 0 1} facets. *Appl. Catal. B*, 162, 27–33, 2015.
14. Le, T.X.H., Bechelany, M., Champavert, J., Cretin, M., A highly active based graphene cathode for the electro-Fenton reaction. *RSC Adv.*, 5, 42536–42539, 2015.
15. George, C., Ammann, M., D'Anna, B., Donaldson, D.J., Nizkorodov, S.A., Heterogeneous Photochemistry in the Atmosphere. *Chem. Rev.*, 115, 4218–4258, 2015.
16. Zou, Z., Ye, J., Sayama, K., Arakawa, H., Direct Splitting of Waterunder Visible Light Irradiation with an Oxide Semiconductor Photocatalyst. *Nature*, 414, 625–627, 2001.
17. Tu, W., Zhou, Y., Zou, Z., Versatile Graphene-Promoting Photocatalytic Performance of Semiconductors: Basic Principles, Synthesis, Solar Energy Conversion, and Environmental Applications. *Adv. Funct. Mater.*, 23, 4996–5008, 2013.
18. Acharya, R. and Parida, K., A review on TiO_2/g-C_3N_4 visible-light- responsive photocatalysts for sustainable energy generation and environmental remediation. *J. Environ. Chem. Eng.*, 8, 103896, 2020.
19. Pati, S.S. and Acharya, R., An overview on g-C_3N_4 as a robust photocatalyst towards the sustainable generation of H_2 energy. *Mater. Today Proc.*, 35, 175–178, 2021.
20. Kim, D., Sakimoto, K.K., Hong, D., Yang, P., Artificial Photosynthesis for Sustainable Fuel and Chemical Production. *Angew. Chem. Int. Ed.*, 54, 3259–3266, 2015.
21. Li, X., Yu, J., Jaroniec, M., Hierarchical photocatalysts. *Chem. Soc Rev.*, 45, 2603–2636, 2016.
22. Li, X., Yu, J., Low, J., Fang, Y., Xiao, J., Chen, X., Engineering heterogeneous semiconductors for solar water splitting. *J. Mater. Chem. A*, 3, 2485–2534, 2015.
23. Yang, J., Wang, D., Han, H., Li, C., Accounts. *Chem. Res.*, 46, 1900–1909, 2013.
24. Chen, S., Shen, S., Liu, G., Qi, Y., Zhang, F., Li, C., Interface Engineering of a CoO_x/Ta_3N_5 Photocatalyst for Unprecedented Water Oxidation Performance under Visible-Light-Irradiation. *Angew. Chem.*, 127, 3090–3094, 2015.
25. Kibria, M.G. and Mi, Z., Artificial Photosynthesis Using Metal/Nonmetal-Nitride Semiconductors: Current Status, Prospects, and Challenges. *J. Mater. Chem. A*, 4, 2801–2820, 2016.
26. Long, R., Li, Y., Song, L., Xiong, Y., Coupling Solar Energy into Reactions: Materials Design for Surface Plasmon-Mediated Catalysis. *Small*, 11, 3873–3889, 2015.
27. Liu, S., Tang, Z.R., Sun, Y., Colmenares, J.C., Xu, Y.J., One-Dimension-Based Spatially Ordered Architectures for Solar Energy Conversion. *Chem. Soc Rev.*, 44, 5053–5075, 2015.

28. Xiang, Q., Cheng, B., Yu, J., Graphene-Based Photocatalysts for Solar-Fuel Generation. *Angew. Chem. Int. Ed.*, 54, 11350–11366, 2015.
29. Perathoner, S., Centi, G., Su, D., Turning Perspective in Photoelectrocatalytic Cells for Solar Fuels. *ChemSusChem*, 9, 345–357, 2016.
30. Zhao, Y., Jia, X., Waterhouse, G., II, Wu, L.Z., Tung, C.H., O'Hare, D., Zhang, T., Layered Double Hydroxide Nanostructured Photocatalysts for Renewable Energy Production. *Adv. Energy Mater.*, 6, 1501974, 2016.
31. Wang, Q., Hisatomi, T., Jia, Q., Tokudome, H., Zhong, M., Wang, C., Pan, Z., Takata, T., Nakabayashi, M., Shibata, N., Li, Y., Scalable water splitting on particulate photocatalyst sheets with a solar-to-hydrogen energy conversion efficiency exceeding 1%. *Nat. Mater.*, 15, 611–615, 2016.
32. Maeda, K., Teramura, K., Lu, D., Takata, T., Saito, N., Inoue, Y., Domen, K., Photocatalyst Releasing Hydrogen from Water. *Nature*, 440, 295–295, 2006.
33. Yuan, Y.J., Lu, H.W., Yu, Z.T., Zou, Z.G., Noble-metal-free olybdenum disulfide cocatalyst for photocatalytic hydrogen production. *ChemSusChem*, 8, 4113–4127, 2015.
34. Kang, D., Kim, T.W., Kubota, S.R., Cardiel, A.C., Cha, H.G., Choi, K.S., Electrochemical synthesis of photoelectrodes and catalysts for use in solar water splitting. *Chem. Rev.*, 115, 12839–12887, 2015.
35. Lu, Q., Yu, Y., Ma, Q., Chen, B., Zhang, H., 2D Transition-metal-dichalcogenide-nanosheet-based composites for photocatalytic and electrocatalytic hydrogen evolution reactions. *Adv. Mater.*, 28, 1917–1933, 2016.
36. Zhang, N., Qu, Y., Pan, K., Wang, G., Li, Y., Synthesis of Pure Phase $Mg_{1.2}Ti_{1.8}O_5$ and $MgTiO_3$ Nanocrystals for Photocatalytic Hydrogen Production. *Nano Res.*, 9, 726–734, 2016.
37. Zhang, K., Wang, L., Sheng, X., Ma, M., Jung, M.S., Kim, W., Lee, H., Park, J.H., Tunable bandgap energy and promotion of H_2O_2 oxidation for overall water splitting from carbon nitride nanowire bundles. *Adv. Energy Mater.*, 6, 1502352, 2016.
38. White, J.L., Baruch, M.F., Pander III, J.E., Hu, Y., Fortmeyer, I.C., Park, J.E., Zhang, T., Liao, K., Gu, J., Yan, Y., Shaw, T.W., Light-driven heterogeneous reduction of carbon dioxide: photocatalysts and photoelectrodes. *Chem. Rev.*, 115, 12888–12935, 2015.
39. Habisreutinger, S.N., Schmidt-Mende, L., Stolarczyk, J.K., Photocatalytic Reduction of CO_2 on TiO_2 and Other Semiconductors. *Angew. Chem. Int. Ed.*, 52, 7372–7408, 2013.
40. Xu, H.Q., Hu, J., Wang, D., Li, Z., Zhang, Q., Luo, Y., Yu, S.H., Jiang, H.L., Visible-Light Photoreduction of CO_2 in a Metal–Organic Framework: Boosting Electron–Hole Separation via ElectronTrap States. *J. Am. Chem. Soc*, 137, 13440–13443, 2015.
41. Tu, W., Zhou, Y., Zou, Z., Photocatalytic conversion of CO_2 into renewable hydrocarbon fuels: state-of-the-art accomplishment, challenges, and prospects. *Adv. Mater.*, 26, 4607–4626, 2014.

42. Li, Q., Mahendra, S., Lyon, D.Y., Brunet, L., Liga, M.V., Li, D., Alvarez, P.J., Antimicrobial nanomaterials for water disinfection and microbial control: potential applications and implications. *Water Res.*, 42, 4591–4602, 2008.
43. Ruales-Lonfat, C., Barona, J.F., Sienkiewicz, A., Bensimon, M., Vélez-Colmenares, J., Benítez, N., Pulgarín, C., Iron oxides semiconductors are efficients for solar water disinfection: a comparison with photo-Fenton processes at neutral pH. *Appl. Catal. B*, 166, 497–508, 2015.
44. Song, H., Son, S.Y., Kim, S.K., Jung, G.Y., A facile synthesis of hierarchical Sn 3 O 4 nanostructures in an acidic aqueous solution and their strong visible-light-driven photocatalytic activity. *Nano Res.*, 8, 3553–3561, 2015.
45. Zhou, K.G., McManus, D., Prestat, E., Zhong, X., Shin, Y., Zhang, H.L., Haigh, S.J., Casiraghi, C., Self-catalytic membrane photo-reactor made of carbon nitride nanosheets. *J. Mater. Chem. A*, 4, 11666–11671, 2016.
46. Prier, C.K., Rankic, D.A., MacMillan, D.W., Visible light photoredox catalysis with transition metal complexes: applications in organic synthesis. *Chem. Rev.*, 113, 5322–5363, 2013.
47. Lang, X., Chen, X., Zhao, J., Heterogeneous visible light photocatalysis for selective organic transformations. *Chem. Soc Rev.*, 43, 473–486, 2014.
48. Chen, X., Li, N., Kong, Z., Ong, W.J., Zhao, X., Photocatalytic fixation of nitrogen to ammonia: state-of-the-art advancements and future prospects. *Mater. Horiz.*, 5, 9–27, 2018.
49. Kudo, A. and Miseki, Y., Heterogeneous photocatalyst materials for water splitting. *Chem. Soc Rev.*, 38, 253–278, 2009.
50. Qu, Y. and Duan, X., Progress, challenge and perspective of heterogeneous photocatalysts. *Chem. Soc Rev.*, 42, 2568–2580, 2013.
51. Tan, L.L., Ong, W.J., Chai, S.P., Mohamed, A.R., Visible-light-activated oxygen-rich TiO2 as next generation photocatalyst: Importance of annealing temperature on the photoactivity toward reduction of carbon dioxide. *Chem. Eng. J.*, 283, 1254–1263, 2016.
52. Tan, L.L., Ong, W.J., Chai, S.P., Mohamed, A.R., Photocatalytic reduction of CO_2 with H_2O over graphene oxide-supported oxygen-rich TiO_2 hybrid photocatalyst under visible light irradiation: process and kinetic studies. *Chem. Eng. J.*, 308, 248–255, 2017.
53. Habisreutinger, S.N., Schmidt-Mende, L., Stolarczyk, J.K., Photocatalytic reduction of CO_2 on TiO_2 and other semiconductors. *Angew. Chem. Int. Ed.*, 52, 7372–7408, 2013.
54. Zhu, S., Liang, S., Tong, Y., An, X., Long, J., Fu, X., Wang, X., Photocatalytic reduction of CO_2 with H_2O to CH_4 on Cu (i) supported TiO_2 nanosheets with defective {001} facets. *Phys. Chem. Chem. Phys.*, 17, 15, 9761–9770, 2015.
55. Liu, Q., Zhou, Y., Kou, J., Chen, X., Tian, Z., Gao, J., Yan, S., Zou, Z., High-yield synthesis of ultralong and ultrathin Zn_2GeO_4 nanoribbons toward improved photocatalytic reduction of CO_2 into renewable hydrocarbon fuel. *J. Am. Chem.*, 132, 14385–14387, 2010.

56. Liu, S., Zhang, N., Xu, Y.J., Core–shell structured nanocomposites for photocatalytic selective organic transformations. *Part. Part. Syst. Charact.*, 31, 540–556, 2014.
57. Zhao, J., Ke, X., Liu, H., Huang, Y., Chen, C., Bo, A., Sheng, X., Zhu, H., Comparing the contribution of visible-light irradiation, gold nanoparticles, and titania supports in photocatalytic nitroaromatic coupling and aromatic alcohol oxidation. *Part. Part. Syst. Charact.*, 33, 628–634, 2016.
58. Prier, C.K., Rankic, D.A., MacMillan, D.W., Visible light photoredox catalysis with transition metal complexes: applications in organic synthesis. *Chem. Rev.*, 113, 5322–5363, 2013.
59. Lang, X., Chen, X., Zhao, J., Heterogeneous visible light photocatalysis for selective organic transformations. *Chem. Soc Rev.*, 43, 473–486, 2014.
60. Hu, Z., Quan, H., Chen, Z., Shao, Y., Li, D., New insight into an efficient visible light-driven photocatalytic organic transformation over CdS/TiO_2 photocatalysts. *Photochem. Photobiol. Sci.*, 17, 51–59, 2018.
61. Frank, S.N. and Bard, A.J., Heterogeneous photocatalytic oxidation of cyanide ion in aqueous solutions at titanium dioxide powder. *J. Am. Chem. Soc*, 99, 303–304, 1977.
62. Frank, S.N. and Bard, A.J., Heterogeneous photocatalytic oxidation of cyanide and sulfite in aqueous solutions at semiconductor powders. *J. Phys. Chem.*, 81, 1484–1488, 1977.
63. Acharya, R., Naik, B., Parida, K.M., Cr (VI) remediation from aqueous environment through modified-TiO_2-mediated photocatalytic reduction. *Beilstein J. Nanotechnol.*, 9, 1448–1470, 2018.
64. Fujishima, A. and Honda, K., Electrochemical Photolysis of Water at a Semiconductor Electrode. *Nature*, 238, 37–38, 1972.
65. Carey, J., Lawrence, J., Tosine, H., Photodechlorination of PCB'sin the Presence of Titanium Dioxide in Aqueous Suspensions. *Bull. Environ. Contam. Toxicol.*, 16, 697–701, 1976.
66. Inoue, T., Fujishima, A., Konishi, S., Honda, K., Photoelectrocatalytic Reduction of Carbon Dioxide in Aqueous Suspensions of Semiconductor Powders. *Nature*, 277, 637–638, 1979.
67. Yu, J., Low, J., Xiao, W., Zhou, P., Jaroniec, M., Enhanced Photocatalytic CO_2-Reduction Activity of Anatase TiO_2 by Coexposed {001} and {101} Facets. *J. Am. Chem. Soc*, 136, 8839–8842, 2014.
68. Fei, J. and Li, J., Controlled Preparation of Porous TiO_2–AgNanostructures through Supramolecular Assembly for Plasmon-Enhanced Photocatalysis. *Adv. Mater.*, 27, 314–319, 2015.
69. Xu, F., Shen, Y., Sun, L., Zeng, H., Lu, Y., Enhanced photocatalytic activity of hierarchical ZnO nanoplate-nanowire architecture as environmentally safe and facilely recyclable photocatalyst. *Nanoscale*, 3, 5020–5025, 2011.
70. Tong, H., Ouyang, S., Bi, Y., Umezawa, N., Oshikiri, M., Ye, J., Nanophotocatalytic materials: possibilities and challenges. *Adv. Mater.*, 24, 229–251, 2012.

71. Shang, L., Tong, B., Yu, H., Waterhouse, G., II, Zhou, C., Zhao, Y., Tahir, M., Wu, L.Z., Tung, C.H., Zhang, T., CdS Nanoparticle-Decorated Cd Nanosheets for Efficient Visible Light-Driven Photocatalytic Hydrogen Evolution. *Adv. Energy Mater.*, 6, 1501241, 2016.
72. Fan, T., Chen, C., Tang, Z., Ni, Y., Lu, C., Synthesis and characterization of g-C_3N_4/$BiFeO_3$ composites with an enhanced visible light photocatalytic activity. *Mater. Sci. Semicond. Process.*, 40, 439–445, 2015.
73. Li, S., Zhang, J., Kibria, M.G., Mi, Z., Chaker, M., Ma, D., Nechache, R., Rosei, F., Remarkably enhanced photocatalytic activity of laser ablated Au nanoparticle decorated $BiFeO_3$ nanowires under visible-light. *Chem. Commun.*, 49, 5856–5858, 2013.
74. Wang, W., Tadé, M.O., Shao, Z., Research progress of perovskite materials in photocatalysis-and photovoltaics-related energy conversion and environmental treatment. *Chem. Soc Rev.*, 44, 5371–5408, 2015.
75. Shi, J. and Guo, L., ABO3-based photocatalysts for water splitting. *Prog. Nat. Sci. Mater. Int.*, 22, 592–615, 2015.
76. Maeda, K., Rhodium-Doped Barium Titanate Perovskite as a Stable p-Type Semiconductor Photocatalyst for Hydrogen Evolution under Visible Light. *ACS Appl. Mater. Interfaces*, 6, 2167–2173, 2014.
77. Li, M., Zhang, J., Dang, W., Cushing, S.K., Guo, D., Wu, N., Yin, P., Photocatalytic hydrogen generation enhanced by band gap narrowing and improved charge carrier mobility in $AgTaO_3$ by compensated co-doping. *Phys. Chem. Chem. Phys.*, 15, 16220–16226, 2013.
78. Layek, S. and Verma, H.C., Magnetic and dielectric properties of multiferroic $BiFeO_3$ nanoparticles synthesized by a novel citrate combustion method. 1502, 05797, 2015.
79. Yang, J., Li, X., Zhou, J., Tang, Y., Zhang, Y., Li, Y., Factors controlling pure-phase magnetic $BiFeO_3$ powders synthesized by solution combustion synthesis. *J. Alloys Compd.*, 509, 9271–9277, 2011.
80. Zeches, R.J., Rossell, M.D., Zhang, J.X., Hatt, A.J., He, Q., Yang, C.H., Kumar, A., Wang, C.H., Melville, A., Adamo, C., Sheng, G., A strain-driven morphotropic phase boundary in BiFeO3. *Science*, 326, 977–980, 2009.
81. Bowen, C.R., Kim, H.A., Weaver, P.M., Dunn, S., Piezoelectric and ferroelectric materials and structures for energy harvesting applications. *Energy Environ. Sci.*, 7, 25–44, 2014.
82. Sando, D., Agbelele, A., Rahmedov, D., Liu, J., Rovillain, P., Toulouse, C., Infante, I.C., Pyatakov, A.P., Fusil, S., Jacquet, E., Carrétéro, C., Crafting the magnonic and spintronic response of $BiFeO_3$ films by epitaxial strain. *Nat. Mater.*, 12, 7, 641–646, 2013.
83. Tang, P., Kuang, D., Yang, S., Zhang, Y., The structural, optical and enhanced magnetic properties of $Bi_{1-x}Gd_xFe_{1-y}Mn_yO_3$ nanoparticles synthesized by sol–gel. *J. Alloys Compd.*, 622, 194–199, 2015.

84. Han, Y.L., Liu, W.F., Wu, P., Xu, X.L., Guo, M.C., Rao, G.H., Wang, S.Y., Effect of aliovalent Pd substitution on multiferroic properties in $BiFeO_3$ nanoparticles. *J. Alloys Compd.*, 661, 115–121, 2016.
85. Wang, J., Wei, Y., Zhang, J., Ji, L., Huang, Y., Chen, Z., Synthesis of pure-phase $BiFeO_3$ nanopowder by nitric acid-assisted gel. *Mater. Lett.*, 124, 242–244, 2014.
86. Tsai, C.J., Yang, C.Y., Liao, Y.C., Chueh, Y.L., Hydrothermally grown bismuth ferrites: controllable phases and morphologies in a mixed KOH/NaOH mineralizer. *J. Mater. Chem.*, 22, 17432–17436, 2012.
87. Rashad, M.M., Effect of synthesis conditions on the preparation of $BiFeO_3$ nanopowders using two different methods. *J. Mater. Sci. Mater. El.*, 23, 882–888, 2012.
88. Wang, Y.P., Zhou, L., Zhang, M.F., Chen, X.Y., Liu, J.M., Liu, Z.G., Room-temperature saturated ferroelectric polarization in $BiFeO_3$ ceramics synthesized by rapid liquid phase sintering. *Appl. Phys. Lett.*, 84, 1731–1733, 2004.
89. Kumar, M., Palkar, V.R., Srinivas, K., Suryanarayana, S.V., Ferroelectricity in a pure $BiFeO_3$ ceramic. *Appl. Phys. Lett.*, 76, 2764–2766, 2000.
90. Gao, T., Chen, Z., Niu, F., Zhou, D., Huang, Q., Zhu, Y., Qin, L., Sun, X., Huang, Y., Shape-controlled preparation of bismuth ferrite by hydrothermal method and their visible-light degradation properties. *J. Alloys Compd.*, 648, 564–570, 2015.
91. Wang, X.R., Yang, C.W., Zhou, D., Wang, Z.Y., Jin, M.L., Chemical coprecipitation synthesis and properties of pure-phase $BiFeO_3$. *Chem. Phys. Lett.*, 713, 185–188, 2018.
92. Li, Y., Wang, X.T., Zhang, X.Q., Li, X., Wang, J., Wang, C.W., New hydrothermal synthesis strategy of nano-sized $BiFeO_3$ for high-efficient photocatalytic applications. *Physica E Low Dimens. Syst. Nanostruct.*, 118, 113865, 2020.
93. Pattnaik, S.P., Behera, A., Martha, S., Acharya, R., Parida, K., Synthesis, photoelectrochemical properties and solar light-induced photocatalytic activity of bismuth ferrite nanoparticles. *J. Nanopart. Res.*, 20, 10, 2018.
94. Fatima, S., Ali, S.I., Iqbal, M.Z., Rizwan, S., The high photocatalytic activity and reduced band gap energy of La and Mn co-doped $BiFeO_3$/graphene nanoplatelet (GNP) nanohybrids. *RSC Adv.*, 7, 35928–35937, 2017.
95. Sando, D., Barthélémy, A., Bibes, M., $BiFeO_3$ epitaxial thin films and devices: past, present and future. *J. Phys. Condens. Matter.*, 26, 473201, 2014.
96. Betancourt-Cantera, L.G., Fuentes, K.M., Bolarín-Miró, A.M., Aldabe-Bilmes, S., Cortés-Escobedo, C.A., Sánchez-De Jesús, F., Enhanced photocatalytic activity of $BiFeO_3$ for water remediation via the addition of Ni^{2+}. *Mater. Res. Bull.*, 132, 111012, 2020.
97. Humayun, M., Zada, A., Li, Z., Xie, M., Zhang, X., Qu, Y., Raziq, F., Jing, L., Enhanced visible-light activities of porous $BiFeO_3$ by coupling with nanocrystalline TiO_2 and mechanism. *Appl. Catal. B*, 180, 219–226, 2016.

98. Putri, L.K., Ng, B.J., Ong, W.J., Lee, H.W., Chang, W.S., Chai, S.P., Engineering nanoscale p–n junction via the synergetic dual-doping of p-type boron-doped graphene hybridized with n-type oxygen-doped carbon nitride for enhanced photocatalytic hydrogen evolution. *J. Mater. Chem. A*, 6, 3181–3194, 2018.
99. Jaffari, Z.H., Lam, S.M., Sin, J.C., Mohamed, A.R., Constructing magnetic Pt-loaded $BiFeO_3$ nanocomposite for boosted visible light photocatalytic and antibacterial activities. *Environ. Sci. Pollut. Res.*, 26, 10204–10218, 2019.
100. Acharya, L., Nayak, S., Pattnaik, S.P., Acharya, R., Parida, K., Resurrection of boron nitride in pn type-II boron nitride/B-doped-g-C_3N_4 nanocomposite during solid-state Z-scheme charge transfer path for the degradation of tetracycline hydrochloride. *J. Colloid Interface Sci.*, 566, 211–223, 2020.
101. Jaffari, Z.H., Lam, S.M., Sin, J.C., Zeng, H., Mohamed, A.R., Magnetically recoverable Pd-loaded $BiFeO_3$ microcomposite with enhanced visible light photocatalytic performance for pollutant, bacterial and fungal elimination. *Sep. Purif. Technol.*, 236, 116195, 2020.
102. Guo, R., Fang, L., Dong, W., Zheng, F., Shen, M., Enhanced photocatalytic activity and ferromagnetism in Gd doped $BiFeO_3$ nanoparticles. *J. Phys. Chem. C*, 114, 21390–21396, 2010.
103. Irfan, S., Li, L., Saleemi, A.S., Nan, C.W., Enhanced photocatalytic activity of La^{3+} and Se^{4+} co-doped bismuth ferrite nanostructures. *J. Mater. Chem. A*, 5, 11143–11151, 2017.
104. Anand, P., Jaihindh, D.P., Chang, W.K., Fu, Y.P., Tailoring the Ca-doped bismuth ferrite for electrochemical oxygen evolution reaction and photocatalytic activity. *Appl. Surf. Sci.*, 540, 148387, 2021.
105. Kumar, P., Chand, P., Singh, V., La^{3+} substituted $BiFeO_3$-a proficient nano ferrite photo-catalyst under the application of visible light. *Chem. Phys. Lett.*, *754*, 137715, 2020.
106. Ponraj, C., Vinitha, G., Daniel, J., Visible light photocatalytic activity of Mn-doped $BiFeO_3$ nanoparticles. *Int. J. Green Energy*, 17, 71–83, 2020.
107. Iorgu, A.I., Maxim, F., Matei, C., Ferreira, L.P., Ferreira, P., Cruz, M.M., Berger, D., Fast synthesis of rare-earth (Pr^{3+}, Sm^{3+}, Eu^{3+} and Gd^{3+}) doped bismuth ferrite powders with enhanced magnetic properties. *J. Alloys Compd.*, 629, 62–68, 2015.
108. Rhaman, M.M., Matin, M.A., Hakim, M.A., Islam, M.F., Bandgap tuning of samarium and cobalt co-doped bismuth ferrite nanoparticles. *Mater. Sci. Eng. B*, 263, 114842, 2021.
109. Zhang, N., Chen, D., Niu, F., Wang, S., Qin, L., Huang, Y., Enhanced visible light photocatalytic activity of Gd-doped $BiFeO_3$ nanoparticles and mechanism insight. *Sci. Rep.*, 6, 1–11, 2016.
110. Nasr, M., Eid, C., Habchi, R., Miele, P., Bechelany, M., Recent progress on titanium dioxide nanomaterials for photocatalytic applications. *ChemSusChem*, 11, 3023–3047, 2018.

111. Wang, H., Zhang, L., Chen, Z., Hu, J., Li, S., Wang, Z., Liu, J., Wang, X., Semiconductor heterojunction photocatalysts: design, construction, and photocatalytic performances. *Chem. Soc Rev.*, 43, 5234–5244, 2014.
112. Robert, D., Photosensitization of TiO2 by MxOy and MxSy nanoparticles for heterogeneous photocatalysis applications. *Catal. Today*, 122, 1-2, 20–26, 2007.
113. Wen, J., Xie, J., Chen, X., Li, X., A review on g-C3N4-based photocatalysts. *Appl. Surf. Sci.*, 391, 72–123, 2017.
114. Xie, M., Fu, X., Jing, L., Luan, P., Feng, Y., Fu, H., Long-lived, visible-light-excited charge carriers of TiO2/BiVO4 nanocomposites and their unexpected photoactivity for water splitting. *Adv. Energy Mater.*, 4, 5, 1300995, 2014.
115. Pelaez, M., Nolan, N.T., Pillai, S.C., Seery, M.K., Falaras, P., Kontos, A.G., Dunlop, P.S., Hamilton, J.W., Byrne, J.A., O'shea, K., Entezari, M.H., A review on the visible light active titanium dioxide photocatalysts for environmental applications. *Appl. Catal. B*, 125, 331–349, 2012.
116. Daghrir, R., Drogui, P., Robert, D., Modified TiO_2 for environmental photocatalytic applications: a review. *Ind. Eng. Chem. Res.*, 52, 10, 3581–3599, 2013.
117. Asahi, R., Morikawa, T., Irie, H., Ohwaki, T., Nitrogen-doped titanium dioxide as visible-light-sensitive photocatalyst: designs, developments, and prospects. *Chem. Rev.*, 114, 19, 9824–9852, 2014.
118. Liu, C., Hu, J., Liu, W., Zhu, F., Wang, G., Tung, C.H., Wang, Y., Binding Modes of Salicylic Acids to Titanium Oxide Molecular Surfaces. *Chem. Eur. J.*, 26, 12, 2666–2674, 2020.
119. Wang, R., Hashimoto, K., Fujishima, A., Chikuni, M., Kojima, E., Kitamura, A., Shimohigoshi, M., Watanabe, T., Light-induced amphiphilic surfaces. *Nature*, 388, 6641, 431–432, 1997.
120. Wang, J., Wang, P., Cao, Y., Chen, J., Li, W., Shao, Y., Zheng, Y., Li, D., A high efficient photocatalyst Ag_3VO_4/TiO_2/graphene nanocomposite with wide spectral response. *Appl. Catal. B*, 136, 94–102, 2013.
121. Kim, D.H., Hong, H.S., Kim, S.J., Song, J.S., Lee, K.S., Photocatalytic behaviors and structural characterization of nanocrystalline Fe-doped TiO_2 synthesized by mechanical alloying. *J. Alloys Compd.*, 375, 1-2, 259–264, 2004.
122. Formo, E., Lee, E., Campbell, D., Xia, Y., Functionalization of electrospun TiO2 nanofibers with Pt nanoparticles and nanowires for catalytic applications. *Nano Lett.*, 8, 2, 668–672, 2008.
123. Zhang, Y., Schultz, A.M., Salvador, P.A., Rohrer, G.S., Spatially selective visible light photocatalytic activity of TiO2/BiFeO3 heterostructures. *J. Mater. Chem.*, 21, 12, 4168–4174, 2011.
124. Yang, Y.C., Liu, Y., Wei, J.H., Pan, C.X., Xiong, R., Shi, J., Electrospun nanofibers of p-type $BiFeO_3$/n-type TiO_2 hetero-junctions with enhanced visible-light photocatalytic activity. *RSC Adv.*, 4, 60, 31941–31947, 2014.

125. Zhu, A., Zhao, Q., Li, X., Shi, Y., $BiFeO_3/TiO_2$ nanotube arrays composite electrode: construction, characterization, and enhanced photoelectrochemical properties. *ACS Appl. Mater. Interfaces*, 6, 1, 671–679, 2014.
126. Xu, J., Liu, J., Zhao, Z., Zheng, J., Zhang, G., Duan, A., Jiang, G., Three-dimensionally ordered macroporous $LaCoxFe_{1-x}O_3$ perovskite-type complex oxide catalysts for diesel soot combustion. *Catal. Today*, 153, 3-4, 136–142, 2010.
127. Gao, B., Deng, J., Liu, Y., Zhao, Z., Li, X., Wang, Y., Dai, H., Mesoporous $LaFeO_3$ catalysts for the oxidation of toluene and carbon monoxide. *Chin. J. Catal.*, 34, 12, 2223–2229, 2013.
128. Liu, Y., Dai, H., Du, Y., Deng, J., Zhang, L., Zhao, Z., Au, C.T., Controlled preparation and high catalytic performance of three-dimensionally ordered macroporous $LaMnO_3$ with nanovoid skeletons for the combustion of toluene. *J. Catal.*, 287, 149–160, 2012.
129. Humayun, M., Zada, A., Li, Z., Xie, M., Zhang, X., Qu, Y., Raziq, F., Jing, L., Enhanced visible-light activities of porous BiFeO3 by coupling with nanocrystalline TiO_2 and mechanism. *Appl. Catal. B*, 180, 219–226, 2016.
130. Jing, L., Zhou, W., Tian, G., Fu, H., Surface tuning for oxide-based nanomaterials as efficient photocatalysts. *Chem. Soc Rev.*, 42, 24, 9509–9549, 2013.
131. Mandal, S.K., Rakshit, T., Ray, S.K., Mishra, S.K., Krishna, P.S.R., Chandra, A., Nanostructures of Sr^{2+} doped $BiFeO_3$ multifunctional ceramics with tunable photoluminescence and magnetic properties. *J. Phys. Condens. Matter*, 25, 5, 055303, 2013.
132. Das, R., Khan, G.G., Varma, S., Mukherjee, G.D., Mandal, K., Effect of quantum confinement on optical and magnetic properties of Pr–Cr-codoped bismuth ferrite nanowires. *J. Phys. Chem. C*, 117, 39, 20209–20216, 2013.
133. Xie, S., Wang, Z., Cheng, F., Zhang, P., Mai, W., Tong, Y., Ceria and ceria-based nanostructured materials for photoenergy applications. *Nano Energy*, 34, 313–337, 2017.
134. Arul, N.S., Mangalaraj, D., Ramachandran, R., Grace, A.N., Han, J.I., Fabrication of CeO_2/Fe_2O_3 composite nanospindles for enhanced visible light driven photocatalysts and super capacitor electrodes. *J. Mater. Chem. A*, 3, 15248–15258, 2015.
135. Han, J., Meeprasert, J., Maitarad, P., Nammuangruk, S., Shi, L., Zhang, D., Investigation of the facet-dependent catalytic performance of Fe_2O_3/CeO_2 for the selective catalytic reduction of NO with NH_3. *J. Phys. Chem. C*, 120, 1523–1533, 2016.
136. Sreeremya, T.S., Krishnan, A., Remani, K.C., Patil, K.R., Brougham, D.F., Ghosh, S., Shape-selective oriented cerium oxide nanocrystals permit assessment of the effect of the exposed facets oncatalytic activity and oxygen storage capacity. *ACS Appl. Mater. Interfaces*, 7, 8545–8555, 2015.
137. Paier, J., Penschke, C., Sauer, J., Oxygen defects and surface chemistry of ceria: quantum chemical studies compared to experiment. *Chem. Rev.*, 113, 3949–3985, 2013.

138. Sultana, S., Mansingh, S., Parida, K.M., Rational design of light induced self-healed Fe based oxygen vacancy rich CeO_2 (CeO_2NS–FeOOH/Fe_2O_3) nanostructure materials for photocatalytic water oxidation and Cr(VI) reduction. *J. Mater. Chem. A*, 6, 11377–11389, 2018.
139. Mansingh, S., Sultana, S., Acharya, R., Ghosh, M.K., Parida, K.M., Efficient Photon Conversion via Double Charge Dynamics CeO_2–$BiFeO_3$ p–n Heterojunction Photocatalyst Promising toward N_2 Fixation and Phenol–Cr (VI) Detoxification. *Inorg. Chem.*, 59, 3856–3873, 2020.
140. Niu, F., Chen, D., Qin, L., Zhang, N., Wang, J., Chen, Z., Huang, Y., Facile synthesis of highly efficient p–n heterojunction CuO/$BiFeO_3$ composite photocatalysts with enhanced visible-light photocatalytic activity. *ChemCatChem*, 7, 3279–3289, 2015.
141. Zhang, T., Shen, Y., Qiu, Y., Liu, Y., Xiong, R., Shi, J., Wei, J., Facial synthesis and photoreaction mechanism of $BiFeO_3$/$Bi_2Fe_4O_9$ heterojunction nanofibers. *ACS Sustain. Chem. Eng.*, 5, 4630–4636, 2017.
142. Li, L., Wang, S., Mu, G., Yin, X., Ou, K., Yi, L., A novel violet/blue light emitting device based on Ce2Si2O7. *Sci. Rep.*, 5, 951– 16659, 2015.
143. Xu, J., Harmer, J., Li, G., Chapman, T., Collier, P., Longworth, S., Tsang, S.C., Size dependent oxygen buffering capacity of ceria nanocrystals. *Chem. Commun.*, 46, 1887–1889, 2010.
144. Lv, Z., Zhou, H., Liu, H., Liu, B., Liang, M., Guo, H., Controlled assemble of oxygen vacant CeO_2@Bi_2WO_6 hollow magnetic microcapsule heterostructures for visible-light photocatalytic activity. *Chem. Eng. J.*, 330, 1297–1305, 2017.
145. Li, Y., Liu, Z., Wu, Y., Chen, J., Zhao, J., Jin, F., Na, P., Carbon dots-TiO_2 nanosheets composites for photoreduction of Cr (VI) under sunlight illumination: Favorable role of carbon dots. *Appl. Catal. B*, 224, 508–517, 2018.
146. Zhao, L., Zhang, Y., Wang, F., Hu, S., Wang, X., Ma, B., Liu, H., Wang, Z.L., Sang, Y., $BaTiO_3$ nanocrystal-mediated micro pseudo-electrochemical cells with ultrasound-driven piezotronic enhancement for polymerization. *Nano Energy*, 39, 461–469, 2017.
147. Gao, F., Chen, X.Y., Yin, K.B., Dong, S., Ren, Z.F., Yuan, F., Yu, T., Zou, Z.G., Liu, J.M., Visible-light photocatalytic properties of weak magnetic $BiFeO_3$ nanoparticles. *Adv. Mater.*, *19*, 2889–2892, 2007.
148. Shi, W., Guo, F., Yuan, S., *In situ* synthesis of Z-scheme Ag_3PO_4/$CuBi_2O_4$ photocatalysts and enhanced photocatalytic performance for the degradation of tetracycline under visible light irradiation. *Appl. Catal. B*, 209, 720–728, 2017.
149. Wang, R.X., Tang, J.H., Zhang, X.Y., Wang, D., Wang, X., Xue, S., Zhang, Z.H., Dionysiou, D.D., Construction of novel Z-scheme Ag/$ZnFe_2O_4$/Ag/$BiTa_{1-x}VxO_4$system with enhanced electron transfer capacity for visible light photocatalytic degradation of sulfanilamide. *J. Hazard. Mater.*, 375, 161–173, 2019.
150. Zhang, X., Wang, X., Chai, J., Xue, S., Wang, R., Jiang, L., Wang, J., Zhang, Z., Dionysiou, D.D., Construction of novel symmetric double Z-scheme

$BiFeO_3/CuBi_2O_4/BaTiO_3$ photocatalyst with enhanced solar-light-driven photocatalytic performance for degradation of norfloxacin. *Appl. Catal. B*, 272, 119017–119027, 2020.

151. Zhao, P., Wang, L., Bian, L., Xu, J., Chang, A., Xiong, X., Xu, F., Zhang, J., Growth mechanism, modified morphology and optical properties of coral-like $BaTiO_3$ architecture through CTAB assisted synthesis. *J. Mater. Sci. Technol.*, 31, 2, 223–228, 2015.
152. Zhao, W.J., Zhang, J., Pan, J.Q., Zhao, W.H., Niu, J.T., Qiu, J.F., Li, C.R., Fabrication and photocatalytic performance of one-dimensional Ag_3PO_4 sensitized $SrTiO_3$ nanowire. *Ceram. Int.*, 44, 18337–18343, 2018.
153. Qiao, J., Zhang, H., Li, G., Li, S., Qu, Z., Zhang, M., Wang, J., Song, Y., Fabrication of a novel Z-scheme $SrTiO_3/Ag_2S/CoWO_4$ composite and its application in sonocatalytic degradation of tetracyclines. *Sep. Purif. Technol.*, 211, 843–856, 2019.
154. Chang, Y., Yu, K., Zhang, C.X., Yang, Z.Q., Feng, Y.J., Hao, H., Jiang, Y.Z., Lou, L.L., Zhou, W.Z., Liu, S.X., Ternary CdS/Au/3DOM-$SrTiO_3$ composites with synergistic enhancement for hydrogen production from visible-light photocatalytic water splitting. *Appl. Catal. B*, 215, 74–84, 2017.
155. Xian, T., Yang, H., Dai, J.F., Wei, Z.Q., Ma, J.Y., Feng, W.J., Photocatalytic properties ofBiFeO3 nanoparticles with different sizes. *Mater. Lett.*, 65, 1573–1575, 2011.
156. Huang, G.Q., Zhang, G.H., Gao, Z.P., Cao, J.W., Li, D.Z., Yun, H., Zeng, T., Enhanced visible-light-driven photocatalytic activity of $BiFeO_3$ via electric-field control of spontaneous polarization. *J. Alloys Compd.*, 783, 943–951, 2019.
157. Hu, X.J., Wang, W.X., Xie, G.Y., Wang, H., Tan, X.F., Jin, Q., Zhou, D.X., Zhao, Y.L., Ternary assembly of g-C_3N_4/graphene oxide sheets/$BiFeO_3$ heterojunction with enhanced photoreduction of Cr (VI) under visible-light irradiation. *Chemosphere*, 216, 733–741, 2019.
158. Zou, Z.M., Yang, X.Y., Zhang, P., Zhang, Y.M., Yan, X.X., Zhou, R.M., Liu, D., Xu, L., Gui, J.Z., Trace carbon-hybridized ZnS/ZnO hollow nanospheres with multi-enhanced visible-light photocatalytic performance. *J. Alloys Compd.*, 775, 481–489, 2019.
159. Qu, Z., Liu, Z., Wu, A., Piao, C., Li, S., Wang, J., Song, Y., Preparation of a coated Z-scheme and H-type $SrTiO_3$/($BiFeO_3$@ZnS)composite photocatalyst and application in degradation of 2,4-dichlorophenol with simultaneous conversion of Cr (VI). *Sep. Purif. Technol.*, 240, 116653–116667, 2020.
160. Kong, L., Ji, Y., Dang, Z., Yan, J., Li, P., Li, Y., Liu, S., g-C3N4 Loading Black Phosphorus Quantum Dot for Efficient and Stable Photocatalytic H2 Generation under Visible Light. *Adv. Funct. Mater.*, 28, 22, 1800668, 2018.
161. Sahoo, S. and Acharya, R., An overview on recent developments in synthesis and molecular level structure of visible-light responsive g-C_3N_4 photocatalyst towards environmental remediation. *Mater. Today Proc.*, 35, 150–155, 2021.

162. Pattnaik, S.P., Behera, A., Acharya, R., Parida, K., Green exfoliation of graphitic carbon nitride towards decolourization of Congo-Red under solar irradiation. *J. Environ. Chem. Eng.*, 7, 6, 103456, 2019.
163. Acharya, R., Naik, B., Parida, K., Visible-Light-Induced Photocatalytic Degradation of Textile Dyes over Plasmonic Silver-Modified TiO_2, in: *Advanced textile engineering materials*, pp. 389–418, 2018.
164. Li, J., Pei, Q., Wang, R., Zhou, Y., Zhang, Z., Cao, Q., Wang, D., Mi, W., Du, Y., Enhanced photocatalytic performance through magnetic field boosting carrier transport. *ACS Nano*, 12, 4, 3351–3359, 2018.
165. Pattnaik, S.P., Behera, A., Martha, S., Acharya, R., Parida, K., Facile synthesis of exfoliated graphitic carbon nitride for photocatalytic degradation of ciprofloxacin under solar irradiation. *J. Mater. Sci.*, 54, 7, 5726–5742, 2019.
166. Sepahvand, H. and Sharifnia, S., Photocatalytic overall water splitting by Z-scheme g-C_3N_4/$BiFeO_3$ heterojunction. *Int. J. Hydrogen Energy*, 44, 42, 23658–23668, 2019.
167. Niu, F., Chen, D., Qin, L., Zhang, N., Wang, J., Chen, Z., Huang, Y., Facile Synthesis of Highly Efficient p–n Heterojunction CuO/$BiFeO_3$ Composite Photocatalysts with Enhanced Visible-Light Photocatalytic Activity. *ChemCatChem*, 7, 20, 3279–3289, 2015.
168. Tao, R., Shao, C., Li, X., Li, X., Liu, S., Yang, S., Zhao, C., Liu, Y., Bi_2MoO_6/BiFeO3 heterojunction nanofibers: enhanced photocatalytic activity, charge separation mechanism and magnetic separability. *J. Colloid Interface Sci.*, 529, 404–414, 2018.
169. Li, C.M., Chen, G., Sun, J.X., Dong, H.J., Wang, Y., Lv, C.D., Construction of Bi2WO6 Homojunction Via QDs Self-Decoration And Its Improved Separation Efficiency of Charge Carriers and Photocatalytic Ability. *Appl. Catal. B*, 160-161, 383–389, 2014.
170. Wang, D., Guo, L., Zhen, Y., Yue, L., Xue, G., Fu, F., AgBr quantum dots decorated mesoporous Bi_2WO_6 architectures with enhanced photocatalytic activities for methylene blue. *J. Mater. Chem. A*, 2, 30, 11716–11727, 2014.
171. Zhang, N., Ciriminna, R., Pagliaro, M., Xu, Y.J., Nanochemistry-derived Bi_2WO_6 nanostructures: towards production of sustainable chemicals and fuels induced by visible light. *Chem. Soc Rev.*, 43, 15, 5276–5287, 2014.
172. Tian, Y., Chang, B., Lu, J., Fu, J., Xi, F., Dong, X., Hydrothermal synthesis of graphitic carbon nitride–Bi_2WO_6 heterojunctions with enhanced visible light photocatalytic activities. *ACS Appl. Mater. Interfaces*, 5, 15, 7079–7085, 2013.
173. Chaiwichian, S., Wetchakun, K., Kangwansupamonkon, W., Wetchakun, N., Novel visible-light-driven $BiFeO_3$-Bi_2WO_6 nanocomposites toward degradation of dyes. *J. Photochem. Photobiol. A*, 349, 183–192, 2017.
174. Tao, R., Li, X., Li, X., Liu, S., Shao, C., Liu, Y., Discrete heterojunction nanofibers of $BiFeO_3$/Bi_2WO_6:Novel architecture for effective charge separation and enhanced photocatalytic performance. *J. Colloid Interface Sci.*, 572, 257–268, 2020.

175. Hao, W., Gao, Y., Jing, X., Zou, W., Chen, Y., Wang, T., Visible Light Photocatalytic Properties of Metastable γ-Bi_2O_3 with Different Morphologies. *J. Mater. Sci. Technol.*, 30, 192–196, 2014.
176. Liu, G., Li, S., Lu, Y., Zhang, J., Feng, Z., Li, C., Controllable synthesis of α-Bi_2O_3 and γ-Bi_2O_3 with high photocatalytic activity by α- Bi_2O_3→γ-Bi_2O_3→α-Bi_2O_3 transformation in a facile precipitation method. *J. Alloys Compd.*, 689, 787–799, 2016.
177. Wang, Y., Li, Z., Yu, H., Guo, L., Controllable synthesis of metastable γ-Bi_2O_3 and optical properties. *Mater. Sci. Semicond. Process.*, 64, 55–62, 2017.
178. Bera, K.K., Majumdar, R., Chakraborty, M., Bhattacharya, S.K., Phase control synthesis of alpha, beta and alpha/beta Bi_2O_3 heterojunction with enhanced and synergistic photocatalytic activity on degradation of toxic dye, Rhodamine-B under natural sunlight. *J. Hazard. Mater.*, 352, 182–191, 2018.
179. Hou, J., Yang, C., Wang, Z., Zhou, W., Jiao, S., Zhu, H., *In situ* synthesis of α–β phase heterojunction on Bi_2O_3 nanowires with exceptional visible-light photocatalytic performance. *Appl. Catal. B*, 142-143, 504–511, 2013.
180. Margha, F.H., Radwan, E.K., Badawy, M.I., Gad-Allah, T.A., Bi_2O_3–$BiFeO_3$ Glass-Ceramic: Controllable β-/γ-Bi_2O_3 Transformation and Application as Magnetic Solar-Driven Photocatalyst for Water Decontamination. *ACS Omega*, 5, 14625–14634, 2020.
181. Walsh, A., Yan, Y., Huda, M.N., Al-Jassim, M.M., Wei, S.H., Band edge electronic structure of $BiVO_4$: elucidating the role of the Bi s and V d orbitals. *Chem. Mater.*, 21, 3, 547–551, 2009.
182. Cao, S.W., Yin, Z., Barber, J., Boey, F.Y., Loo, S.C.J., Xue, C., Preparation of Au-$BiVO_4$ heterogeneous nanostructures as highly efficient visible-light photocatalysts. *ACS Appl. Mater. Interfaces*, 4, 1, 418–423, 2012.
183. Chatchai, P., Kishioka, S.Y., Murakami, Y., Nosaka, A.Y., Nosaka, Y., Enhanced photoelectrocatalytic activity of FTO/WO_3/$BiVO_4$ electrode modified with gold nanoparticles for water oxidation under visible light irradiation. *Electrochim. Acta*, 55, 3, 592–596, 2010.
184. Chen, C.J., Liao, C.H., Hsu, K.C., Wu, Y.T., Wu, J.C., P–N junction mechanism on improved NiO/TiO_2 photocatalyst. *Catal. Commun.*, 12, 14, 1307–1310, 2011.
185. Fan, T., Chen, C., Tang, Z., Hydrothermal synthesis of novel $BiFeO_3$/$BiVO_4$ heterojunctions with enhanced photocatalytic activities under visible light irradiation. *RSC Adv.*, 6, 12, 9994–10000, 2016.
186. Soltani, T., Tayyebi, A., Lee, B.K., $BiFeO_3$/$BiVO_4$ p–n heterojunction for efficient and stable photocatalytic and photoelectrochemical water splitting under visible-light irradiation. *Catal. Today*, 340, 188–196, 2020.
187. Hu, Q., Di, J., Wang, B., Ji, M., Chen, Y., Xia, J., Li, H., Zhao, Y., *In-situ* preparation of NH_2-MIL-125 (Ti)/BiOCl composite with accelerating charge carriers for boosting visible light photocatalytic activity. *Appl. Surf. Sci.*, 466, 525–534, 2019.

188. Tong, X., Yang, Z., Feng, J., Li, Y., Zhang, H., BiOCl/UiO-66 composite with enhanced performance for photo-assisted degradation of dye from water. *Appl. Organomet. Chem.*, 32, 2, 4049, 2018.
189. Tripathy, S.P., Subudhi, S., Acharya, R., Acharya, R., Das, M., Parida, K., Adsorptive removal of Cr (VI) onto UiO-66-NH_2 and its determination by radioanalytical techniques. *J. Radioanal. Nucl. Ch.*, 322, 2, 983–992, 2019.
190. Li, Y., Xu, H., Ouyang, S., Ye, J., Metal–organic frameworks for photocatalysis. *Phys. Chem. Chem. Phys.*, 18, 11, 7563–7572, 2016.
191. Qiu, J., Zhang, X., Xie, K., Zhang, X.F., Feng, Y., Jia, M., Yao, J., Noble metal nanoparticle-functionalized Zr-metal organic frameworks with excellent photocatalytic performance. *J. Colloid Interface Sci.*, 538, 569–577, 2019.
192. Zhang, M., Wang, L., Zeng, T., Shang, Q., Zhou, H., Pan, Z., Cheng, Q., Two pure MOF-photocatalysts readily prepared for the degradation of methylene blue dye under visible light. *Dalton Trans.*, 47, 12, 4251–4258, 2018.
193. Bargozideh, S., Tasviri, M., Shekarabi, S., Daneshgar, H., Magnetic $BiFeO_3$ decorated UiO-66 as a p–n heterojunction photocatalyst for simultaneous degradation of a binary mixture of anionic and cationic dyes. *New J. Chem.*, 44, 30, 13083–13092, 2020.
194. Luo, S., Chai, F., Zhang, L., Facile and fast synthesis of urchin-shaped Fe_3O_4@ Bi_2S_3 core-shell hierarchical structures and their magnetically recyclable photocatalytic activity. *J. Mater. Chem. C.*, 22, 4832–4836, 2012.
195. Bhoi, Y.P., Nayak, A.K., Gouda, S.K., Mishra, B.G., Photocatalytic mineralization of carbendazim pesticide by a visible light active novel type-II Bi_2S_3/ $BiFeO_3$ heterojunction photocatalyst. *Catal. Commun.*, 114, 114–119, 2018.
196. Ma, Y., Lv, P., Duan, F., Sheng, J., Lu, S., Zhu, H., Du, M., Chen, M., Direct Z-scheme Bi_2S_3/$BiFeO_3$ heterojunction nanofibers with enhanced photocatalytic activity. *J. Alloys Compd.*, 834, 155158, 2020.
197. Deng, S., Yang, Z., Lv, G., Zhu, Y., Li, H., Wang, F., Zhang, X., WO_3 nanosheets/g- C_3N_4 nanosheets' nanocomposite as an effective photocatalyst for degradation of rhodamine B. *Appl. Phys. A*, 125, 44, 2019.
198. Lee, T., Lee, Y., Jang, W., Soon, A., Understanding the advantage of hexagonal WO_3 as an efficient photoanode for solar water splitting: a first-principles perspective. *J. Mater. Chem. A*, 4, 11498–11506, 2016.
199. Zhao, Z.G. and Miyauchi, M., Nanoporous-walled tungsten oxide nanotubes as highly active visible-light-driven photocatalysts. *Angew. Chem. Int. Ed.*, 47, 7051–7055, 2008.
200. Huang, L., Xu, H., Li, Y., Li, H., Cheng, X., Xia, J., Xu, Y., Cai, G., Visible-lightinducedWO_3/gC_3N_4 composites with enhanced photocatalytic activity. *Dalton Trans.*, 42, 8606–8616, 2013.
201. Xu, H., Liu, L., She, X., Mo, Z., Xu, Y., Huang, L., Song, Y., Li, H., WO_3 nanorod photocatalysts decorated with few-layer g-C_3N_4 nanosheets: controllable synthesis and photocatalytic mechanism research. *RSC Adv.*, 6, 80193–80200, 2016.

202. Ali, S., Humayun, M., Pi, W., Yuan, Y., Wang, M., Khan, A., Yue, P., Shu, L., Zheng, Z., Fu, Q., Luo, W., Fabrication of $BiFeO_3$-g-C_3N_4-WO_3 Z-scheme heterojunction as highly efficient visible-light photocatalyst for water reduction and 2, 4-dichlorphenol degradation: Insight mechanism. *J. Hazard. Mater.*, 397, 122708, 2020.
203. Bargozideh, S., Tasviri, M., Kianifar, M., Construction of novel magnetic $BiFeO_3$/MoS_2 composite for enhanced visible-light photocatalytic performance towards purification of dye pollutants. *Int. J. Environ. Anal. Chem.*, 2020.
204. Camacho-Escobara, L., Palma-Goyesb, R.E., Ortiz-Landerosa, J., Romero-Ibarrac, I., Gamba-Vásquezb, O.A., Vazquez-Arenasd, J., Unraveling the structural and composition properties associated with the enhancement of the photocatalytic activity under visible light of Ag_2O/$BiFeO_3$-Ag synthesized by microwave-assisted hydrothermal method. *Appl. Surf. Sci.*, 521, 146357–146366, 2020.

8

Photocatalytic Degradation of Pollutants Using $ZnTiO_3$-Based Semiconductor

Waseem Raza[1]* and Khursheed Ahmad[2]

[1]Department of Chemistry, Indian Institute of Technology Delhi, Hauz Khas New Delhi, India
[2]Discipline of Chemistry, Indian Institute of Technology Indore, Simrol, Khandwa Road, India

Abstract

Recently, removal of industrial organic wastes and harmful pollutants from water has been widely investigated for better environmental sustainability and protection. These hazardous pollutants not only lead to serious contamination but also problematic to human and aquatic life worldwide. Photocatalysis is an important chemical approach which has been established one of the most promising and "green" machinery to remediate the environmental related problem. Photocatalysis is a catalytic procedure, taking place on the surface of semiconducting material in the presence of illumination of light. Hence, seeking for a favorable catalysts with promising efficiency has been receiving a great deal of attention for enhanced photocatalytic activity. Various heterogeneous semiconductor photocatalysts, such as metal oxides, organic complexes and chalcogenides, have been used for photo degradation of pollutants. Recently, most dynamic oxides perovskite-type mixed metal oxides have been extensively examined and applied for environmental remediation because of their good physical and chemical properties with uncertainty of arrangement and structure. Among all, zinc titanate ($ZnTiO_3$) is one of the most fascinating photocatalytic material and has been broadly explored for environmental protection. Therefore, this chapter compiles the recent advancement in perovskite ($ZnTiO_3$) oxides as advanced photocatalysts for environmental remediation.

***Keywords*:** Photocatalysis, $ZnTiO_3$, perovskite, heterogeneous, environmental protection

**Corresponding author*: waseemrazambd@gmail.com

Khursheed Ahmad and Waseem Raza (eds.) Perovskite Materials for Energy and Environmental Applications, (217–240) © 2022 Scrivener Publishing LLC

8.1 Introduction

Water is an essential and one of the most crucial need for life. The presence of water is important for ecosystem, moreover life cannot be sustained without water. However, addition of any unwanted component can make it polluted [1]. Fast growing population and rapid development of industries are the most serious source for water pollution [2, 3]. There are various types of pollutants regularly added to water system, such as organic, inorganic, heavy metals, and so on. The existence of such pollutants in water not only harmful for human health but also dangerous for ecosystem, food web, and environment. Therefore, it is highly essential to control or treat such hazardous pollutants for human and aquatic life and to avoid the accumulation in environment. So, water pollution is a hot topic nowadays and gaining much more attention for discharge of sewage and remediation of toxic pollutants [4]. A lot of different techniques such as physical adsorption, chemical and biological methods have been reported in literature [5, 6]. However, these traditional methods are always not so effective for eliminating such toxic, persistent organic, and highly hazardous metal ions pollutants [1]. Traditional physical methods, such as coagulation, filtration, flocculation, and adsorption, cannot effectively treat pollutants, but transfer one form to other [7]. Biological methods are also not so good for all pollutants such as organic dyes cannot be decomposed by microorganism [8]. Hence, evolution of efficient, green, and economical method for treatment of wastewater is highly needed. Among all chemical methods, advanced oxidation processes (AOPs) have been considered as a promisisng approach for wastewater mitigation, because AOPs effectively degrade the pollutants into CO_2, H_2O, and mineral acid. [9] During the degradtion process, highly reactive species, such as hydroxyl radical, ${}^{\bullet}$OOH radical, and superoxide radical anion, take part for minerlization of wide range of organic and inorganic pollutants [10]. Among various AOPs, semiconductor-based heterogeneous photocatalysis has became one of the most promisisng approach and is considered as green technology for wastewater treatment. Moroever, heterogenous photocatalysis may completely degrade the hazardous waste, dyes, metals, and toxic chemicals directly into CO_2 and H_2O [7, 11, 12]. Heterogenous photocatalysis being an AOPs has some advanteges over tradional methods, such as required a semiconductor (low price, simple preparation and high photochemical stability, recylced) that interacts with photons (endlees sun light) to start the chemical reaction, works at ambient pressure and temperature and completlety mineralized the pollutnats [13]. Photocatalysis is a catalysis process that takes palce on the surface of semiconductor in the presence of light, which produces reactive species for degradation of pollutants. Various metal oxides semiconductors have been widely investigated for the degradation of pollutnats [14]. Among all TiO_2 and ZnO have widely investigted semiconductors for wastewater treatment due to their unique propeties such as nontoxic nature, low cost, chemically stable easily avialability and efficiently production of reactive species [15–18]. Moroever, these semiconductors show eco-friendly nature for photocatalytic degradation of pollutants [19]. However, the industrial application of TiO_2 and ZnO are limited due to their wide band gap, low quantum efficiency, and fast recombination of photogenrated charge carriers. On the other hand ZnO also have several drawbacks, such as poor photosensitization, low quantum yield, and high photocorrosion [20]. Therfore, scientists have applied different methods to modified TiO_2 and ZnO into a heterostructure to improve the activity, stabiltiy, and durability by coating ZnO with TiO_2 photocatalyst [21]. The composite materials made up of the addition of TiO_2 and ZnO are expected to be a novel material having the quality of both materials. Moroever, the simultaneous preparation of both ZnO-TiO_2 heterojunction gives three different compunds, such as Zn_2TiO_4 (zinc orthotitanate) with cubic and tertragonal strucuture, $Zn_2Ti_3O_8$ (zinc polytitanate) with a cubic defect spinel structure and $ZnTiO_3$ (zinc metatitanate or zinc titanate) with cubic and hexagonal structure [20, 22–24]. However, Zn_2TiO_4 possess fcc lattice, in which octahedral sites are occupied by Zn and Ti cations, whereas tetrahedral sites are occupied by Zn cations conataining 56 atoms per unit and most stable one (≤ 418°C) [25], whereas $Zn_2Ti_3O_8$ is a metastable, low temperature form of $ZnTiO_3$, which exist below 800°C [26]. The preparation of pure/pristine form of $ZnTiO_3$ is not easy because it decomposes to Zn_2TiO_4 and rutile TiO_2 above 945°C [26]. However, cubic $ZnTiO_3$ is thermodynamically more stable at low temperature and also converts to hexagonal at T ≥ 800°C [27]. Among all forms of ZnO-TiO_2 composite, $ZnTiO_3$ shows favorable catalytic properties and considered as a promising photocatalytic material for environmental applications. $ZnTiO_3$ is a perovskite-type oxide. The perovskite oxides possess a general formula of ABO_3, where A generally stands for cations form rare-earth or alkaline-earth metals, which are placed on the 12-fold coordinated cub-octahedral cages of the oxygen sublattice of the oxygen sublattice, and B usually a metal cation element stand for transition-metal cations with 3d, 4d, and 5d configuration (e.g., Mn, Fe, Co, Ni, Cu, Ti, etc.) surrounded by six oxygen atoms in an octahedral coordination as shown in Figure 8.1 [28–30].

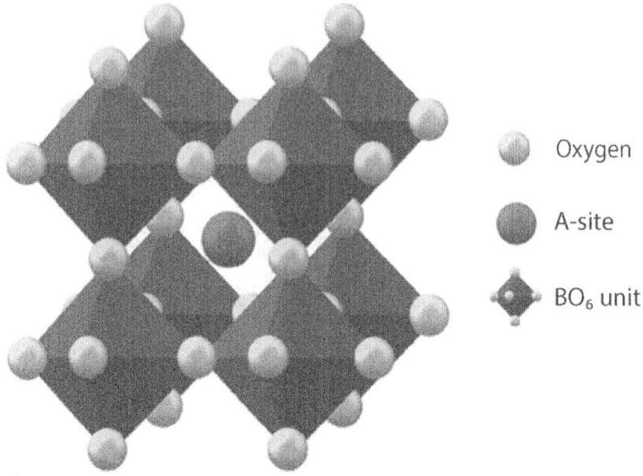

Figure 8.1 Schematic illustration of a perovskite with the general chemical formula ABO_3 structure [30].

One of the great advantage of perovskite structure that distortions widely exists resulting in a huge family of such materials with different composition by changing either A or B cation or substituted by same or other cations with different valences [31]. However, the size of A^{2+} is larger than B^{4+} in ABO_3 crystal and oxygen exists as O^{2-} and oxygen can be present in excess and deficiency as nonstoichiometry which gives the formula of $A_{1-X}A'XB_{1-y}B'O_3\pm\delta$ [32, 33]. Moreover, perovskite oxides offer enormous opportunity in redox active sites, physiochemical properties, oxygen vacancies and surface properties due to different possible substitutions at either A and B sites. Based on crystal structure and composition, perovskite oxides can also form layered or double perovskites as well as morphology of the perovskite oxides also have an enormous impact on their physiochemical properties [34]. The above factors basically effect the excitation and transfer of photogenerated electron/hole pairs and greatly modify the photocatalytic activity [35]. To check the deviation from the ideal perovskite structure, Goldschmidt gives a postulate in terms of tolerance factor (t) as given in equation 8.1 [36].

$$t = (r_A + r_O)/\sqrt{2}\,(r_B + r_O) \qquad (8.1)$$

where r_A, r_B, and r_O are ionic radii of elements A, B, and O ions, respectively. Moreover, t value gives the idea of symmetry in perovskite oxides, if equal to 1 means symmetry will be perfect cubic and usually produced at high temperature [37]. If t value less one ($t < 1$) then symmetry will be distorted to rhombohedral or orthorhombic, whereas higher t value than one (t > 1) gives hexagonal symmetry [38]. However, doping in perovskite oxides is an easy task, can be possible by replacing of both cations either A or B sites, which can change the symmetry and composition, resulting creation of oxygen or cation vacancies [39]. These changes will highly alter the light absorption, band structure, adsorption, and photocatalytic activity of perovskite oxide materials. Moreover, optical properties and band gap of the perovskite oxides can also be alter by doping at O sites, resulting tailoring of photocatalytic activity easily [37]. The periodic table shows the different A site and B site cations, which can be used for preparation of various photocatalytically active perovskite oxides as displayed in Figure 8.2. Among the various functional materials and titanium based perovskite oxides, zinc titanate oxide has been widely investigated with both hexagonal and cubic structures in photocatalytic degradation due their unique properties [40]. Moreover, $ZnTiO_3$ is a promising material for photocatalysis due to its low oxidation potential and high reduction potential, as well as its low recombination rate of photogenerated charge carriers.

However, $ZnTiO_3$ possesses two forms, namely cubic and hexagonal with wide band gaps, such as 3.8 and 3.6, respectively [41]. The cubic $ZnTiO_3$ has half vacant octahedral sites with defect spinel structure [42]. However, among both two crystal structures, hexagonal $ZnTiO_3$ is a well-known photocatalyst for potential photocatalytic reactions.

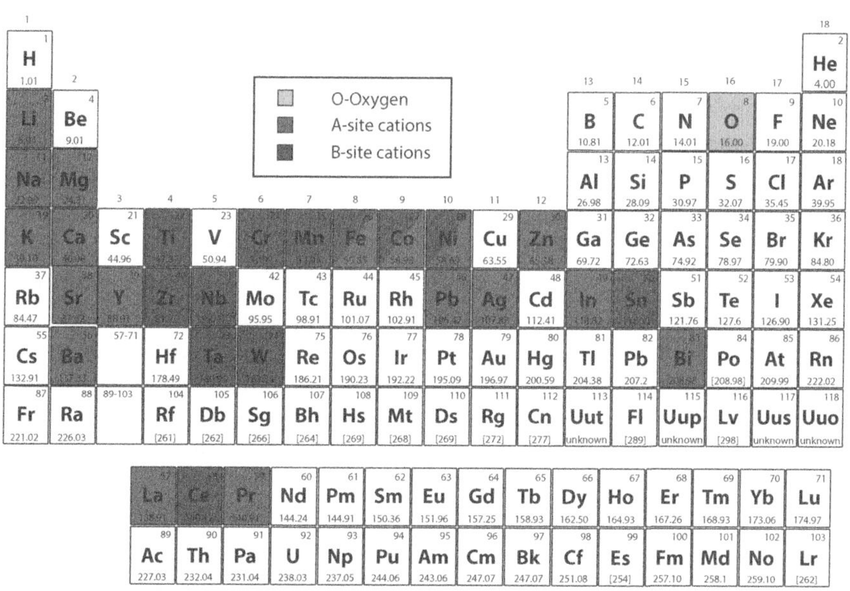

Figure 8.2 Periodic table showing different cations and oxygen used for A and B sites for the synthesis of photocatalytically active perovskite oxides [30].

8.2 Synthesis of $ZnTiO_3$

There are different reports for the synthesis of $ZnTiO_3$ perovskite material, such as solid state method, ball milling method, sputtering method, coprecipitation method, hydrothermal method, sol-gel method, combustion method, molten salt method, electrospinning method, sonochemical method, pechini method, atomic layer deposition method, and so on.

The standard solid state reaction has been applied for the synthesis of pristine $ZnTiO_3$, in which stoichiometric amounts of ZnO and TiO_2 powders were mixed as the starting materials and grounded together in ethanol using ball milling. However, the drying of chemicals was done in nitrogen atmosphere. The mixtures were calcined at 700°C to 900°C for 24 hours [43]. The ball milling method has been used for the synthesis of $ZnTiO_3$. In a typical procedure, precisely weight analytical grade ZnO and TiO_2 powders were grounded in a mortar pestle in acetone to obtain homogenous mixture. In a planetary ball mill, high energy ball milling was applied, in which a rotating disk rotate in opposite direction at a speed of 300 rpm and the vials around at 450 rpm. After the ball milling, the powder was collected [44]. Nikam *et al.* applied combustion method for the synthesis of $ZnTiO_3$ photocatalyst. Typically, zinc acetate (1 mol), titanium tetraisopropoxide (1 mol), citric acid (0.5 mol), and ammonium nitrate (0.5 mol) were taken in flask to prepare a homogenous mixture in isopropanol and methanol mixture. The obtained paste was kept into a tubular furnace at 650°C and 750°C for 3 hours [45]. The hexagonal pristine $ZnTiO_3$ powder was prepared using molten salt method, stoichiometric amount of ZnO, TiO_2 as raw materials were mixed by ball milling method in ethanol for 6 hours. After that, powder was dried at 60°C for 12 hours, and molten salt synthesis was done in a corundum crucible at different temperature for 2 hours [46]. The nanopowder $ZnTiO_3$ was synthesized using sol-gel assist electrospinning method. The zinc acetate dihydrtae and titanium tetrabutoxide (1:0.9) were mixed in acetic acid on the magnetic stirrer followed by heating at 50°C, the obtained product was calcined at 700°C to 1000°C. After that, electrospinning method was applied to synthesize the $ZnTiO_3$ fibers [47]. Han *et al.* used a facile self-assembly method for the synthesis of $ZnTiO_3$. In a typical procedure, raw materials of zinc acetate (2.19 g) and polyvinylpyrrolidone (3.0 g) and tetrabutyl titanate (3.40 g) were added into water and ethylene glycol and heated at 50°C for 24 hours. The obtained product was calcined at 600°C for 3 hours to get final product as shown in Figure 8.3 [48].

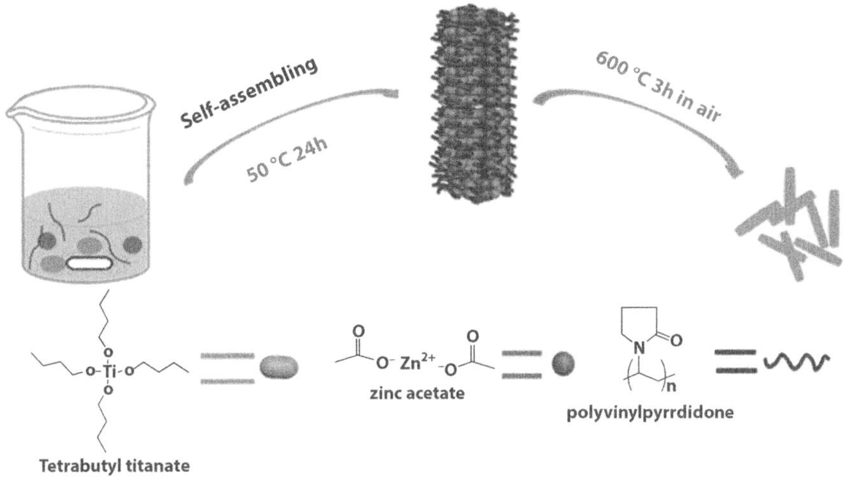

Figure 8.3 Synthesis of $ZnTiO_3$ using self-assembly method [48].

8.3 Fundamental Need and Basic Mechanism for Photocatalytic Degradation of Pollutants

The basic step in the photocatalytic degradation of pollutants is the excitation of electron form valence band (VB) to the conduction band (CB) leaving behind a hole, when light shines with correct energy falls on the semiconducting materials ($ZnTiO_3$) as shown in equation 8.2.

$$ZnTiO_3 + h\nu \rightarrow e^- + h^+ \tag{8.2}$$

However, the applied shine light should be equal to or greater than the band gap of the semiconductor, so that the generated electron/holes pairs can participate in the redox reactions. The generated charge carries will move to the surface of semiconductor to participate the redox reaction, where they will act as an electron donor or acceptor for generating the active radicle species, if charge separation is maintained. If the pollutants degradation reaction will take place in aqueous solution in the presence of oxygen, then the hole will react with water and produce the hydroxyl radical and the separated electron in the CB band will participate to generate the superoxide radicle after reaction with oxygen [1]. However, the reaction of electron with oxygen and hole with water during surface reaction depends on the potential of CB and VB of semiconductor. The CB and VB of $ZnTiO_3$ is reported as −0.203 eV and +2.897 eV respectively. So, photogenerated electrons have highly negative potential than the reduction potential of oxygen $O_2^{\bullet -}/O_2$ (−0.13 eV vs. NHE), so they can easily produce the superoxide radicles [49]. The production of hydroxyl radicles are also possible by the holes generated by $ZnTiO_3$ because the oxidation potential of holes are higher than the oxidation potential of hydroxyl anion $^{\bullet}OH/^{-}OH$ (+1.99 eV vs. NHE) [5]. So, generated electrons/holes can produce the reactive species for pollutants degradation as shown in equations 8.3 to 8.5.

$$e^- + O_2 \rightarrow ^{\bullet -} \tag{8.3}$$

$$h^+_{vb} + H_2O \rightarrow H^+ + ^{\bullet}OH \tag{8.4}$$

$$h^+_{vb} + ^{-}OH \rightarrow ^{\bullet}OH \tag{8.5}$$

The photogenerated species act as reactive species for the degradation of pollutants. The generated superoxide radicle species act as strong reducing agent and generated hydroxyl radicles acts as a strong oxidizing agent for the degradation of pollutnats as shown in equation 8.6 [1]. The general mechanism of photocatalytic production of charge carriers is given in Figure 8.4.

$$\text{Pollutants molecules} + \text{reactive species} \rightarrow \text{degraded products} \tag{8.6}$$

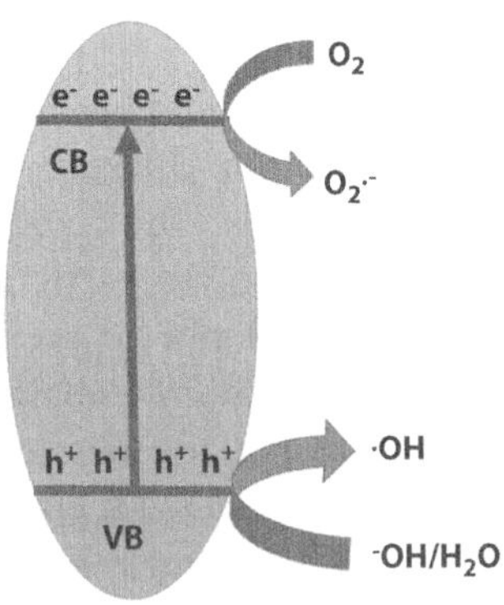

Figure 8.4 Mechanism for photocatalytic production of charge carrier at the surface of $ZnTiO_3$ semicoponductor.

So, perovskite oxide-based $ZnTiO_3$ is good photocatalytic material for pollutant degradation. The photocatalytic degradation of pollutants using $ZnTiO_3$ semiconductor gives complete mineralization or low molecular weight products because it uses advanced oxidation methods for the degradation of pollutants [50]. Moreover, the reusability of the catalyst is easy because of the heterogenous nature of reaction at the surface of semiconductor photocatalyst, which gives degraded products at the end. So, this method is considered as an eco-friendly and green approach for waste water treatment [51].

8.4 Photocatalytic Degradation of Pollutants Based on $ZnTiO_3$

Considerable information is given in the literature on the degradation of pollutants using perovskite photocatalytic materials in terms of mechanism, reactions, and performances [52, 53]. Perovskite-type photocatalysts have shown attractive photocatalytic properties because of different significant advantages corresponding to binary oxides, such as vast capacity to tune the band structure, suitable band gap edge potential, and other photophysical properties [54, 55]. However, designing of photocatalyst for a targeted application is very deceiving and is a paramount step to achieve good results. Therefore, the selection of semiconductor photocatalyst should be done by keeping in mind the band bap edge potential and feasible reaction conditions of the targated application [56]. So, $ZnTiO_3$ is a typical perovskite-type oxide photocatalyst and has been extensively investigated for degradation of pollutants due to its unique properties. However, $ZnTiO_3$ possesses a wide band gap (3.1–3.65 eV), and therefore, can absorbed only UV light from the solar spectrum. The natural sunlight contains mostly visible light and near-infrared (NIR) regions. On the other hand, fast recombination rate of the photogenerated electrons/holes also limit the practical application of $ZnTiO_3$ photocatalyst. Therefore, it is an urgent need to develop the visible light-induced photocatalyst to achieve the higher quantum yield. There are different methods to enhance the photocatalytic activity of $ZnTiO_3$ into visible region, among all band gap engineering is promising approach. The band gap engineering grants the band gap tuning of $ZnTiO_3$ photocatalyst by doping with hetero atoms in their crystal lattice. As disused earlier, doping can be possible at any three different sites, such as A, B, and O sites, which may enhance the photocatalytic activity by absorbing more visible light by band gap alteration. The CB minimum of $ZnTiO_3$ is mostly made up of d orbitals and lies at a slightly higher negative value than zero potential, whereas the VB maximum is made up of 2p orbitals and having more positive potentials but below than 3 eV as shown in Figure 8.5 (a). So, we can say that $ZnTiO_3$ have satisfactory CB and VB band edges potential for photocatalytic degradation of pollutants. Therefore, alternation in the band gap of pristine $ZnTiO_3$ semiconductor photocatalyst is possible by the addition of deep and shallow intra-band energy levels between CB and VB, without affecting the band gap but improves the light absorption ability of semiconductor by add-on shoulder as shown in Figure 8.5 (b). However, the shift toward longer wavelength (red shift) is also possible by doping and narrowing the band gap of $ZnTiO_3$ semiconductor as shown in Figure 8.5 (c) [57, 58].

The cations of second and third series of transitions metals are d-electron-rich cations with high atomic energy, secure good electron mobility, which can provide donor impurities in $ZnTiO_3$ semiconductor without changing the CB minimum. However, the VB maximum of the host material will be altered if the doping will be carried out at O-sites with higher atomic energies than the oxygen atom, resulting reduction in the band gap. Therefore, if doping of nitrogen is performed in the host material, then a new band N-2p orbital get introduced above the VB maximum due to its higher potential energy than O-2p orbitals, resulting in enhancement in the visible light absorption [37]. However, the reduction in the band gap after nitrogen doping depends on the interaction between N-2p and O-2p orbitals. Therefore, improvement in the absorption band or altering the absorption band can be achieved by adding the new energy levels with in the band gap of $ZnTiO_3$ material. The newly formed levels between VB and CB levels of $ZnTiO_3$ provides transitions, which improves the light absorptions intensity toward longer wavelength. However, the reduction in the band gap upon introduction of dopant level may be due to hybridization between heteroatom and host orbitals. Overall, doping of heteroatoms at different sites of $ZnTiO_3$ (ABO_3) will improve the light absorption intensity by introducing the impurity level or add-on shoulder via band tailing. Abirami *et al.* reported the synthesis of Ag-doped $ZnTiO_3$ using sol-gel method with phase transition from cubic to hexagonal at 600°C to 800°C for enhanced photocatalytic activity [27]. Kong *et al.* reported the synthesis of $ZnTiO_3$ photocatalyst using modified alcoholysis procedure with different phases for degradation of azo and methyl violet (MV) dyes under solar light illumination [59]. The hexagonal $ZnTiO_{3,}$ which was calcined at 800°C for 3 hours, exhibits the highest degradation of all dyes among all phases. In the modified alcoholysis procedure, ethylene glycol was used

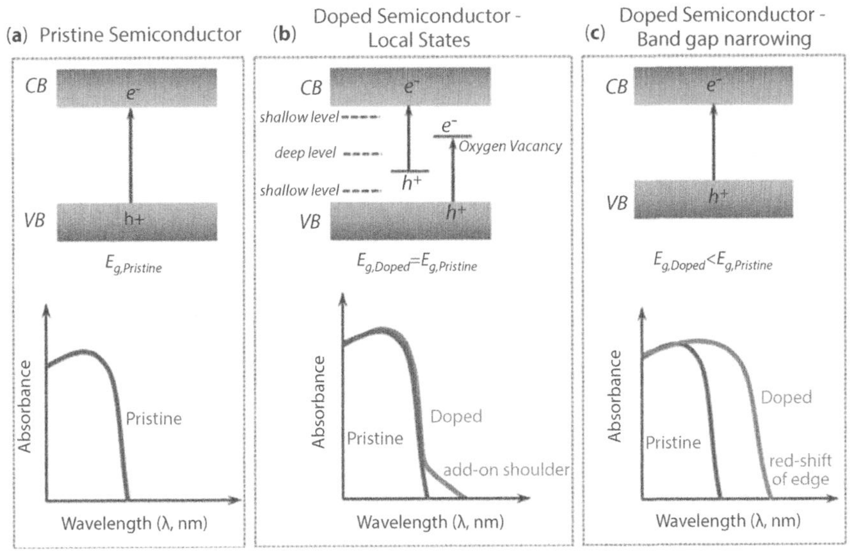

Figure 8.5 Presentation of designing, engineering of electronic band structure and absorption of (a) pristine semiconductor, (b) formation of shallow, deep and add-on shoulder levels, and (c) reduction in band gap and red shift due to doping [58].

as a reagent, solvent and acetylacetone were also used as stabilizer and chelating agent for the synthesis of $ZnTiO_3$. After the complete synthesis, final powder was calcined at 600°C, 700°C, 800°C, and 900°C for 2 and 3 hours. The scanning electron microscopy (SEM) of the as-prepared sample with calcination at 800°C for 3 hours was carried out to known the morphology. Figure 8.6 (a) shows the SEM image of $ZnTiO_3$ calcined at 800°C for 3 hours that particles are nearly spherical with uniform size. They also checked the photocatalytic activity for degradation of MV for different calcined temperature for different hours. It could be seen from Figure 8.6 (b) that sample calcined at 800°C for 3 hours exhibits the highest photocatalytic activity for the degradation of MV dye. It could be seen from Figure 8.6 (b) that the photocatalytic activity increases with calcination temperature and time may be due to change in phase from cubic to hexagonal. The photocatalytic activity decreases with calcination temperature exceeding above 800°C for 3 hours may be due to change in phase again from hexagonal $ZnTiO_3$ to Zn_2TiO_4. Therefore, hexagonal $ZnTiO_3$ calcined at 800°C for 3 hours, exhibiting highest activity than other samples.

Figure 8.6 (c) shows the degradation of MV dye in the presence and absence of $ZnTiO_3$ calcined at 800°C for 3 hours with or without solar light. For comparative purpose, figure also exhibits the MV dye

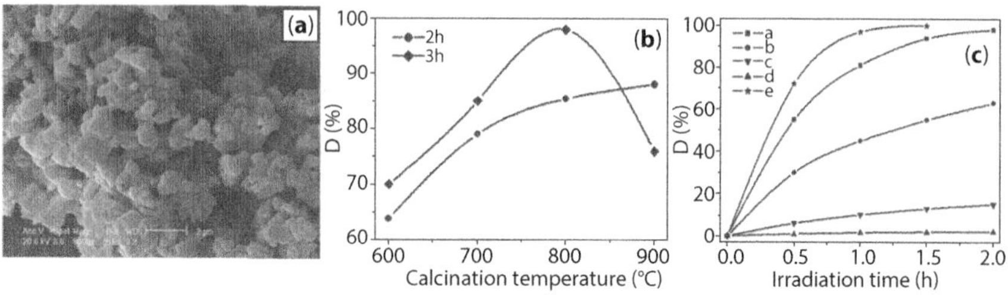

Figure 8.6 (a) SEM image of hexagonal $ZnTiO_3$ calcined at 800°C for 3 hours, (b) Photocatalytic catalytic degradation of MV dye as a function of different calcined temperatures and time. (c) Photocatalytic degradation of MV dye as a function of time (a) in the presence of $ZnTiO_3$ calcined at 800°C for 3 hours under solar light, (b) in the presence of TiO_2 under solar light, (c) in the presence of solar light without $ZnTiO_3$ (d) in the presence of $ZnTiO_3$ calcined at 800°C for 3 hours without solar light, and (e) in the presence of P25 with solar light [59].

degradation in the presence of P25 under solar light illumination. The sample calcined at 800°C for 3 hours ($ZnTiO_3$) shows the higher activity than TiO_2 calcined at 800°C for 2 hours. However, P25 exhibits the higher activity than $ZnTiO_3$ calcined at 800°C for 3 hours due to less specific surface area. Surendar *et al.* synthesized the pristine and La-doped $ZnTiO_3$ photocatalyst using sol-gel method for the degradation of rhodamine B (RhB) under visible light illumination. They also observed the effect of La on the crystal and physiochemical properties of $ZnTiO_3$ photocatalyst [60]. They synthesized pristine and different (1-5) atom% $ZnTiO_3$ photocatalysts and calcined at different temperature for 3 hours and named as ZT (pristine) ZLT-1, ZLT-2, ZLT-3, ZIT-4, and ZLT-5, respectively. The surface morphology was evaluated using transmission electron microscopy (TEM), which shows hexagonal image for pristine and cubic for 2% La-$ZnTiO_3$ as shown in Figure 8.7 (a & b). The efficiency of the as prepared photocatalysts were evaluated by degrading the RhB dye under visible light illumination.

Figure 8.7 (c) exhibits the photocatalytic degradation of RhB dye in the presence of pristine and different % of La-doped $ZnTiO_3$ and commercial N-TiO_2. It could be seen from the figure that La-doped $ZnTiO_3$ photocatalyst exhibits higher activity than pristine and commercial N-TiO_2. The 2% La-doped $ZnTiO_3$ photocatalyst exhibiting highest activity toward RhB degradation among all photocatalysts. Figure 8.7 (c) shows that first photocatalytic activity increases till 2 atom % of La then decreases. The decrease in the activity may be due to increase in the amount of oxygen vaccines, which act as recombination sites for charge carriers. Another reason may

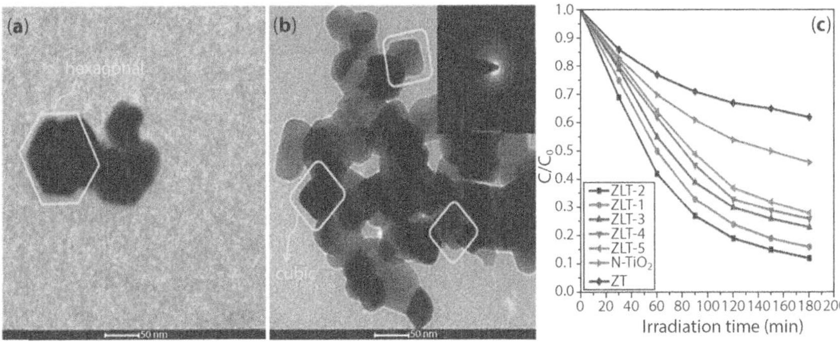

Figure 8.7 TEM images of (a) pristine $ZnTiO_3$, (b) 2% La-doped $ZnTiO_3$ and (c) Photocatalytic degradation of RhB dye in the presence of pristine and different % of La-doped $ZnTiO_3$ and N-TiO_2 photocatalyst [60].

be due to agglomeration of La-nanoparticles at surface of $ZnTiO_3$ photocatalyst, which covers the active sites of photocatalyst. They also proposed a mechanism for the degradation of RhB in the presence of ZLT-2 as shown in Figure 8.8. The $ZnTiO_3$ has a band gap around 3.8 eV with a VB edge potential 0f +3.0 eV which is mainly made up of O 2p orbital, therefore, holes are able to oxidize the H2O to $^{\bullet}OH(E(^{\bullet}OH/H_2O)$ =

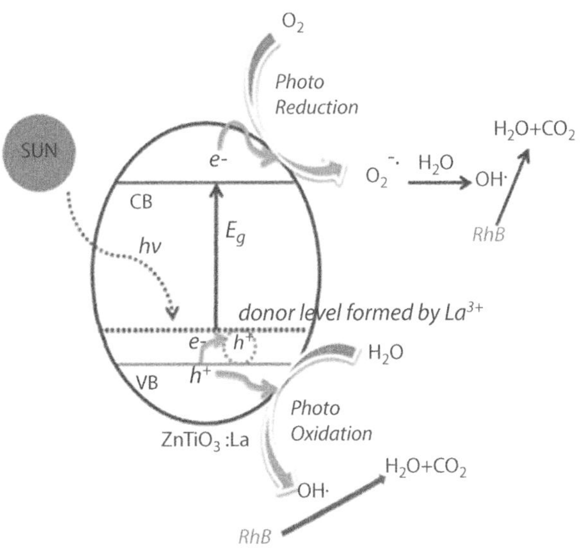

Figure 8.8 The photocatalytic degradation of RhB dye in the presence of 2% La-doped $ZnTiO_3$ photocatalyst under solar light illumination [61].

+2.68 eV vs. NHE). However, the CB edge potential is −0.8 eV made up of Ti 3d orbitals, the electrons are so energetic that can reduced the oxygen to superoxide radicle anion (E ($O_2/O_2^{\bullet-}$) = −0.3 eV vs NHE.

Dutta *et al.* synthesized the nano-sized pristine and Ag-doped/dispersed $ZnTiO_3$ photocatalysts using facile microwave assisted method for degradation of RhB dye in aqueous solution under visible light irradiation [61]. The synthesized powder was calcined at different temperatures (300°C, 400°C, and 500°C) for 5 hours. The morphology of the prepared un-doped and Ag-doped $ZnTiO_3$ were evaluated using TEM and HR-TEM, exhibiting nanorod like morphology as shown in Figure 8.9 (a, c, and d). The HR-TEM image of pristine shows lattice fringe spacing around 0.25 nm, which is corresponding to (110) plane of $ZnTiO_3$ as shown in Figure 8.9 (b). It could be seen from Figure 8.9 (c) that the size of nanorods increased with Ag doping and indicates the presence of Ag nanoparticles on the surface of $ZnTiO_3$ photocatalyst as shown in Figure 8.9 (d).

They choose RhB dye to evaluate the photocatalytic activity of three different samples, such as pristine and different percentage of (2% and 5%) Ag-doped $ZnTiO_3$ photocatalysts under UV and visible light illumination. Figure 8.10 (a & b) present the photocatalytic activity of pristine $ZnTiO_3$

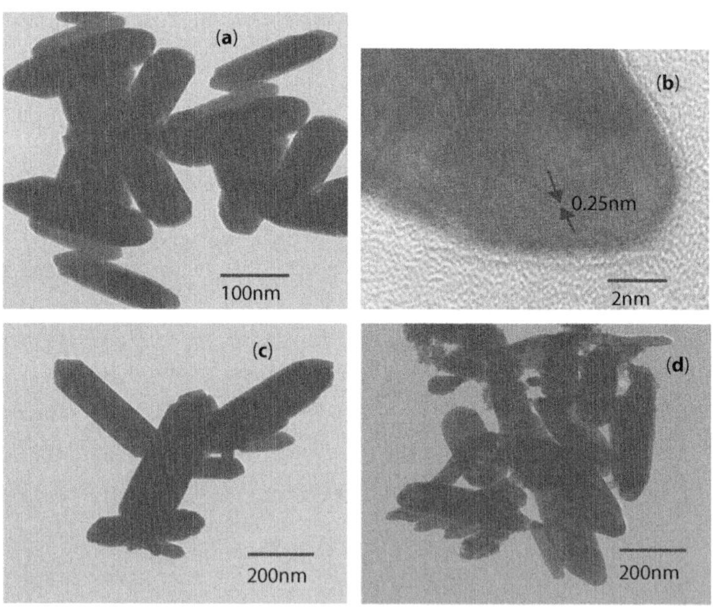

Figure 8.9 (a and b) TEM and HR-TEM images of pristine $ZnTiO_3$, (c) TEM image of Ag-doped $ZnTiO_3$, and (d) TEM image of Ag-doped $ZnTiO_3$ [61].

under UV and visible light illumination respectively. The best results is presented by sample calcined at 500°C, which shows 100% degradation of RhB dye after 150 min under UV light. The pristine $ZnTiO_3$ calcined at different temperature also exhibits degradation of RhB dye. It could be seen from Figure 8.10 (b) that sample calcined at 300°C shows poor activity under visible light illumination as compared with sample calcined at 500°C. However, sample calcined at 500°C exhibiting higher activity among all calcined samples may be due to its reduced band gap of 3.02 eV. They also did experiments in the presence of sunny days but the degradation of RhB was better in the presence of Xe lamp than sunny days. This may be due to sunny days provide less intensity as compared with Xe lamp intensity. Figure 8.10 (c) shows change in absorbance as function of time for RhB dye degradation in the presence of pristine $ZnTiO_3$

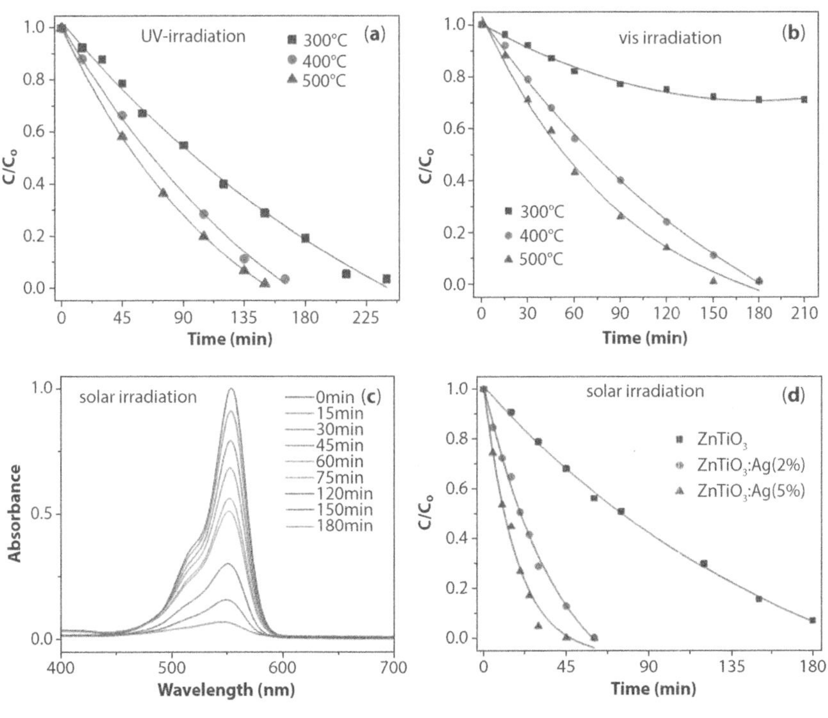

Figure 8.10 Photocatalytic degradation of RhB dye at different calcined temperature in the presence of pristine $ZnTiO_3$ photocatalyst (a) under UV light, (b) under visible light, (c) change in the absorbance of RhB as a function of time in the presence of pristine $ZnTiO_3$ calcined at 500°C under visible light illumination, and (d) change in the concentration of RhB in the presence of pristine and Ag doped (2 & 5%) $ZnTiO_3$ calcined at 500°C under visible light illumination [61].

calcined at 500°C, which exhibiting 93% degradation after 180 min under solar light irradiation. Figure 8.10 (d) presents the change in concentration as a function of time in the presence of pristine and different % of Ag (2 and 5)–doped $ZnTiO_3$ photocatalysts calcined at 500°C under visible light illumination. It could be seen from the that 5% Ag-doped $ZnTiO_3$ shows highest degradation of RhB dye under solar light with around 100% degradation of RhB after 45 min. However, pristine and 2% Ag-doped $ZnTiO_3$ also exhibiting 36 and 87% degradation of RhB dye after 45 min under solar light illumination. The improvement of degradation of RhB dye after doping may be due to the deposition of plasmonic metal Ag NPs on the $ZnTiO_3$, which can efficiently improve the charge separation of photogenerated electron-hole pairs and extend the light absorption ability of $ZnTiO_3$, resulting in enhancement in the activity of photocatalyst under sunlight.

Azar *et al.* reported the synthesis of pristine and different % of (5%, 10%, and 20%) S-doped $ZnTiO_3$ photocatalysts using hydrothermal method and calcined at 750°C for 3 hours for the degradation of Crystal violet (CV) and RhB under natural sunlight [62]. The photocatalytic activity of the semiconductor photocatalyst depends on different factors, such as morphology, surface area, particle size structural defects, and degree of crystallinity. Figure 8.11 (a & b) exhibit the degradation of CV and RhB in the presence and absence of pristine and different % of (5-20) S-doped $ZnTiO_3$ photocatalysts as well as in the absence of photocatalyst respectively. It could be seen from Figure 8.11 (a & b) that highest degradation of CV and RhB was found to be 93% and 95% after 180 minute in the presence 20% S-doped $ZnTiO_3$. Figure 8.11 (c & d) show the change in the absorbance of aqueous solution of CV in the presence of pristine and 20% S-doped $ZnTiO_3$ photocatalyst under solar light illumination. It is clearly seen from the figures that decrease in the absorbance of CV is much higher in the presence of S-doped $ZnTiO_3$ as compared with pristine $ZnTiO_3$ photocatalyst. The enhanced photocatalytic activity for S-doped $ZnTiO_3$ photocatalyst may be because the absorption of more visible light due to smaller band gap, improvement of pore size, and surface area. The results show that degradation of CV is more than RhB in the given time this may be due to stability of RhB and resistant against degradation process.

Figure 8.12 (a & b) exhibit the degradation of RhB dye in terms of change in absorbance of aqueous solution of RhB in the presence of pristine and 20% S-doped $ZnTiO_3$ photocatalyst. The degradation of RhB is higher in the presence of S-doped $ZnTiO_3$ than pristine one as confirm from Figure 8.12 (a & b). The higher photocatalytic activity for 20% S-doped $ZnTiO_3$ photocatalyst may be due to formation of doping level above the

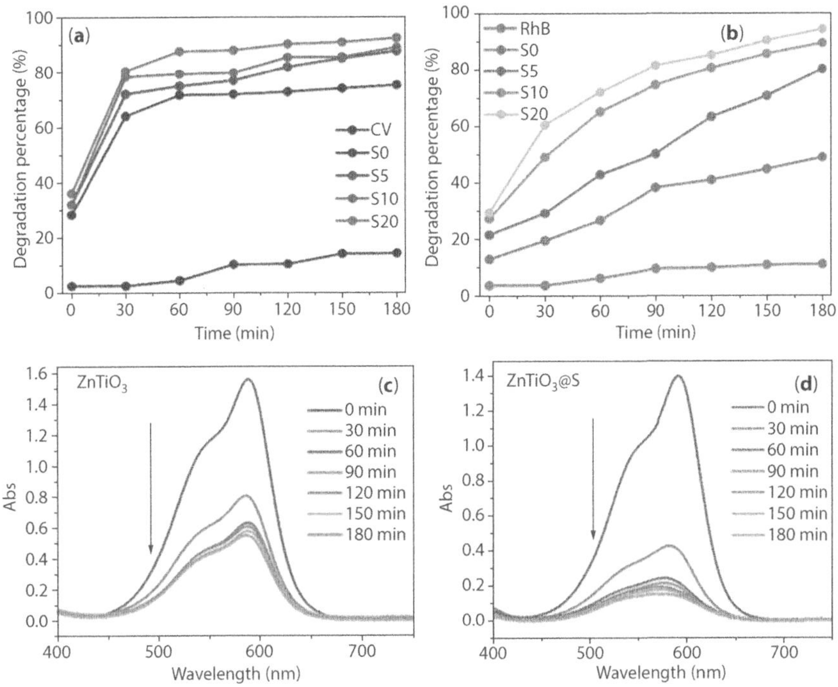

Figure 8.11 Photocatalytic degradation of CV in the presence of (a) pristine and (b) different % of S-doped $ZnTiO_3$ photocatalyst changes in the absorbance of aqueous solution of CV in the presence of (c) pristine $ZnTiO_3$ and (d) 20% S-doped $ZnTiO_3$ photocatalyst [62].

VB of $ZnTiO_3$. As the 3p orbital of sulfur is higher in energy than 2p orbital of oxygen, therefore, sulfur penetrates the crystal lattice which expand the VB of $ZnTiO_3$, resulting reduction in band gap of $ZnTiO_3$. The photocatalytic degradation of dyes in the presence of 20% S-doped $ZnTiO_3$ semiconductor photocatalyst is shown in Figure 8.12 (c). Upon shining light equal to or more than the band gap of S-doped $ZnTiO_3$ photocatalyst, electrons are excited to CB leaving holes in VB. These photogenerated charge carriers take part in the reaction for the formation of reactive species, such as superoxide radicle anion and hydroxyl radicles as shown in equations 8.3 to 8.5. These reactive species are responsible for degradation of dyes. However, superoxide radicle anions not only take part for the degradation of dyes but also inhibit the recombination of charge carries as shown in the equations below.

$$O_2 + H^+ \rightarrow HOO^\bullet \tag{8.7}$$

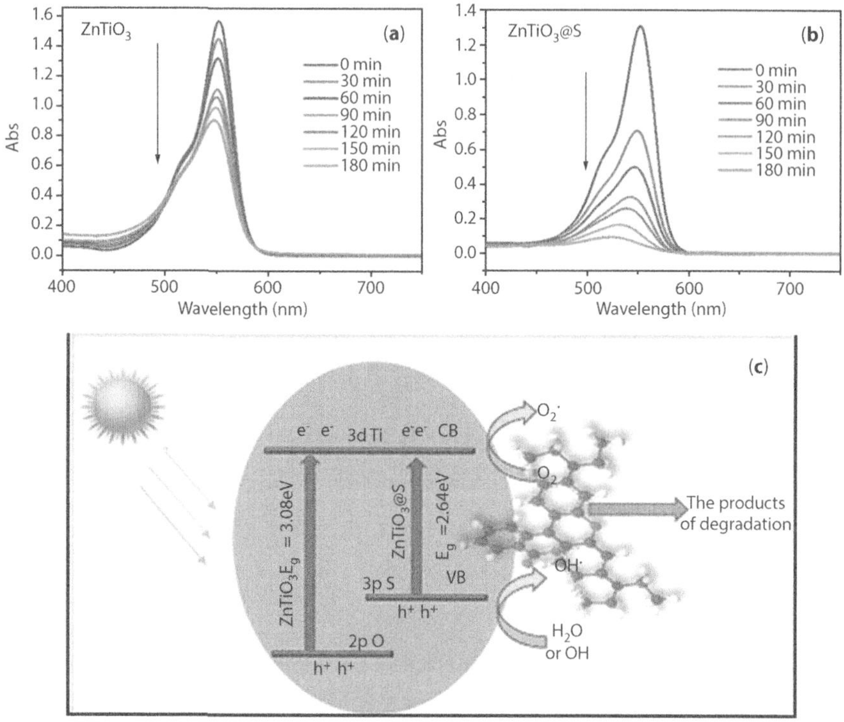

Figure 8.12 Photocatalytic degradation of RhB in the presence of (a) pristine and (b) different % of S-doped $ZnTiO_3$ photocatalyst, and (c) schematic illustration for the degradation of dyes in the presence of 20% S-doped $ZnTiO_3$ under visible light illumination [62].

$$2HOO^{\bullet} \rightarrow H_2O_2 + O_2 \tag{8.8}$$

$$H_2O_2 \rightarrow 2^{\bullet}OH \tag{8.9}$$

$$\text{CV or RhB} + {}^{\bullet}OH \rightarrow CO_2 + H_2O \tag{8.10}$$

8.5 Conclusion

To address the energy and environmental issues, photocatalysis is an alternative promising and green approach by harvesting solar energy. This chapter presents the recent progress on the study and improvement of perovskite ($ZnTiO_3$) oxides for waste water treatment with fundamental information. The $ZnTiO_3$ has wide band gap can harvest only UV light which is 3% to 5% of solar spectrum. Therefore, the development of visible light active $ZnTiO_3$ photocatalyst is the main aim to improve the photocatalytic activity of pristine $ZnTiO_3$. Many efforts have been carried out for the development and designing of visible light active $ZnTiO_3$ photocatalyst to utilize the maximum part of the solar spectrum. Therefore, different strategies have been carried out, among all band gap, engineering has been extensively investigated to improve the light absorption and to reduce the electron/hole pairs recombination for the enhancement of photocatalytic activity for the degradation of pollutants. The tuning of band gap in the case of perovskite (ABO_3) can be done at three different alteration sites, such as A, B, and O-site, which can tailor their optical, chemical, and physical properties, resulting improvement in the photocatalytic activity. Therefore, band gap engineering at different sites can help in the design and architecture of next-generation excellent photocatalyst for efficient targets and can frame the road for future research.

References

1. Raza, W. and Faraz, M., Advanced Functional Textiles and Polymers: Fabrication, Processing and Applications, John Wiley & Sons, Inc. 171–199, 2019.
2. Hu, J.S., Ren, L.L., Guo, Y.G., Liang, H.P., Cao, A.M., Wan, L.J., Bai, C.L., Mass production and high photocatalytic activity of ZnS nanoporous nanoparticles. *Angew. Chem. Int. Ed.*, 44, 1269–1273, 2005. https://doi.org/10.1002/anie.200462057.
3. Li, Y.-F., Xu, D., Il Oh, J., Shen, W., Li, X., Yu, Y., Mechanistic Study of Codoped Titania with Nonmetal and Metal Ions: A Case of C + Mo Codoped TiO 2. *ACS Catal.*, 2, 391–398, 2012. https://doi.org/10.1021/cs2006668.
4. Raza, W., Haque, M.M., Muneer, M., Bahnemann, D., Synthesis of visible light driven TiO2 coated carbon nanospheres for degradation of dyes. *Arab. J. Chem.*, 12, 3534–3545, 2019. https://doi.org/10.1016/j.arabjc.2015.09.002.
5. Raza, W. and Ahmad, K., Visible Light-Driven Photocatalysts for Environmental Applications Based on Graphitic Carbon Nitride, *Handbook of Nanomaterials and Nanocomposites for Energy and Environmental Applications*, pp. 1–25, Springer International Publishing, 2020.
6. Raza, W. and Faraz, M., Novel g-C3N4/Fe-ZnO/RGO nanocomposites with boosting visible light photocatalytic activity for MB, Cr (VI), and outstanding catalytic activity toward para-nitrophenol reduction. *Nanotechnology*, 31, 31, 1–16, 2020.
7. Gaya, U.I. and Abdullah, A.H., Heterogeneous photocatalytic degradation of organic contaminants over titanium dioxide: A review of fundamentals, progress and problems. *J. Photoch. Photobio. C*, 9, 1–12, 2008. https://doi.org/10.1016/j.jphotochemrev.2007.12.003.

8. Vieno, N.M., Tuhkanen, T., Kronberg, L., Seasonal variation in the occurrence of pharmaceuticals in effluents from a sewage treatment plant and in the recipient water. *Environ. Sci. Technol.*, 39, 8220–8226, 2005. https://doi.org/10.1021/es051124k.
9. Vieno, N.M., Härkki, H., Tuhkanen, T., Kronberg, L., Occurrence of pharmaceuticals in river water and their elimination in a pilot-scale drinking water treatment plant. *Environ. Sci. Technol.*, 41, 5077–5084, 2007. https://doi.org/10.1021/es062720x.
10. Daghrir, R., Drogui, P., Robert, D., Modified TiO2 for environmental photocatalytic applications: A review. *Ind. Eng. Chem. Res.*, 52, 3581–3599, 2013. https://doi.org/10.1021/ie303468t.
11. Rasalingam, S., Wu, C.M., Koodali, R.T., Modulation of pore sizes of titanium dioxide photocatalysts by a facile template free hydrothermal synthesis method: Implications for photocatalytic degradation of rhodamine B. *ACS Appl. Mater. Interfaces*, 7, 4368–4380, 2015. https://doi.org/10.1021/am508883f.
12. Reddy, K.H., Martha, S., Parida, K.M., Fabrication of novel p-BiOI/n-ZnTiO3 heterojunction for degradation of rhodamine 6G under visible light irradiation. *Inorg. Chem.*, 52, 6390–6401, 2013. https://doi.org/10.1021/ic400159m.
13. Surendar, T., Kumar, S., Shanker, V., Influence of La-doping on phase transformation and photocatalytic properties of $ZnTiO_3$ nanoparticles synthesized via modified sol-gel method. *Phys. Chem. Chem. Phys.*, 16, 728–735, 2014. https://doi.org/10.1039/c3cp53855a.
14. Ahmad, K. and Raza, W., Graphene-Based Nanocomposites for Photocatalytic Dye Degradation Applications, *Environmental Nanotechnology for Water Purification*, Scrivener Publishing LLC, 101–122, 2020.
15. Raza, W., Haque, M.M., Muneer, M., Synthesis of visible light driven ZnO: Characterization and photocatalytic performance. *Appl. Surf. Sci.*, 322, 215–224, 2014. https://doi.org/10.1016/j.apsusc.2014.10.067.
16. Raza, W., Haque, M.M., Muneer, M., Fleisch, M., Hakki, A., Bahnemann, D., Photocatalytic degradation of different chromophoric dyes in aqueous phase using la and Mo doped TiO2 hybrid carbon spheres. *J. Alloys Compd.*, 632, 837–844, 2015. https://doi.org/10.1016/j.jallcom.2015.01.222.
17. Haque, M.M., Raza, W., Khan, A., Muneer, M., Heterogeneous Photocatalyzed Degradation of Barbituric Acid and Matrinidazole Under Visible Light Induced Ni, Mn, Mo and La-Doped TiO2. *J. Nanoeng. Nanomanuf.*, 4, 135–139, 2014. https://doi.org/10.1166/jnan.2014.1182.
18. Raza, W., Faisal, S.M., Owais, M., Bahnemann, D., Muneer, M., Facile fabrication of highly efficient modified ZnO photocatalyst with enhanced photocatalytic, antibacterial and anticancer activity. *RSC Adv.*, 6, 78335–78350, 2016. https://doi.org/10.1039/c6ra06774c.

19. Kuo, Y.L., Chen, H.W., Ku, Y., Analysis of silver particles incorporated on TiO2 coatings for the photodecomposition of o-cresol. *Thin Solid Films*, 515, 3461–3468, 2007. https://doi.org/10.1016/j.tsf.2006.10.085.
20. Li, Z.X., Shi, F.B., Ding, Y., Zhang, T., Yan, C.H., Facile synthesis of highly ordered mesoporous $ZnTiO_3$ with crystalline walls by self-adjusting method. *Langmuir*, 27, 14589–14593, 2011. https://doi.org/10.1021/la2034615.
21. Surendar, T., Kumar, S., Shanker, V., Influence of La-doping on phase transformation and photocatalytic properties of $ZnTiO_3$ nanoparticles synthesized via modified sol-gel method. *Phys. Chem. Chem. Phys.*, 16, 728–735, 2014. https://doi.org/10.1039/c3cp53855a.
22. Eskandarloo, H., Badiei, A., Behnajady, M.A., Tavakoli, A., Ziarani, G.M., Ultrasonic-assisted synthesis of Ce doped cubic-hexagonal $ZnTiO_3$ with highly efficient sonocatalytic activity. *Ultrason. Sonochem.*, 29, 258–269, 2016. https://doi.org/10.1016/j.ultsonch.2015.10.004.
23. Reddy, K.H., Martha, S., Parida, K.M., Erratic charge transfer dynamics of Au/$ZnTiO_3$ nanocomposites under UV and visible light irradiation and their related photocatalytic activities. *Nanoscale*, 10, 18540–18554, 2018. https://doi.org/10.1039/c8nr06158k.
24. Nolan, N.T., Seery, M.K., Pillai, S.C., Crystallization and phase-transition characteristics of sol-gel-synthesized zinc titanates. *Chem. Mater.*, 23, 1496–1504, 2011. https://doi.org/10.1021/cm1031688.
25. Sharma, U. and Jeevanandam, P., Synthesis Temperature Dependent Morphological Evolution in Zinc Titanate Heteronanostructures and Their Application in Environmental Remediation. *ChemistrySelect*, 1, 6382–6395, 2016. https://doi.org/10.1002/slct.201601269.
26. Jose, M., Elakiya, M., Dhas, S.A.M.B., Structural and optical properties of nanosized ZnO/$ZnTiO_3$ composite materials synthesized by a facile hydrothermal technique. *J. Mater. Sci. Mater. Electron.*, 28, 13649–13658, 2017. https://doi.org/10.1007/s10854-017-7207-9.
27. Abirami, R., Kalaiselvi, C.R., Kungumadevi, L., Senthil, T.S., Kang, M., Synthesis and characterization of $ZnTiO_3$ and Ag doped $ZnTiO_3$ perovskite nanoparticles and their enhanced photocatalytic and antibacterial activity. *J. Solid State Chem.*, 281, 121019, 2020. https://doi.org/10.1016/j.jssc.2019.121019.
28. Arandiyan, H., Wang, Y., Sun, H., Rezaei, M., Dai, H., Ordered meso- and macroporous perovskite oxide catalysts for emerging applications. *Chem. Commun.*, 54, 6484–6502, 2018. https://doi.org/10.1039/c8cc01239c.
29. Peña, M.A. and Fierro, J.L.G., Chemical structures and performance of perovskite oxides. *Chem. Rev.*, 101, 1981–2017, 2001. https://doi.org/10.1021/cr980129f.
30. Kumar, A., Kumar, A., Krishnan, V., Perovskite Oxide Based Materials for Energy and Environment-Oriented Photocatalysis. *ACS Catal.*, 10, 10253–10315, 2020. https://doi.org/10.1021/acscatal.0c02947.

31. Huang, X., Zhao, G., Wang, G., Irvine, J.T.S., Synthesis and applications of nanoporous perovskite metal oxides. *Chem. Sci.*, 9, 3623–3637, 2018. https://doi.org/10.1039/c7sc03920d.
32. Arandiyan, H., Wang, Y., Sun, H., Rezaei, M., Dai, H., Ordered meso- and macroporous perovskite oxide catalysts for emerging applications. *Chem. Commun.*, 54, 6484–6502, 2018. https://doi.org/10.1039/c8cc01239c.
33. Peña, M.A. and Fierro, J.L.G., Chemical structures and performance of perovskite oxides. *Chem. Rev.*, 101, 1981–2017, 2001. https://doi.org/10.1021/cr980129f.
34. Huang, Y.H., Dass, R.I., King, Z.L., Goodenough, J.B., Double perovskites as anode meterials for solid-oxide fuel cells. *Science*, 312, 254–257, 2006. https://doi.org/10.1126/science.1125877.
35. Shi, J. and Guo, L., ABO3-based photocatalysts for water splitting. *Prog. Nat. Sci. Mater. Int.*, 22, 592–615, 2012. https://doi.org/10.1016/j.pnsc.2012.12.002.
36. Goldschmidt, V.M., The Laws of Crystal. *Naturwissenschaften*, 14, 477–485, 1926. https://link.springer.com/content/pdf/10.1007/BF01507527.pdf (accessed November 29, 2020).
37. Wang, W., Tadé, M.O., Shao, Z., Research progress of perovskite materials in photocatalysis- and photovoltaics-related energy conversion and environmental treatment. *Chem. Soc Rev.*, 44, 5371–5408, 2015. https://doi.org/10.1039/c5cs00113g.
38. Labhasetwar, N., Saravanan, G., Kumar Megarajan, S., Manwar, N., Khobragade, R., Doggali, P., Grasset, F., Perovskite-type catalytic materials for environmental applications. *Sci. Technol. Adv. Mater.*, 16, 1–13, 2015. https://doi.org/10.1088/1468-6996/16/3/036002.
39. Grabowska, E., Selected perovskite oxides: Characterization, preparation and photocatalytic properties-A review. *Appl. Catal. B*, 186, 97–126, 2016. https://doi.org/10.1016/j.apcatb.2015.12.035.
40. Wu, H., Min, Y., Zhang, Q., Li, W., Yuan, J., Wu, Z., Wang, S., Low-temperature synthesis of mesoporous $ZnTiO_3$-graphene composite for the removal of norfloxacin in aqueous solution. *RSC Adv.*, 6, 103822–103829, 2016. https://doi.org/10.1039/c6ra17556b.
41. Mrázek, J., Spanhel, L., Chadeyron, G., Matějec, V., Evolution and Eu3+ doping of sol-gel derived ternary Zn xTiyOz - Nanocrystals. *J. Phys. Chem. C*, 114, 2843–2852, 2010. https://doi.org/10.1021/jp9036217.
42. Qian, D., Gerward, L., Jiang, J.Z., Deformation-induced reactions of ZnO and TiO2. *J. Mater. Sci.*, 39, 5389–5392, Springer, 2004. https://doi.org/10.1023/B:JMSC.0000039251.24875.ee.
43. Chang, Y.S., Chang, Y.H., Chen, I.G., Chen, G.J., Chai, Y.L., Fang, T.H., Wu, S., Synthesis, formation and characterization of $ZnTiO_3$ ceramics. *Ceram. Int.*, 30, 2183–2189, 2004. https://doi.org/10.1016/j.ceramint.2004.01.002.
44. Manik, S.K., Bose, P., Pradhan, S.K., Microstructure characterization and phase transformation kinetics of ball-milled prepared nanocrystalline

Zn2TiO4 by Rietveld method. *Mater. Chem. Phys.*, 82, 837–847, 2003. https://doi.org/10.1016/j.matchemphys.2003.07.012.
45. Nikam, L., Panmand, R., Kadam, S., Naik, S., Kale, B., Enhanced hydrogen production under a visible light source and dye degradation under natural sunlight using nanostructured doped zinc orthotitanates. *New J. Chem.*, 39, 3821–3834, 2015. https://doi.org/10.1039/c4nj01995d.
46. Liu, X., Molten salt synthesis of ZnTiO 3 powders with around 100 nm grain size crystalline morphology. *Mater. Lett.*, 80, 69–71, 2012. https://doi.org/10.1016/j.matlet.2012.04.048.
47. Perween, S. and Ranjan, A., Improved visible-light photocatalytic activity in $ZnTiO_3$ nanopowder prepared by sol-electrospinning. *Sol. Energy Mater. Sol. Cells*, 163, 148–156, 2017. https://doi.org/10.1016/j.solmat.2017.01.020.
48. Han, M.C., Zhang, J.H., Cui, P., Yi, T.F., Li, X., Porous $ZnTiO_3$ rods as a novel lithium storage material for Li-ion batteries. *Ceram. Int.*, 46, 14030–14037, 2020. https://doi.org/10.1016/j.ceramint.2020.02.202.
49. Wan, Z., Zhang, G., Wu, X., Yin, S., Novel visible-light-driven Z-scheme Bi12GeO20/g-C3N4 photocatalyst: Oxygen-induced pathway of organic pollutants degradation and proton assisted electron transfer mechanism of Cr(VI) reduction. *Appl. Catal. B*, 207, 17–26, 2017. https://doi.org/10.1016/j.apcatb.2017.02.014.
50. Singh, M., Kumar, A., Krishnan, V., Influence of different bismuth oxyhalides on the photocatalytic activity of graphitic carbon nitride: a comparative study under natural sunlight. *Mater. Adv.*, 1, 1262–1272, 2020. https://doi.org/10.1039/d0ma00294a.
51. Kumar, A., Kumar, A., Krishnan, V., Perovskite Oxide Based Materials for Energy and Environment-Oriented Photocatalysis. *ACS Catal.*, 10, 10253–10315, 2020. https://doi.org/10.1021/acscatal.0c02947.
52. Turchi, C.S. and Ollis, D.F., Photocatalytic degradation of organic water contaminants: Mechanisms involving hydroxyl radical attack. *J. Catal.*, 122, 178–192, 1990. https://doi.org/10.1016/0021-9517(90)90269-P.
53. Bhatkhande, D.S., Pangarkar, V.G., Beenackers, A.A.C.M., Photocatalytic degradation for environmental applications - A review. *J. Chem. Technol. Biotechnol.*, 77, 102–116, 2002. https://doi.org/10.1002/jctb.532.
54. Kanhere, P. and Chen, Z., A review on visible light active perovskite-based photocatalysts. *Molecules*, 19, 19995–20022, 2014. https://doi.org/10.3390/molecules191219995.
55. Zhang, G., Liu, G., Wang, L., Irvine, J.T.S., Inorganic perovskite photocatalysts for solar energy utilization. *Chem. Soc Rev.*, 45, 5951–5984, 2016. https://doi.org/10.1039/c5cs00769k.
56. Melchionna, M. and Fornasiero, P., Updates on the Roadmap for Photocatalysis. *ACS Catal.*, 10, 5493–5501, 2020. https://doi.org/10.1021/acscatal.0c01204.
57. Gao, M., Zhu, L., Peh, C.K., Ho, G.W., Solar absorber material and system designs for photothermal water vaporization towards clean water and energy

production. *Energy Environ. Sci.*, 12, 841–864, 2019. https://doi.org/10.1039/c8ee01146j.

58. Li, J. and Wu, N., Semiconductor-based photocatalysts and photoelectrochemical cells for solar fuel generation: A review. *Catal. Sci. Technol.*, 5, 1360–1384, 2015. https://doi.org/10.1039/c4cy00974f.
59. Kong, J.Z., Li, A.D., Zhai, H.F., Li, H., Yan, Q.Y., Ma, J., Wu, D., Preparation, characterization and photocatalytic properties of $ZnTiO_3$ powders. *J. Hazard. Mater.*, 171, 918–923, 2009. https://doi.org/10.1016/j.jhazmat.2009.06.092.
60. Surendar, T., Kumar, S., Shanker, V., Influence of La-doping on phase transformation and photocatalytic properties of $ZnTiO_3$ nanoparticles synthesized via modified sol-gel method. *Phys. Chem. Chem. Phys.*, 16, 728–735, 2014. https://doi.org/10.1039/c3cp53855a.
61. Dutta, D.P., Singh, A., Tyagi, A.K., Ag doped and Ag dispersed nano $ZnTiO_3$: Improved photocatalytic organic pollutant degradation under solar irradiation and antibacterial activity. *J. Environ. Chem. Eng.*, 2, 2177–2187, 2014. https://doi.org/10.1016/j.jece.2014.09.015.
62. Tavakoli-Azar, T., Mahjoub, A.R., Sadjadi, M.S., Farhadyar, N., Sadr, M.H., "Improving the photocatalytic performance of a perovskite $ZnTiO_3$ through $ZnTiO_3$@S nanocomposites for degradation of Crystal violet and Rhodamine B pollutants under sunlight". 119, 1–11, 108091, 2020. https://doi.org/10.1016/j.inoche.2020.108091.

9

Types of Perovskite Materials

Faria Khatoon Naqvi*, Yashfeen Khan, Saba Beg and Anees Ahmad

Department of Chemistry, Aligarh Muslim University, Aligarh, (U.P.) India

Abstract

In this chapter, we will be discussing various types of perovskites materials. The perovskite is an enormous family of materials possessing variety of structure. The compounds of perovskites family having a crystal structures associated with the mineral perovskite, $CaTiO_3$. The general perovskite structure can be characterized by the general formula ABX_3 (X = O,F), where A is a large metal cation close-packed in layers with oxygen ions and B is a smaller metal ion situated in an octahedrally coordinated hole between the close-packed layers. We will discribe types of pervoskite materials and their examples.

Keywords: Perovskite, $CaTiO_3$

Abbreviations

$CaTiO_3$	Calcium Titanate Oxide
$BaTiO_3$	Barium Titanate Oxide
BX_6	Corner sharing octahedra with the A cation
$GdFeO_3$	Gadolium Orthoferrite
$SrTiO_3$	Stroncium Titanate
$GdBaCo_2O_{5.5}$	Cobaltite
$CsCa_2Nb_3O_{10}$	Dion-Jacobson perovskites phase
$K_2La_2Ti_3O_{10}$	Ruddlesden-Popper perovskites phase
$BaBi_4Ti_4O_{15}$	Aurivillius phase
t	Tolerance factor

**Corresponding author*: naqvifaria@gmail.com

Khursheed Ahmad and Waseem Raza (eds.) Perovskite Materials for Energy and Environmental Applications, (241–256) © 2022 Scrivener Publishing LLC

9.1 Introduction

The family structure of perovskites is a huge family of compounds holding crystal structures related to the mineral perovskite $CaTiO_3$. Perovskites took their name from the ($CaTiO_3$) structure, first discovered in the Ural Mountains of Russia by Gustav Rose in 1839 and is named after a Russian mineralogist Count Lev Aleksevich Von Perovski (1792–1856) (represented in Figure 9.1) [1].

The ideal cubic perovskite structure is not very common and also the mineral perovskite itself is slightly distorted. Distorted perovskites have reduced symmetry, which is important for their magnetic and electric properties. Because of these properties, perovskites have great industrial importance, especially the ferroelectric tetragonal form of $BaTiO_3$. In the ideal form, the crystal structure of cubic ABX3 perovskite can be described as consisting of corner sharing [BX6] octahedra with the A cation occupying the 12-fold coordination site formed in the middle of the cube of eight of such octahedra.

The perovskite family of oxides is probably the best-studied family of oxides. The interest in compounds belonging to this family of crystal structures arise in the large and ever surprising variety of properties exhibited and the flexibility to accommodate almost all of the elements in the periodic system.

A large number of perovskite type oxides have been studied because of their interesting properties, including superconductivity, insulator-metal transition, ionic conduction characteristics, dielectric properties, and ferroelectricity [2].

Perovskite is one of the most frequently encountered structures in solid-state physics, and it accommodates most of the metal ions in the periodic

Figure 9.1 Pictorial representation of (a) Gustav Rose, discoverer (b) perovskite mineral species ($CaTiO_3$) and, (c) Lev Aleksevich von Perovski, mineralogist.

table with a significant number of different anions. During the last few years, many experimental and theoretical investigations were devoted to the study of perovskite solids typically ABO_3.

These solids are currently gaining considerable importance in the field of electrical ceramics, refractories, geophysics, material science, astrophysics, particle accelerators, fission-fusion reactors, heterogeneous catalysis, environment, and so on. Perovskite-structured oxides can accept considerable substitutions in one or both cationic sites (i.e., A and B sites) while retaining their original crystal structures. Such a feature provides the chemical tailoring of the materials via partial replacement of the cationic site(s) with foreign metal ions, hence, 29 modifying their structural, microstructural, electrical, and magnetic properties [3–7].

The physicochemical properties of these materials are dependent on the crystal structure, lattice defect, exposed lattice plane, surface morphology, particle size, and specific surface area, as well as the pore structure [8–10]. Up to now, a large number of perovskite-type oxides and perovskite-like oxides have been generated and investigated for clarifying their physicochemical properties.

The cubic perovskite is called the ideal perovskite. This class of materials has great potential for a variety of device applications due to their simple crystal structures and unique ferroelectric and dielectric properties. As one of the most abundant and widely investigated minerals, perovskites solids are widely studied as the candidate materials of substrate materials [11, 12].

Lately, perovskite-structured ceramics are considered as one of the most preferred materials. This may be because of their peculiar properties, like dielectric properties, ferroelectric, optical properties, pyroelectric, and thermoelectric. Depending on these mentioned properties, perovskite ceramics have variegated applications, such as random access memories, tunable microwave devices, capacitors, displays, piezoelectric devices, sensors, actuators, transducers, and wireless communications [13–16].

9.1.2 Types of Perovskite

Innumerable oxide compounds belong to a few perovskite-based homologous series. Variegated types of perovskite materials are shown in Figure 9.2.

The perovskite structures are existsing in different forms and can be classified into various categories, depending on their arrangements as follows:

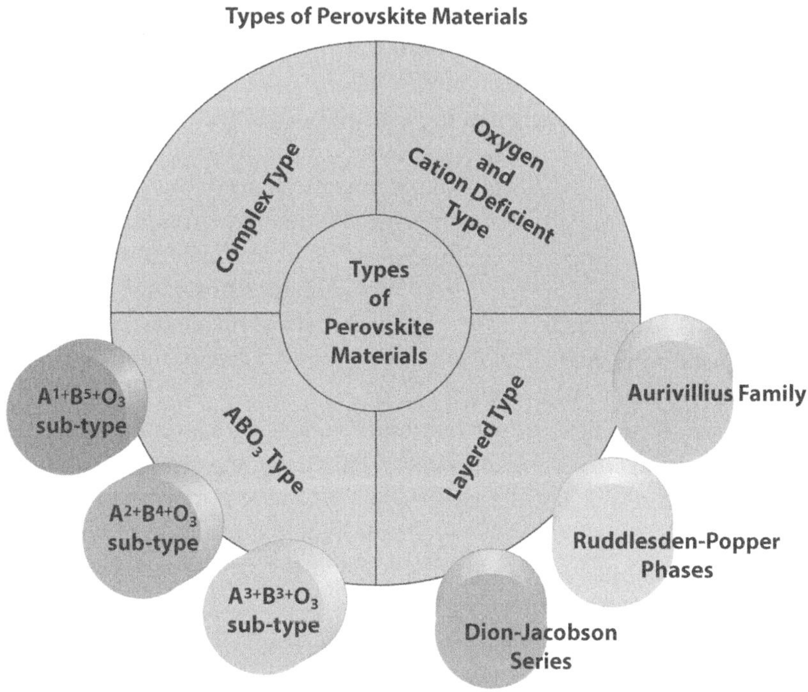

Figure 9.2 Types of perovskites materials.

9.1.2.1 ABO_3 Type of Perovskite Materials

ABO_3-type oxides form an important class of materials of great technical value in several device applications. Such oxides are known to exist with a wide range of A and B ions. Most of these oxides have the relatively simple structure of the mineral perovskite ($CaTiO_3$) (Figure 9.3). The ideal perovskite structure has a cubic unit cell [17–19].

However, distortions are common and depend on the value of the tolerance factor "*t*" given by Goldschmidt as,

$$t = r_A + r_O / \sqrt{2}\,(r_B + r_O)$$

where r_A and r_B are ionic radii of A and B cations and r_O is ionic radius of oxygen anion (in Å).

Rhombohedral structure or orthorhombic $GdFeO_3$-type structure (Figure 9.4) is found when *t* is less than unity. Polytypic structures are observed when *t* is greater than unity. Other structures, such as ilmenite and pyrochlore, also occur. ABO_3-type oxides display a wide spectrum of

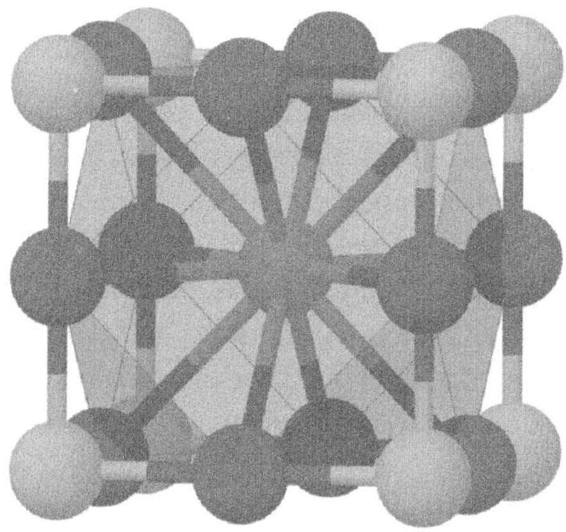

Figure 9.3 Ideal structure of $CaTiO_3$ (designed by materials project software).

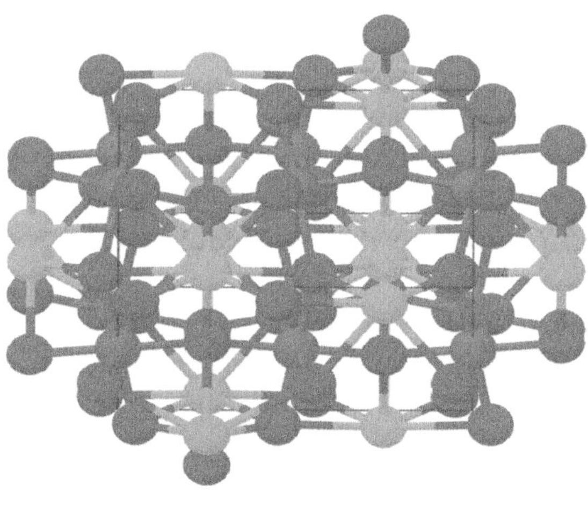

Figure 9.4 Orthorhombic $GdFeO_3$-type structure when "t" is less than unity.

physical properties of technical importance, such as ferroelectricity, antiferroelectricity, piezoelectricity, insulating behavior, semiconductivity, metallic conductivity, superconductivity, ferromagnetism, antiferromagnetism, and so on. The ferroelectric properties are mostly controlled by the relative sizes of the ions. The electrical and magnetic properties are mainly dependent on the electronic configuration of the ions. Dopants and substituents have a profound influence on the properties of the ABO_3 oxides.

ABO_3 types of perovskite structures are adopted by many oxides. The idealized form is a cubic structure (Pm_3m) which is rarely encountered. The orthorhombic (Amm_2) and tetragonal (P_4mm) phases are the most common non-cubic variants. $SrTiO_3$ and $CaRbF_3$ are examples of cubic perovskites. Barium titanate is an example of a perovskite which can take on the rhombohedral (R_3m), orthorhombic, tetragonal, and cubic forms, depending on the temperature. ABO_3 type of perovskites is further classified into various types according to their valencies on A atom or on B atom [20].

a. **$A^{1+}B^{5+}O_3$ Type of Perovskite,** e.g.: $KNbO_3$, $CsIO_3$, $AgTaO_3$

b. **$A^{2+}B^{4+}O_3$ Type of Perovskite,** e.g.: $CaTiO_3$, $BaTiO_3$, $CaMnO_3$

c. **$A^{3+}B^{3+}O_3$ Type of Perovskite,** e.g.: $LaFeO_3$, $BiMnO_3$, $NdFeO_3$

9.1.2.2 Oxygen and Cation-Deficient Perovskites

Perovskites of the type ACuO3-δ, which have Cu atoms in the octahedral sites, are deficient in oxygen, alkaline earth, and trivalent ions (Y^{3+}, lanthanoids, Bi^{3+}, Tl^{3+}) occupy the A site. A typical composition is $YBa_2Cu_3O_{7-x}$

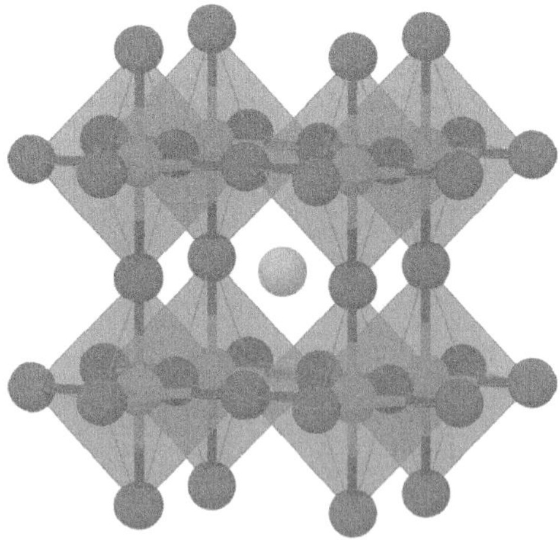

Figure 9.5 $BaTiO_3$ structure (drawn by Materials Project software).

with x = 0.04. These compounds are high-temperature superconductors. The structure is a superstructure of perovskite, but with approximately 2/9 of the oxygen positions vacant, in such way that 2/3 of the Cu atoms have square pyramidal coordination and 1/3 have square-planar coordination, see Figure 9.5. The cobaltite $GdBaCo_2O_{5.5}$ is another example of an oxygen-deficient perovskite related structure where the Co^{3+} ions have octahedral and square pyramidal coordination [21].

Also, the Brownmillerite structure is an oxygen-deficient superstructure of cubic perovskite with an ordering of oxygen vacancies. Ruddlesden and Popper designed a series of homologous compounds with the general formula $AO(ABO_3)_n$ where AO represent a rock salt structure layer separating blocks of perovskite layers characterized by n = 1,2,3…∞.

9.1.2.3 Complex Perovskites

The complex perovskite type compounds, $A(Bx'By'')O_3$ can be divided into compounds which contain twice as much lower valence state element as a higher valence state element, $A(B'_{0.67}B'_{0.33})O_3$, those which contain twice as much of the higher valence state element as the lower valence state element, $A(B'_{0.33}B'_{0.67})O_3$, those which contain the two B elements in equal amounts, A(B′0.5B″0.5)O3 and oxygen-deficient phases $A(Bx'By'')O_{3-z}$ [22]. The valences of the A and B cations are generally close to 2+ and 4+, respectively, but in some special cases their valences can be 3+ and 3+ if the B^{3+} cation has six coordinations. The valence variation at the A cation position can cause distortion or displacement of the oxygen anion array, resulting in the buckling of the $(AO_3)^{4-}$ layers (Figure 9.6).

This buckling may bring distortion of the octahedra with B cations at the centers. The B cation must have the flexibility to tolerate this effect,

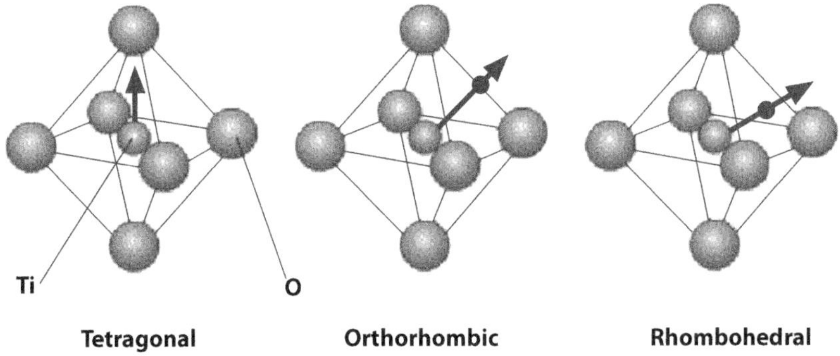

Figure 9.6 Showing three phases of $BaTiO_3$ structure where O_6 octahedra surrounded the Ti atom.

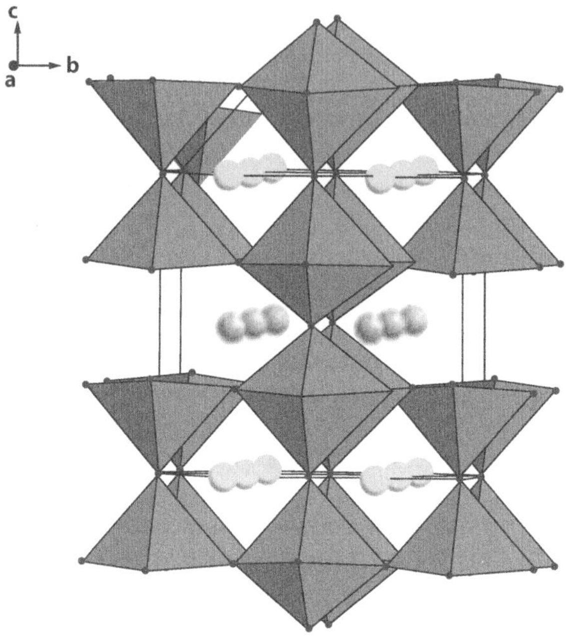

Figure 9.7 The cobaltite $GdBaCo_2O_{5.5}$ has a perovskite related structure were Ba atoms are yellow and Gd atoms are gray.

and the transition metal elements are the candidates for filling the B cation position because of its multi valency or the special *3d* and *4d* electron configurations.

This is the reason that transition metal oxides have perovskite-type structures, and they usually have extraordinary physical properties. The oxygen and cation-deficient phases will be regarded as those which contain considerable vacancies and not those phases, which are only slightly nonstoichiometric. Many of these contain B ions of one element in two valence stated and should not be confused with the complex perovskite compounds, which contain different elements in different valence states [23–25] (Figure 9.7).

9.1.2.4 Layered Perovskites

Layered perovskites are intergrowths of perovskite and other structures, and they consist of two-dimensional perovskite slabs interleaved with cations or cationic structural units [26].

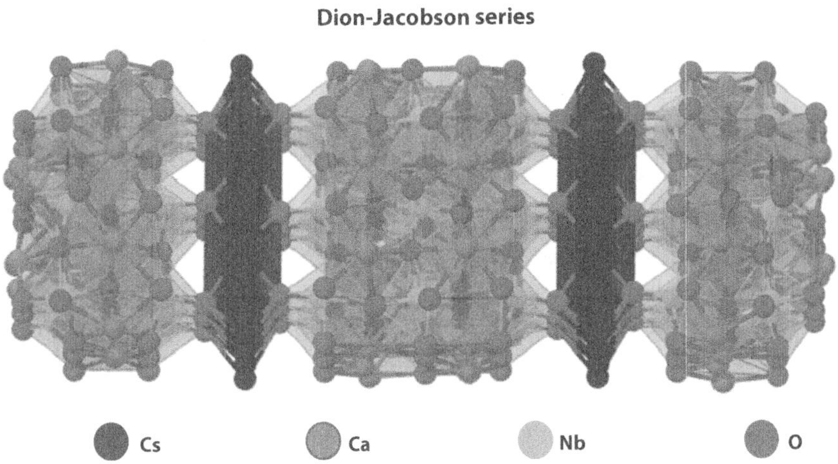

Figure 9.8 Structures of selected Dion-Jacobson perovskites ($CsCa_2Nb_3O_{10}$).

9.1.2.4.1 The Dion-Jacobson Series of Layered Perovskites

A'[$A_{n-1}B_nO_{3n+1}$], typified by the n= 3 phase $CsCa_2Nb_3O_{10}$ in Figure 9.8, has one interlayer cation per formula unit [26–28].

9.1.2.4.2 Ruddlesden-Popper Phases

$A'_2[A_{n-2}B_nO_{3n+1}]$, such as K_2La2-Ti_3O_{10} in Figure 9.9, have two interlayer cations per formula unit and possess twice the interlayer charge density of the Dion-Jacobson phases [28, 29].

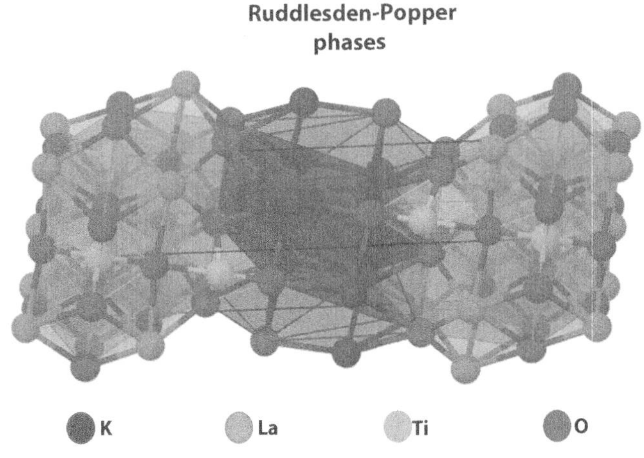

Figure 9.9 Structures of selected Ruddlesden-Popper perovskites ($K_2La_2Ti_3O_{10}$).

9.1.2.4.3 Aurivillius Family

The *Aurivillius family* of layered bismuth oxides was first reported by Bengt Aurivillius in1949 and 1950, who discovered new phases with layered structures [1–3]. The Aurivillius compounds crystallize in a layered structure, which consists of two structural elements, described as intergrowth of fluorite-like $[M_2O_2]^{2+}$ slabs alternating with perovskite-like $[A_{n-1}B_nO_{3n+1}]^{2-}$ blocks where thickness of perovskite layers increases as n increases, as shown in Figures 9.10 and 9.11.

Aurivillius phases, including $Bi_2W_2O_9$ (or $Bi_2O_2[W_2O_7]$ to emphasize the location of the bismuth oxide in the interlayer) shown in Figure 9.1d, are intergrowths of perovskite and bismuth oxide and have a covalent network of $Bi_2O_2^{2+}$ between the two-dimensional perovskite slabs. Depending on the choice and stoichiometry of the A and B-site cations, perovskites and layered perovskites can possess a wide variety of interesting properties including superconductivity, colossal magneto-resistance, ferroelectricity, and catalytic activity [30].

The cation site in the $[M_2O_2]^{2+}$ slabs are almost exclusively occupied by Bi^{3+} cation forming $[Bi_2O_2]^{2+}$ slabs. The general chemical formula of these

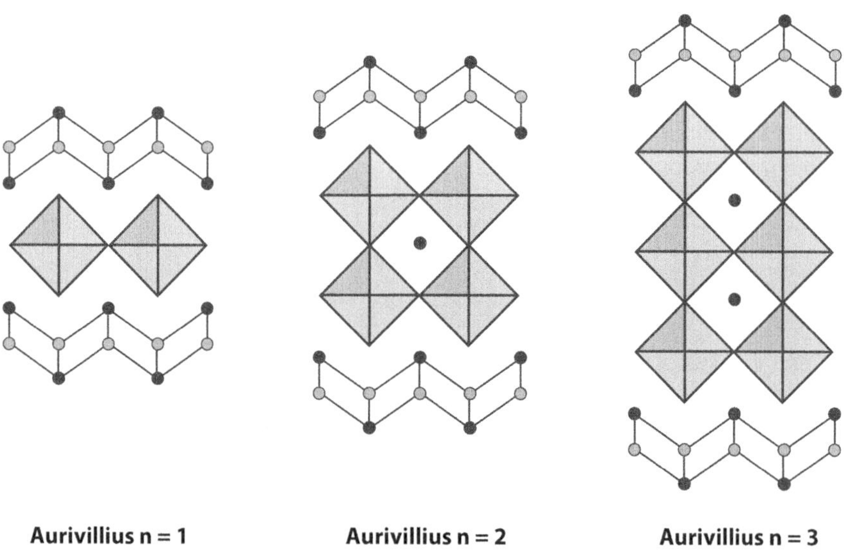

Figure 9.10 Schematic representation of the structure of the Aurivillius compounds with n =1, 2, and 3.

Figure 9.11 Crystal structure of the n = 4 Aurivillius phase $BaBi_4Ti_4O_{15}$.

compounds can be written as $Bi_2An_{-1}B_nO_{3n+3}$ or $(Bi_2O_2)^{2+}$ $(A_{n-1}B_nO_{3n+1})^{2-}$ where n refers to the number of perovskite blocks and n = 1 to 8. The bismuthate layer $[Bi_2O_2]^{2+}$ is inviolate to substitution, with the exception of other lone pair cations such as Pb^{2+} [31] and Tl^{+} [32], which could occupy the $[Bi_2O_2]^{2+}$ layers since they do not possess a stereochemically active lone pair. In the $[Bi_2O_2]^{2+}$ square-pyramidal units, Bi^{3+} is present at the apex and is coordinated to four oxygen atoms at the corners of the base. These units are linked by sharing edges to form infinite Bi_2O_2 sheets [33]. As a result, the bismuth oxide layer acts as a backbone for the structure of Aurivillius phases and permits the substitution of metallic ion in the perovskite layer [34].

Perovskite blocks are connected through B-O-B linkages and consist of an infinite two-dimensional array of corner-linked BO6 octahedra. The A site can be occupied by large ionic radius cation, such as Na^{+}, K^{+}, Ca^{2+}, Sr^{2+}, Ba^{2+}, Pb^{2+}, and Bi^{3+}, which are 12-fold coordinated. The B site cation can be occupied generally by a cation with small ionic radius, higher charge, and are octahedrally coordinated, such as Ti^{4+}, Nb^{5+}, Ta^{5+}, V^{5+}, and W^{6+}. The B-site cation is inside an oxygen octahedron sharing corners to form oxygen-metal-oxygen chains. Typical examples of layered Aurivillius compounds are Bi_2WO_6 (n = 1), $Bi_2SrM_2O_9$ (M = Nb, Ta) (n = 2), and $Bi_4Ti_3O_{12}$ (n = 3). The cation-deficient Aurivillius phase $(Bi_2O_2)(W_2O_7)$ is known

Figure 9.12 Schematic illustration of n = 1, 2, n = 2, 3 and n = 3, 4 mixed layers structures formed by Aurivillius compounds.

[10] with a general formula $(Bi_2O_2)(B_n\ VIO_{3n+1})$. A new family of mixed layered type bismuth compounds having frequent intergrowth structure formed by two Aurivillius compounds is presented in Figure 9.12, with intergrowth of n and (n+1) members of Aurivillius family. The general formula can be written as:

$(Bi_2O_2)(A_{n-1}B_nO_{3n+1})$ + (Bi_2O_2) $(A_{(n+1)-1}B_{n+1}O_{3(n+1)+1}$ or $Bi_4A_{2n-1}B_{2n+1}O_{6n+9}$,

where n = 1, 2, and 3 [35].

References

1. Cheng, Z. and Lin, J., Layered organic–inorganic hybrid perovskites: structure, optical properties, film preparation, patterning and templating engineering. *CrystEngComm*, 12, 2646–2662, 2010.
2. Warner, T.E., *Synthesis, properties and mineralogy of important inorganic materials*, Wiley, Chichester, West Sussex, U.K., 2011.
3. Schaak, R.E. and Mallouk, T.E., Prying apart Ruddlesden-Popper phases: Exfoliation into sheets and nanotubes for assembly of perovskite thin films. *Chem. Mater.*, 12, 3427–3434, 2000.
4. Uma, S., Raju, A.R., Gopalakrishnan, J., Bridging the Ruddlesden–Popper and the Dion–Jacobson series of layered perovskites: synthesis of layered oxides, A2-xLa2Ti 3-xNbxO 10 (A= K, Rb), exhibiting ion exchange. *J. Mater. Chem.*, 3, 709–713, 1993.
5. Asano, H., Hayakawa, J., Matsui, M., Preparation and properties of triple perovskite La{3-3x}Ca{1+ 3x} Mn3O10 ferromagnetic thin films. *Appl. Phys. Lett.*, 71, 844–846, 1997.
6. Samara, G., Pressure and temperature dependence of the dielectric properties and phase transitions of the ferroelectric perovskites: PbTiO3 and BaTiO3. *Ferroelectrics*, 2, 277–289, 1971.
7. Waser, R., Baiatu, T., Härdtl, K.H., dc Electrical Degradation of Perovskite-Type Titanates: I, Ceramics. *J. Am. Ceram. Soc*, 73, 1645–1653, 1990.
8. Steele, B., Powell, B., Moody, P., Anionic conduction in refractory oxide solid solutions possessing the fluorite, pyrochlore and perovskite structures, in: *Proc. Brit. Ceram. Soc*, pp. 87–102, 1968.
9. Wentzcovitch, R., Karki, B., Karato, S., Da Silva, C., High pressure elastic anisotropy of MgSiO 3 perovskite and geophysical implications. *Earth Planetary Sci. Lett.*, 164, 371–378, 1998.
10. Navrotsky, A. and Weidner, D.J., Perovskite: a structure of great interest to geophysics and materials science, in: *American Geophysical Union Geophysical Monograph Series*, Wiley, p. 45, 1989.

11. Osako, M. and Ito, E., Thermal diffusivity of MgSiO3 perovskite. *Geophys. Res. Lett.*, 18, 239–242, 1991.
12. Piel, H., High T c superconductors for accelerator cavities. *Nucl. Instrum. Methods Phys. Res. A*, 287, 294–305, 1990.
13. De, A.K., Luckscheiter, B., Lutze, W., Malow, G., Schiewer, E., Development of glass ceramics for the incorporation of fission products. Radioactive Waste Processing and Disposal, Volume 1 By U.S. Nuclear Regulatory Commission, 1980.
14. Nagata, S., Katsui, H., Hoshi, K., Tsuchiya, B., Toh, K., Zhao, M., Shikama, T., Hodgson, E., Recent research activities on functional ceramics for insulator, breeder and optical sensing systems in fusion reactors. *J. Nucl. Mater.*, 442, S501–S507, 2013.
15. Zhu, J., Li, H., Zhong, L., Xiao, P., Xu, X., Yang, X., Zhao, Z., Li, J., Perovskite oxides: preparation, characterizations, and applications in heterogeneous catalysis. *ACS Catal.*, 4, 2917–2940, 2014.
16. Seyfi, B., Baghalha, M., Kazemian, H., Modified LaCoO3 nano-perovskite catalysts for the environmental application of automotive CO oxidation. *Chem. Eng. J.*, 148, 306–311, 2009.
17. Goldschmidt, V.M., Geochemische Verteilungsgesetze der Elemente. Skrifter Norske Videnskaps, Akad, Oslo, 8, 1926.
18. Tsunoda, Y., Sugimoto, W., Sugahara, Y., Intercalation behavior of n-alkylamines into a protonated form of a layered perovskite derived from aurivillius phase Bi2SrTa2O9. *Chem. Mater.*, 15, 632–635, 2003.
19. Merz, W.J., Switching time in ferroelectric BaTiO3 and its dependence on crystal thickness. *J. Appl. Phys.*, 27, 938–943, 1956.
20. Roth, R.S., Classification of perovskite and other ABO3-type compounds. *J. Res. Natl. Bur. Stand.*, 58, 75–88, 1957.
21. Modeshia, D.R. and Walton, R.I., Solvothermal synthesis of perovskites and pyrochlores: crystallisation of functional oxides under mild conditions. *Chem. Soc Rev.*, 39, 4303–4325, 2010.
22. Teraoka, Y., Nobunaga, T., Yamazoe, N., Effect of cation substitution on the oxygen semipermeability of perovskite-type oxides. *Chem. Lett.*, 17, 3, 503–506, 1988.
23. Radaelli, P., Iannone, G., Marezio, M., Hwang, H., Cheong, S., Jorgensen, J., Argyriou, D., Structural effects on the magnetic and transport properties of perovskite A1– xAx′MnO3 (x= 0.25, 0.30). *Phys. Rev. B*, 56, 8265, 1997.
24. Müller, K. A. and Kool, T. W., *Properties of perovskites and other oxides*, Default Book Series., May 2010, https://doi.org/10.1142/7591.
25. Zheludev, I.S., *Physics of Crystalline Dielectrics: Volume 1 Crystallography and Spontaneous Polarization*, Springer-Verlag New York Inc., U.S., 2013.
26. Von Hippel, D., *Dielectric materials and applications*, Artech House on Demand, 1994.
27. Wiesendanger, E., Dielectric, mechanical and optical properties of orthorhombic KNbO3. *Ferroelectrics*, 6, 263–281, 1973.

28. Aurivillius, B., Mixed bismuth oxides with layer lattices. II. Structure of Bi4Ti3O12. Arkiv for Kemi, 1, 54, 463–480, 1949.
29. Frit, B. and Mercurio, J.P., *J. Alloys Compd.188*, 27, 1992.
30. Kendall, K.R., Thomas, J.K., Loye, H.-C.Z., 30. Recent developments in perovskite-based oxide ion conductors. *Solid State Ion.*, 70/71, 221, 1994.
31. Millán, P., Castro, A., Torrance. J.B., The first doping of lead2+ into the bismuth oxide layers of the aurivillius oxides. *Materials Research Bulletin*, 28, 2, 117–122, 1993, ISSN 0025-5408, https://doi.org/10.1016/0025-5408(93)90079-S.
32. Millán, P., Castro, A., Thallium substitution in the bismuth–oxygen layer of aurivillius phases. *Materials Research Bulletin*, 34, 1, 25–34, 1999, ISSN 0025-5408, https://doi.org/10.1016/S0025-5408(98)00213-X.
33. Lee, C.K., Lim, G.S., West, A.R., Phase diagrams and stoichiometries of the solid electrolytes, Bi4V2O11: M, M = Co, Cu, Zn, Ca, Sr. *J. Mater. Chem.*, 4, 9, 1441, 1994.
34. Kendall, K.R., Thomas, J.K., Loye, H.-C.Z., Synthesis and ionic conductivity of a new series of modified Aurivillius phases. *Chem. Mater.*, 7, 50, 1995.
35. Champarnaud-Mesjard, J.C., Frit, B., Watanabe, A., Crystal structure of Bi2W2O9, the n=2 member of the homologous series (Bi2O2)BVInO3n+1 of cation-deficient Aurivillius phases. *J. Mater. Chem.*, 9, 1319, 1999.

10

Effects of Various Additives to $CH_3NH_3PbI_3$ Perovskite Solar Cells

Takeo Oku

Department of Materials Science, The University of Shiga Prefecture, Hikone, Shiga, Japan

Abstract

Effectiveness of various additives to $CH_3NH_3PbI_3$-based organic inorganic hybrid photovoltaic devices was described. Addition of a small amount of SbI_3, AsI_3, CsI, $PbCl_2$, CuBr, NH_4Cl, K, $HC(NH_2)_2$ or polysilane to the perovskite compounds improved the photovoltaic parameters and stabilities of the devices. Elemental substitutions, structural transitions, and homogeneity of the microstructures were investigated. Photoconversion efficiencies were improved by various additives to the perovskite. Highly (100) oriented perovskite grains were also formed by NH_4Cl addition and air blowing. Annealing process at high temperatures at ~200°C using polysilane improved the stability of the perovskite solar cells. The present article showed the effectiveness of various additives to the $CH_3NH_3PbI_3$-based perovskite solar cells.

10.1 Introduction

Perovskite-based photovoltaic devices are now one of the attractive candidates for new generation photovoltaic devices. Solar cells consisting of a $CH_3NH_3PbI_3$ ($MAPbI_3$) have been developed and broadly investigated [1–6], which provided higher conversion efficiencies than those of organic solar cells [7–11]. After achievement of the 15% efficiency [12], higher photoconversion efficiencies have been reported [13–16], and the efficiencies reached 20% [17–22].

Email: oku@mat.usp.ac.jp

Khursheed Ahmad and Waseem Raza (eds.) Perovskite Materials for Energy and Environmental Applications, (257–316) © 2022 Scrivener Publishing LLC

The photoconversion efficiencies of the photovoltaic devices are dependent on the compositions and microstructures of the perovskite halide compounds [23, 24]. Bromine (Br) or chlorine (Cl) doping at the iodine (I) positions have been studied [7, 25]. The doped Cl lengthened diffusion lengths of excitons, leading to increase of the conversion efficiency [7, 14, 25]. In addition, elemental dopings, such as tin (Sn) [26–28], antimony (Sb) [29–34], germanium (Ge) [35–37], thallium (Tl) [36], indium (In) [36], arsenic (As) [38], and copper (Cu) [39–42] at the lead (Pb) positions have been investigated. Wavelength ranges of photoconversion were reported to be expanded by Sn or Tl-doping [26, 36]. Detailed investigations on the elemental substitution at the Pb positions are intriguing for Pb-free devices from the view point of electronic structures [43, 44] and tolerance factors [24, 45]. Adding cesium (Cs) [19, 46], rubidium (Rb) [20, 42], ethylammonium [47], guanidinium [48], potassium, or formamidinium ($NH = CHNH_3$, FA) [17, 49, 50] at methylammonium (CH_3NH_3, MA) sites also increased the efficiencies of the devices. Addition of other compounds [51–54], controls of TiO_2 layers [55, 56], hole transport layers [57], and microstructures of the perovskites [58] also improved the photovoltaic properties.

In the present review, several perovskite photovoltaic devices with substitution elements and additives are described. Effects of SbI_3, AsI_3, $PbCl_2$, $SnBr_2$, CsI, Cu halides, NH_4Cl, KI, and polysilane additions to the perovskite solution were investigated. Information on micostructures and photovoltaic properties is valuable for the realization of practical use of the perovskite solar cells. Electron diffraction and high-resolution transmission electron microscopy (HRTEM) are very useful method to analyze the microstructures of various photovoltaic materials [59–68] and perovskite-based structures in nanoscopic scale [69–73]. In the present review, basic fabrication, photovoltaic properties, and microstructures of perovskite-based photovoltaic devices were described. First, the fundamental crystal structures of the perovskites are described. Then, the basic device configurations are described. After that, various elemental doping and morphology control of the perovskite layers are also described.

10.2 Crystal Structures of Perovskite Halides

$CH_3NH_3PbX_3$ ($MAPbX_3$, X = I, Br, or Cl) compounds have a perovskite-type structure and provide structure transformation by annealing at elevated temperatures [74–76]. Crystal structures of the $MAPbI_3$ are shown in Figure 10.1 [24]. Disordering of $CH_3NH_3^+$ and I^- were reported for the cubic $CH_3NH_3PbI_3$ [77]. On the other hand, hydrogen positions were determined for the orthorhombic $MAPbI_3$ [78]. Transition temperatures and crystal systems of $MAPbX_3$ are listed in Table 10.1.

As well as the $MAPbI_3$, $[HC(NH_2)_2]PbI_3$ ($FAPbI_3$) provided high photo-conversion efficiencies [17]. The Crystal structure of $FAPbI_3$ [79] is shown in Figure 10.2(a).

The $CH_3NH_3^+$ cations could be substituted by alkali elements such as Na^+, K^+, Rb^+, or Cs^+. Structural models of $CsGeI_3$ and $CsPbBr_3$ for the high temperature cubic phase are shown in Figure 10.2(b) and 10.2(c), respectively,

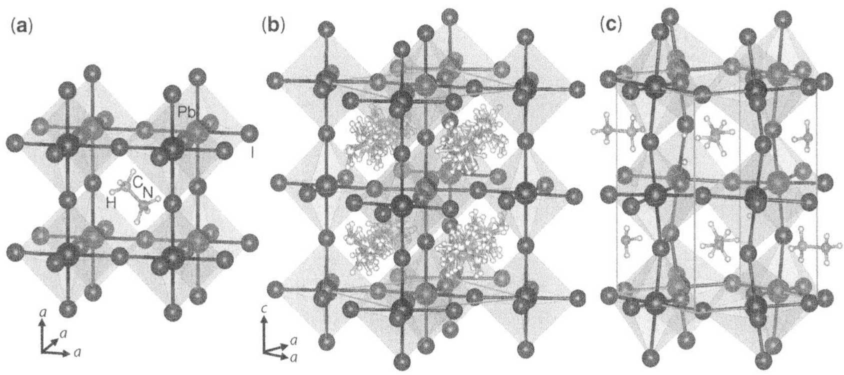

Figure 10.1 Crystal structures of (a) cubic, (b) tetragonal, and (c) orthorhombic $MAPbI_3$.

Table 10.1 Structure transformations of $MAPbX_3$.

Material	**$CH_3NH_3PbCl_3$**	**$CH_3NH_3PbBr_3$**	**$CH_3NH_3PbI_3$**
Crystal system	Cubic	Cubic	Cubic
Transition temperature (K)	177	236	330
Crystal system	Tetragonal	Tetragonal	Tetragonal
Transition temperature (K)	172	149~154	161
Crystal system	Orthorhombic	Orthorhombic	Orthorhombic

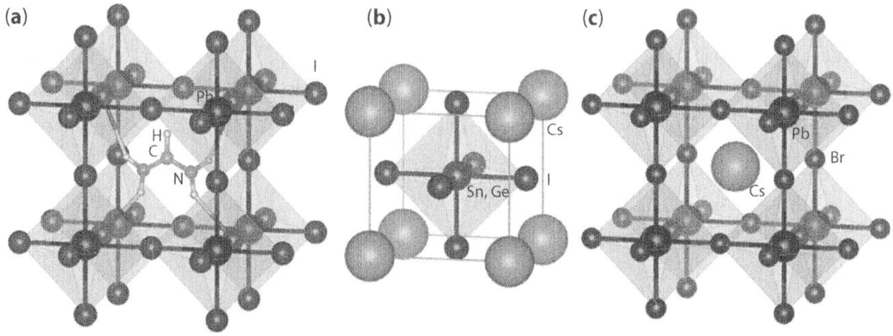

Figure 10.2 Crystal structures of (a) $FAPbI_3$, (b) $CsGeI_3$, and (c) $CsPbBr_3$.

Table 10.2 Crystal systems and temperatures of $CsSnI_3$.

Temperature (K)	**300**	**350**	**478**
Crystal system	Orthorhombic	Tetragonal	Cubic
Space group	*Pnma*	*P4/mbm*	*Pm3m*
Z	4	2	1
Lattice parameters	a = 8.6885 Å b = 12.3775 Å c = 8.3684 Å	a = 8.7182 Å c = 6.1908 Å	a = 6.1057 Å

and crystal system changes and structural parameters of $CsSnI_3$ are listed in Table 10.2 [24, 80–82]. $CsSnI_{2.95}F_{0.05}$ photovoltaic devices provided a photoconversion efficiency of 8.5% [6].

10.3 Basic Configuration of Solar Cells

Three types of device configurations, such as standard, planer, and inverted structures, have been studied for the perovskite solar cells. The most widely fabricated standard cell structures would be the standard type, which has a cell structure of (transparent conducting substrate)/(electron transport layer)/(mesoporous TiO_2)/perovskite/(hole transport layer)/(metal electrode). One of the reasons is that the standard perovskite structures have similar structures to dye-sensitized solar cells with longer research history [83–87].

A representative production procedure of the standard $MAPbI_3$ solar cell is shown in Figure 10.3(a) [12, 88, 89]. Mixture solutions of CH_3NH_3I and $PbCl_2$, PbI_2, and other reagents were spin-coated on the mesoporous TiO_2 and annealed at 100°C to 150°C in ambient air. To obtain the stability of the device under the sun, higher annealing temperatures would be more suitable, and a high temperature annealing process at ~200°C in ambient air was developed recently by using polysilanes [53]. A photograph of the

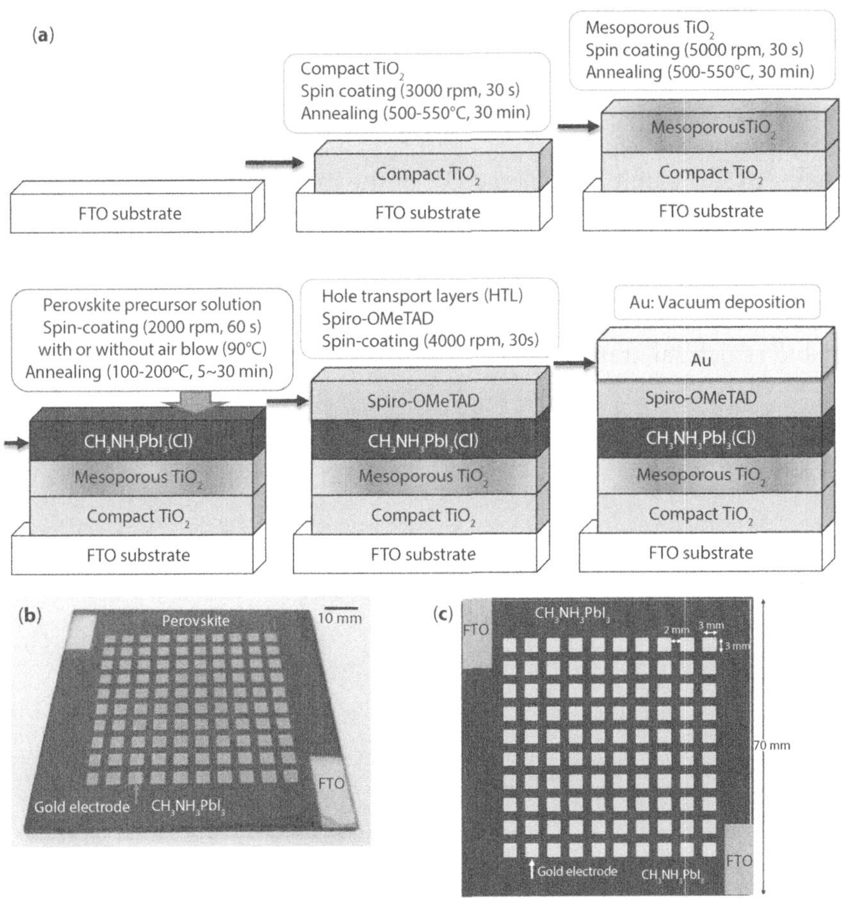

Figure 10.3 (a) Fabrication process of perovskite solar cells. (b) Photograph and (c) schematic plan view of the cell with the size of 70 × 70 mm.

cell with the size of 70 × 70 mm and a schematic plan view are shown in Figure 10.3(b) [90].

The current density–voltage (*J–V*) characteristics of typical perovskite solar cells under solar simulator indicated the annealing effects on the perovskite compounds [88, 89]. The as-deposited $MAPbI_3$ device provided a lower photoconversion efficiency, and better photovoltaic properties were obtained after annealing at 100°C for 15 minutes.

X-ray diffraction (XRD) profiles of $MAPbI_3$ on the ordinary glass substrates are shown in Figure 10.4(a). The reflections could be indexed with cubic and tetragonal structures for annealed and as-deposited thin films, respectively. Although the as-deposited film have a single phase structure, a broad peak of PbI_2 appeared after annealing. Figures 10.4(b) is enlarged X-ray diffraction profiles. Split diffraction reflections of 004–220 for the as-deposited film changed into a single reflection of 200 after annealing, which indicates the structure transition from tetragonal to cubic structures. The $MAPbI_3$ single crystal provides the structure transformation at ~330 K, as listed in Table 10.1. As listed in Table 10.3, the unit cell volume (*V*) of the high temperature cubic phase is 261 Å^3, which is bigger than that of the tetragonal phase (246 Å^3) [88]. Here, *Z* is the number of chemical unit in the unit cell. This result would be due to the disordering of iodine and thermal expansion of the unit cell.

The X-ray diffraction profiles of Figure 10.4 indicated the tetragonal-cubic structural transition of $MAPbI_3$ by partial separation of PbI_2 at

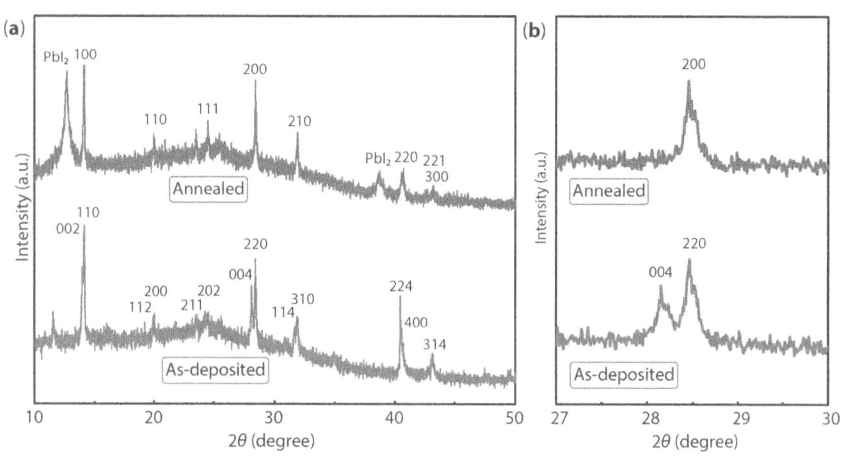

Figure 10.4 (a) X-ray diffraction profiles of as-deposited and annealed thin films of a $CH_3NH_3PbI_3$ compound. (b) Magnified profiles at 2*θ* between 27° and 30°. Copyright (2014) The Japan Society of Applied Physics.

Table 10.3 Measured structure parameters of $CH_3NH_3PbI_3$.

Samples	Crystal system	Lattice constants (Å)	V ($Å_3$)	Z	V/Z ($Å_3$)
As-deposited	Tetragonal	$a = 8.862$, $c = 12.645$	993.1	4	248.3
Annealed	Cubic	$a = 6.272$	246.8	1	246.8
Ref. 114 (220 K)	Tetragonal	$a = 8.800$, $c = 12.685$	982.3	4	245.6
Ref. 99 (330 K)	Cubic	$a = 6.391$	261.0	1	261.0

elevated temperatures, which resulted in the reduction of the unit cell volume from 248.3 to 246.8 $Å^3$ (Table 10.3). This structure transition might be due to the PbI_2 separation, which is different from the ordinary second-order phase transition at 330 K [74, 75, 91].

A cross-sectional TEM image is shown in Figure 10.5(a), and layered structures are observed [92]. Figures 10.5(b) and 10.5(c) are a TEM image and an electron diffraction pattern of the $MAPbI_3$ perovskite and TiO_2 interface, respectively [24, 88]. Polycrystalline $MAPbI_3$ shows dark contrast, which is due to Pb with the largest atomic number in this thin film. Sizes of TiO_2 nanoparticles are ~50 nm, and Debye–Scherrer rings of anatase TiO_2 nanoparticles are observed. Other reflections, apart from the Debye–Scherrer rings due to anatase TiO_2, are due to the perovskite nanoparticles. A structure model and the simulated electron diffraction pattern along the cubic [210] are shown in Figures 10.5(d) and 10.5(f), respectively. The calculated diffraction pattern agrees with the actual pattern of Figure 10.5(c).

Figure 10.6(a) is a HRTEM image of the $CH_3NH_3PbI_3$ compound obtained along the [100] direction [24]. The image taken at the thin crystal indicates projected potential in the crystal directly [69, 71]. The size and darkness of black dots corresponding to Pb sites can be clearly recognized, and the sites of iodine (I) show some darkness, as indicated in the crystal structure of $MAPbI_3$ in Figure 10.6(b). $CH_3NH_3^+$ ions are not represented as black dots in this image because of their smaller atomic number of H, C, and N. Figure 10.6(c) is a HRTEM image of a TiO_2 nanoparticle, and lattice fringes of (100) anatase TiO_2 are observed. The TEM results showed size distributions of TiO_2 nanoparticles, as shown in Figure 10.5(b), and no epitaxial relationships is observed at the TiO_2 and $MAPbI_3$ interface.

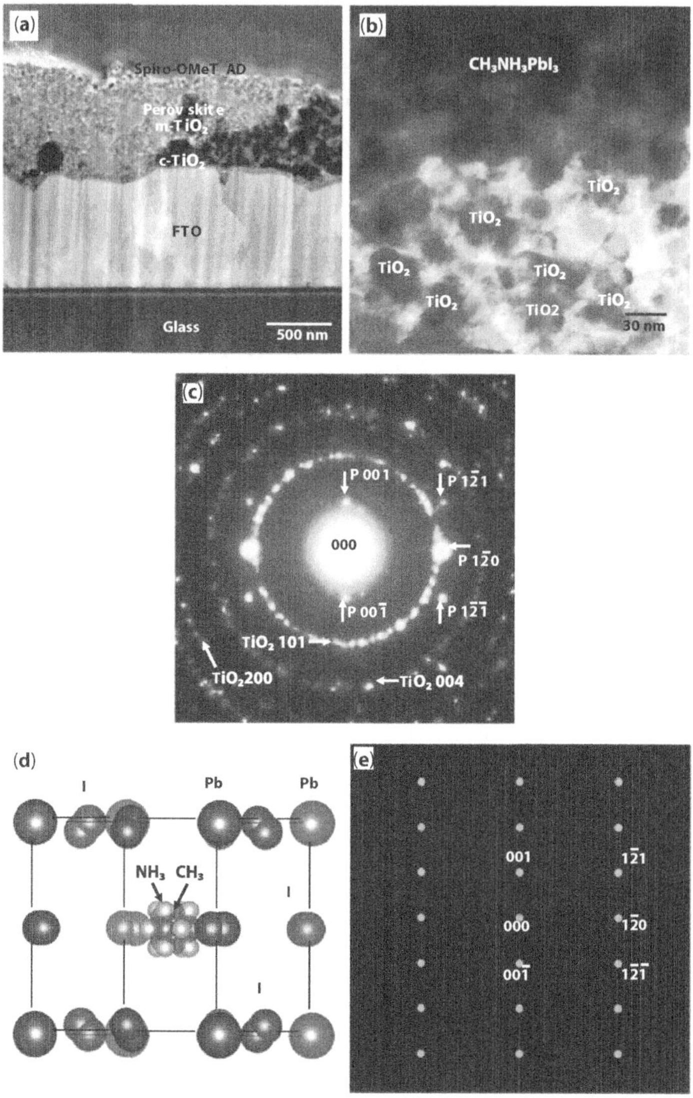

Figure 10.5 (a) Cross-sectional transmission electron microscope image of perovskite solar cell. (b) Transmission electron microscope image and (c) selected area diffraction of the interface of perovskite and TiO_2. The perovskite compound was indicated by "P." (d) Atomic structure model and (e) electron diffraction calculated along the [210] direction of cubic $CH_3NH_3PbI_3$. Copyright (2014) The Japan Society of Applied Physics.

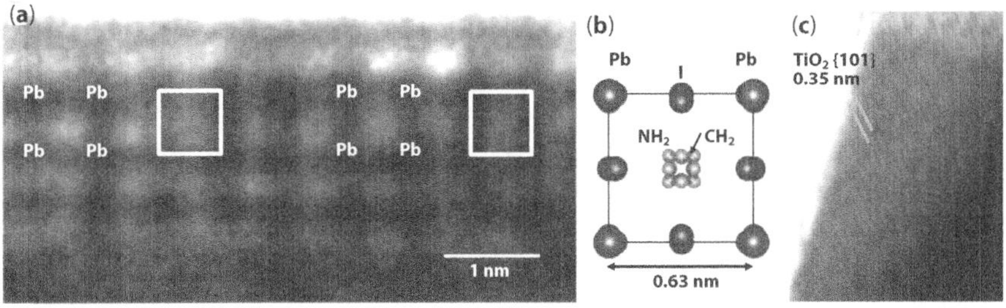

Figure 10.6 (a) HRTEM image and (b) (100)-projected cubic $CH_3NH_3PbI_3$. (c) HRTEM image of TiO_2.

An energy level diagram of the standard $MAPbI_3$ device is shown in Figure 10.7(b). Carriers are generated by the sunlight from the FTO side. The TiO_2 electron transport layer accepts electrons from the perovskite, and these electrons are transported to the FTO electrode. On the other hands, the separated holes are transported to the gold (Au) electrode through the spiro-OMeTAD hole transport layer.

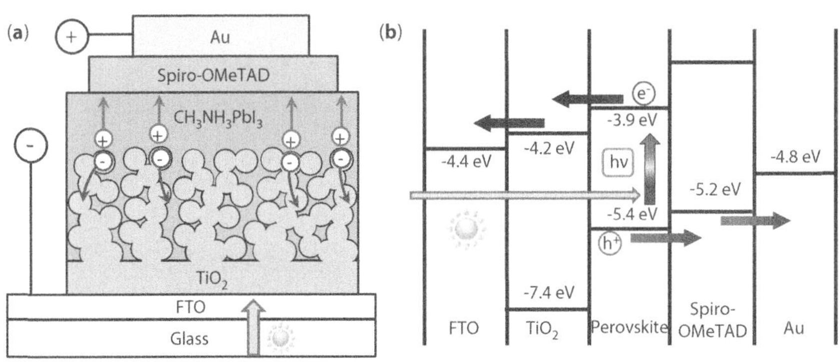

Figure 10.7 (a) Model of interfacial structure. (b) Energy level diagram of perovskite solar cell. Copyright (2014) The Japan Society of Applied Physics.

10.4 Cl Doping to Perovskites

Effects of Cl-doping at iodine sites in $CH_3NH_3PbI_3$ have been investigated [93, 94]. The current density–voltage properties of the $CH_3NH_3PbI_{3-x}Cl_x$ perovskites under the solar simulator are shown in Figure 10.8(a), and summarized cell parameters are listed in Table 10.6. The standard $CH_3NH_3PbI_3$ device provided a photoconversion efficiency (η) of 6.16%, and the averaged η was 5.53%, as shown in Table 10.4. The highest η of 8.16% was attained for the $CH_3NH_3PbI_{2.88}Cl_{0.12}$ device. As the Cl concentration increased, the short-circuit current density (J_{SC}) and open-circuit voltage (V_{OC}) decreased, as observed in Figure 10.8(b) and Table 10.4. Energy gaps (E_g) of $CH_3NH_3PbI_3$, $CH_3NH_3PbI_{2.88}Cl_{0.12}$ and $CH_3NH_3PbI_{1.8}Cl_{1.2}$ were measured to be 1.578, 1.590, and 1.593, respectively, by optical absorption, and this indicates an increase of the energy gap of the standard $MAPbI_3$ by the Cl doping.

X-ray diffraction profiles of the present perovskite cells are shown in Figure 10.9(a). Diffraction reflections are indexed by tetragonal and cubic structures for $CH_3NH_3PbI_{1.8}Cl_{1.2}$ and $CH_3NH_3PbI_3$ perovskites, respectively. A broad peak due to PbI_2 is observed for the standard $MAPbI_3$ perovskite.

Figure 10.9(b) is enlarged X-ray diffraction profiles of Figure 10.9(a) at 2θ of ~28.5°. A cubic 200 reflection split into tetragonal 004/220 reflections for the $CH_3NH_3PbI_{1.8}Cl_{1.2}$, which indicates structure transition from the cubic to tetragonal perovskites by heavy Cl-doping [24]. No $PbCl_2$

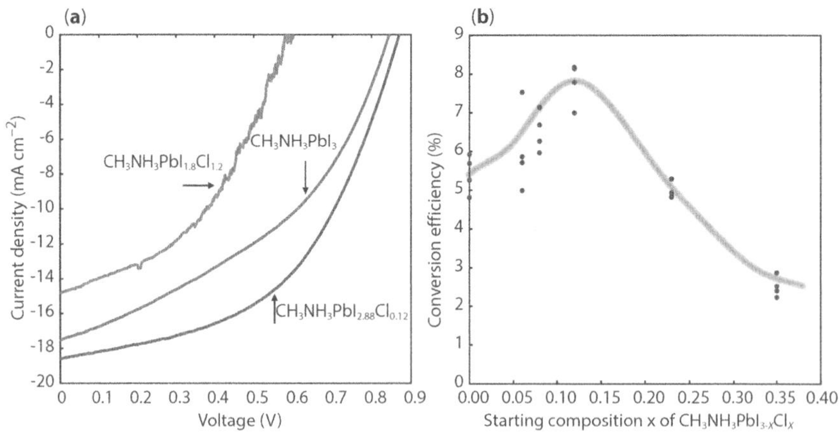

Figure 10.8 (a) Current density–voltage properties of the present devices. (b) Conversion efficiencies of the cells as a function of Cl concentration.

Table 10.4 Measured photovoltaic parameters of $CH_3NH_3PbI_{3-x}Cl_x$ devices.

Preparation composition	J_{SC} (mA cm^{-2})	V_{OC} (V)	FF	η (%)	η_{ave} (%)
$CH_3NH_3PbI_3$	17.5	0.844	0.416	6.16	5.53
$CH_3NH_3PbI_{2.94}Cl_{0.06}$	17.7	0.871	0.487	7.53	6.02
$CH_3NH_3PbI_{2.92}Cl_{0.08}$	18.1	0.825	0.478	7.14	6.52
$CH_3NH_3PbI_{2.88}Cl_{0.12}$	18.6	0.869	0.504	8.16	7.77
$CH_3NH_3PbI_{2.77}Cl_{0.23}$	13.9	0.865	0.440	5.29	4.97
$CH_3NH_3PbI_{2.65}Cl_{0.35}$	11.7	0.709	0.347	2.87	2.51
$CH_3NH_3PbI_{1.80}Cl_{1.20}$	14.8	0.598	0.436	3.87	2.00

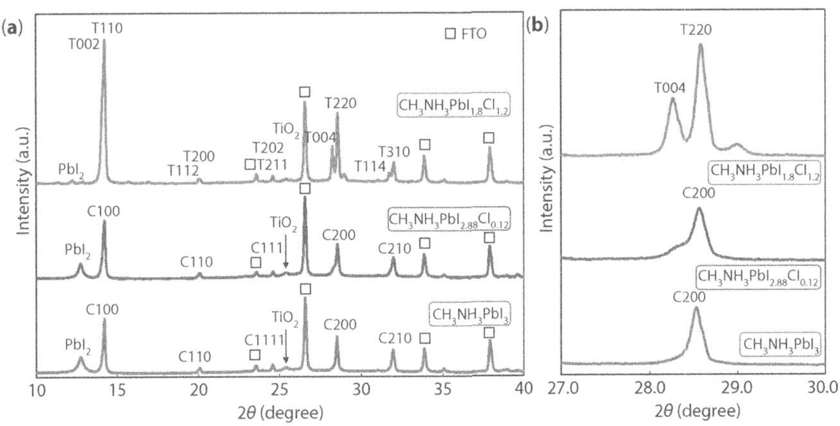

Figure 10.9 (a) X-ray diffraction profiles of $CH_3NH_3PbI_{3-x}Cl_x$. (b) Enlarged profiles of (a).

compound was detected, and the Cl-doping suppressed PbI_2 formation. For the $CH_3NH_3PbI_{2.88}Cl_{0.12}$, a small shoulder is observed just left of the 200 reflection in Figure 10.9(b), which is due to the pseudo-cubic structure between the tetragonal and cubic structures. Table 10.5 shows the measured structure parameters of the present perovskite compounds.

The diffraction profiles in Figure 10.9 indicated structural transition from the tetragonal $MAPbI_3$ to the cubic structure by fractional PbI_2 separation from the $MAPbI_3$ [88], which resulted in the decrease of the unit cell volume of the cubic $MAPbI_3$ from 261 to 244 Å^3, as listed in Table 10.5. From the EDS analysis, the site occupancy of I might be lesser than 1,

Table 10.5 Measured structural parameters of the present perovskite compounds.

Preparation composition	Crystal system	Lattice constant (Å)	V (Å³)	Z	V/Z (Å³)
$CH_3NH_3PbI_3$	Cubic	$a = 6.252$	244.4	1	244.4
$CH_3NH_3PbI_{2.88}Cl_{0.12}$	Pseudo-cubic	$a = 6.245$	243.5	1	243.5
$CH_3NH_3PbI_{1.80}Cl_{1.20}$	Tetragonal	$a = 8.826$, $c = 12.618$	982.8	4	245.7

leading to the decrease of the cell volume. The photoconversion properties were reported to be improved by the tetragonal to cubic transform [88].

Figure 10.10(a) is a scanning electron microscope (SEM) image of the $CH_3NH_3PbI_{2.88}Cl_{0.12}$ crystal, and the particle size is *c. a.* 10 μm. Elemental mapping images of Pb, I, and Cl by energy dispersive X-ray spectroscopy (EDS) are shown in Figure 10.10(b-d), respectively, and the compositions of Pb/I/Cl were measured to be 1.00:2.70:0.11. The iodine anions might be deficient, which lead to an increase of hole concentration of the perovskites.

The splitting of XRD peaks from C200 to T004/T220 was due to the Cl-doping. The reduction of the crystal symmetry might decrease the efficiencies. On the other hand, the cubic perovskite structure was still preserved as the pseudo-cubic structure by adding a small amount of Cl. The Cl doping would lengthen diffusion length of carriers [7, 25], which increased the efficiencies.

External quantum efficiency (EQE) spectra of the present devices is shown in Figure 10.11(a). The ordinary $MAPbI_3$ compound provides photo conversion in the range of 300 to 800 nm. The energy gap of $MAPbI_3$ increased from 1.578 to 1.590 eV by the present Cl-doping, which would lead to the increase of open-circuit voltage. Figure 10.11(b) is internal quantum efficiency (IQE) spectra of the present devices calculated from reflectance and external quantum efficiencies. These IQE spectra increased in the range of 550 to 780 nm, which indicates the possible increase of the photoconversion in that range.

In summary, the effects of $PbCl_2$ addition to $MAPbI_3$ precursors were studied. The cubic perovskite transformed to the tetragonal structure by the heavy Cl-doping. A small amount of Cl-doping improved increased the photoconversion, which would be due to enlargement of band gap and diffusion length of excitons. Both the IQE and EQE increased by adding a small amount of chlorine, and the IQE results designate that further

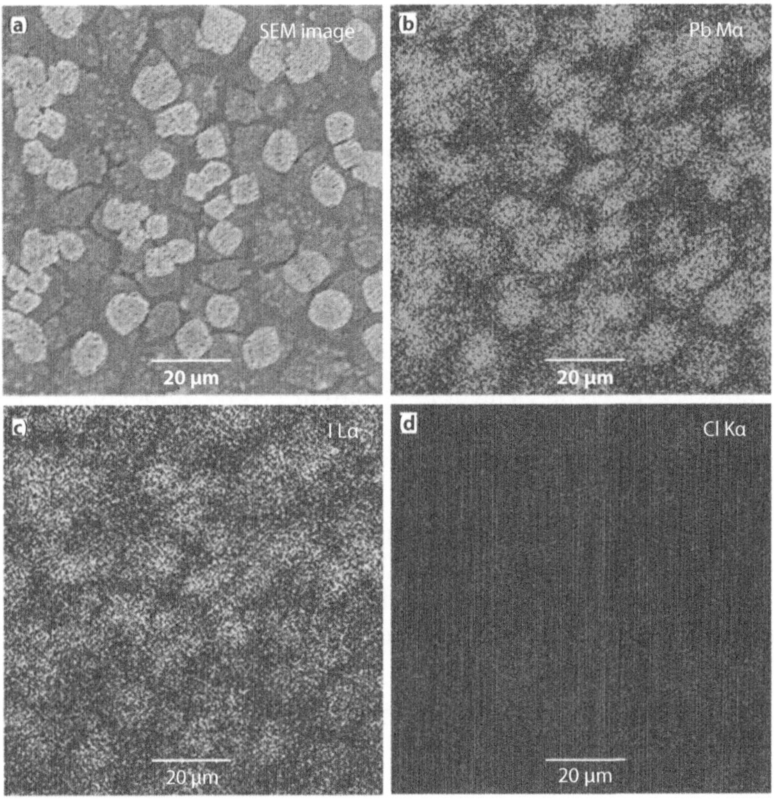

Figure 10.10 (a) SEM image of $CH_3NH_3PbI_{2.88}Cl_{0.12}$ device. EDS mapping images of (b) Pb, (c) I, and (d) Cl.

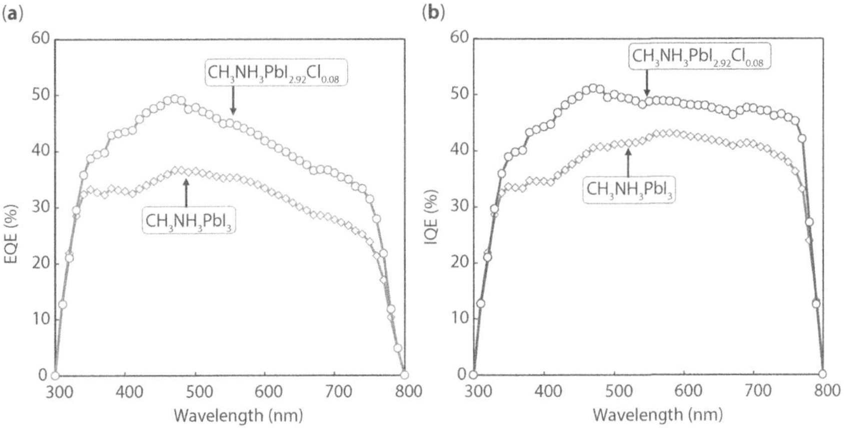

Figure 10.11 (a) External quantum efficiency and (b) internal quantum efficiency spectra of $CH_3NH_3PbI_{2.92}Cl_{0.08}$ and $CH_3NH_3PbI_3$ devices.

improvement of efficiencies might be possible by decreasing the optical reflection in the range of 550 to 780 nm.

10.5 Sb or As Doping to Perovskites

Photovoltaic parameters of devices strongly depend on the perovskite structures. Elemental dopings, such as copper (Cu), germanium (Ge), or tin (Sn), at the Pb atomic positions, have been studied [26–28, 35-37], which leads to Pb-free devices and influences the photoconversion properties [26, 28, 35, 95, 96].

Antimony (Sb) is an element in the group 15, and Sb-doping at the Pb site was expected as electronic carriers. Effects of SbI_3 addition to a standard perovskite solution were described here [29–34].

Current density–voltage properties of the $CH_3NH_3Pb_{1-x}Sb_xI_3$ devices indicated an effect of Sb-doping in the $MAPbI_3$, and the measured parameters of the present devices are listed in Table 10.6.

An ordinary $MAPbI_3$ device provided an averaged photoconversion efficiency of 6.37%, as listed in Table 10.6. The highest efficiency of 9.07% was obtained for the $CH_3NH_3Pb_{0.97}Sb_{0.03}I_3$ device [29]. As the Sb concentration (x) increased, the conversion efficiencies decreased, as shown in Figure 10.12(a). Higher efficiencies were reported by co-addition of NH_4Cl and SbI_3 to $MAPbI_3$ [30]. EQE spectra of the standard and Sb-doped perovskite

Table 10.6 Measured photovoltaic parameters of $CH_3NH_3Pb_{1-x}Sb_xI_3$ devices.

Sb (x)	J_{SC} (mA cm^{-2})	V_{OC} (V)	*FF*	η (%)	η_{ave} (%)
0.00	17.0	0.758	0.509	6.56	6.37
0.01	16.0	0.789	0.534	6.74	6.41
0.02	16.9	0.792	0.518	6.94	6.72
0.03	19.2	0.843	0.560	9.07	8.47
0.05	15.7	0.755	0.575	6.82	5.61
0.07	14.7	0.692	0.502	5.11	4.07
0.10	12.1	0.630	0.476	3.63	3.27
0.15	13.1	0.570	0.402	3.00	2.85

devices are shown in Figure 10.12(b). The EQE was improved in the range of 400 to 700 nm by addition of a small amount of Sb.

X-ray diffraction profiles of the present $CH_3NH_3Pb_{1-x}Sb_xI_3$ devices are shown in Figure 10.13(a). Indices of the reflections are based on the cubic perovskite structure. In addition to the perovskite peaks, reflections of PbI_2 are observed in the profiles. The Sb addition suppressed PbI_2 formation, and little PbI_2 was detected for $CH_3NH_3Pb_{1-x}Sb_xI_3$ devices with $x \geq 0.05$. Figure 10.13(b) is measured lattice constants (a) of Sb-doped $MAPbI_3$ perovskites. A small increase of lattice constants is observed in the x range of 0.03 to 0.05, and further addition of Sb decrease the lattice constants. The X-ray diffraction profile of the standard $MAPbI_3$ in Figure 10.13(a) showed PbI_2 formation. This might indicate the PbI_2 separation from the standard $MAPbI_3$, which would correspond to the smaller lattice constant (a = 6.266 Å) than that (6.391 Å) of single crystal $MAPbI_3$ [77].

Increase of the conversion efficiencies can be described by the following mechanisms. The first is a doping effect of Sb at the Pb sites. Excess charge of Sb^{3+} compared with that of Pb^{2+} would have a role of charge carriers for the perovskite crystal, which lead to increase of J_{SC}. Another mechanism is an attraction effect of I^- by Sb^{3+}, which might prevent PbI_2 elimination from $MAPbI_3$ and increase the lattice constants. The lattice constants decreased by an increase of the amount of Sb with an ionic size smaller than Pb.

Dendritic structures were also observed by adding FA and Sb [24]. Figures 10.14(a) and 14(b) are optical microscopy images of $HC(NH_2)_2PbIBr_2$ and $HC(NH_2)_2Pb_{0.95}Sb_{0.05}I_3$, respectively. They are $FAPbI_3$ doped with Br and Sb at the I and Pb sites, respectively. By adding a small amount of Br and Sb, perovskite crystals with dendritic structures grew densely, and the surface

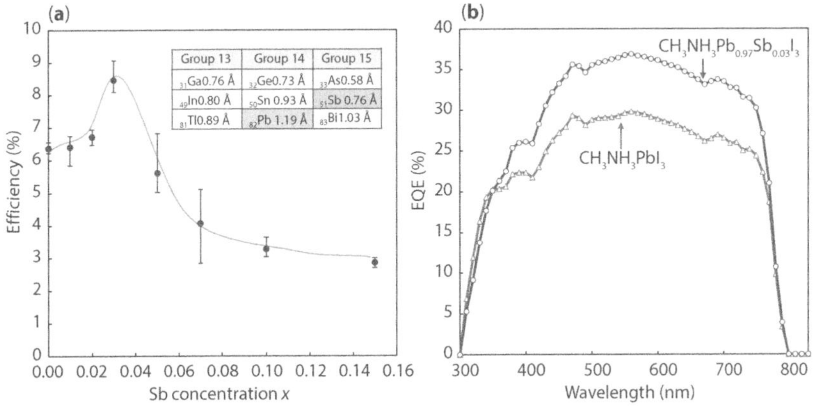

Figure 10.12 (a) Photoconversion efficiencies of $CH_3NH_3Pb_{1-x}Sb_xI_3$ as a function of Sb concentration. (b) EQE spectra of $CH_3NH_3Pb_{0.97}Sb_{0.03}I_3$ and standard $CH_3NH_3PbI_3$ devices.

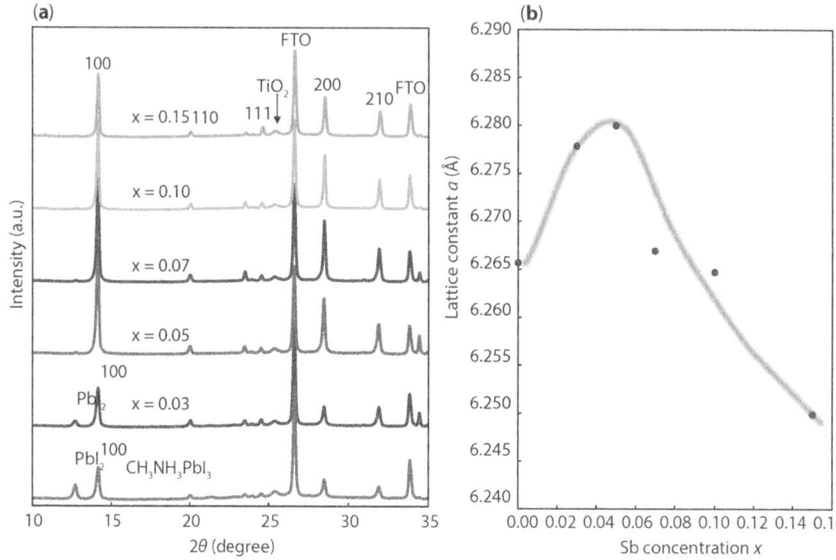

Figure 10.13 (a) X-ray diffraction profiles of $CH_3NH_3Pb_{1-x}Sb_xI_3$ devices. (b) Lattice constants of Sb-doped $MAPbI_3$ perovskites.

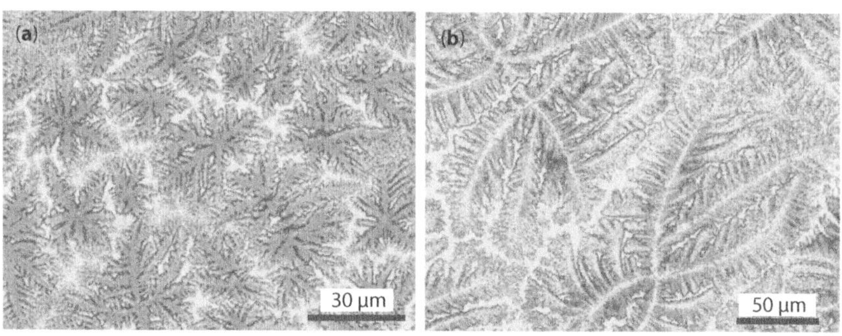

Figure 10.14 Optical microscope images of (a) $HC(NH_2)_2PbIBr_2$ and (d) $HC(NH_2)_2Pb_{0.95}Sb_{0.05}I_3$.

coverage of the perovskite grains on the cell substrate increased. Because of these dendritic structures, short circuit current densities and photo conversion efficiencies increased. Gibbs–Thomson coefficients and liquidus line gradients were estimated, and the dendrites would be formed by satisfying neutral stable conditions on the grain growth rate by increasing the kinds of solute elements. Fractal dimensions of these dendrite structures were calculated to be ~2.8 by using a box-counting method [24]. These dendrite structures would contribute increase of the surface coverage and interfacial area at the interpenetrating *pn* junction, which provided the higher current densities and conversion efficiencies.

In addition to the Sb doping, arsenic (As) doping at the Pb sites was also reported. The effects of AsI_3, $SnBr_2$, and CsI addition to $CH_3NH_3PbI_3$ perovskite solar cells were investigated [38]. Arsenic is also a group 15 element, and an improvement of conversion efficiencies of $MAPbI_3$ was expected as well as Sb-doping. The amounts of additives are 5% AsI_3, 5% AsI_3 5% + 5% CsI, and 5% AsI_3 + 5% $SnBr_2$, and denoted as +AsI_3, + AsI_3 + CsI, and + AsI_3 + $SnBr_2$ devices.

The + $SnBr_2$ device provided an η of 4.47%, as shown in Figure 10.15(a). A J_{SC} for the + $SnBr_2$ device decreased by the $SnBr_2$ addition, although the +AsI_3 device provided a J_{SC} of 11.6 mA cm^{-2}. The highest conversion efficiency was obtained for the $MA_{0.95}Cs_{0.05}Pb_{0.95}As_{0.05}I_3$, which provided an η

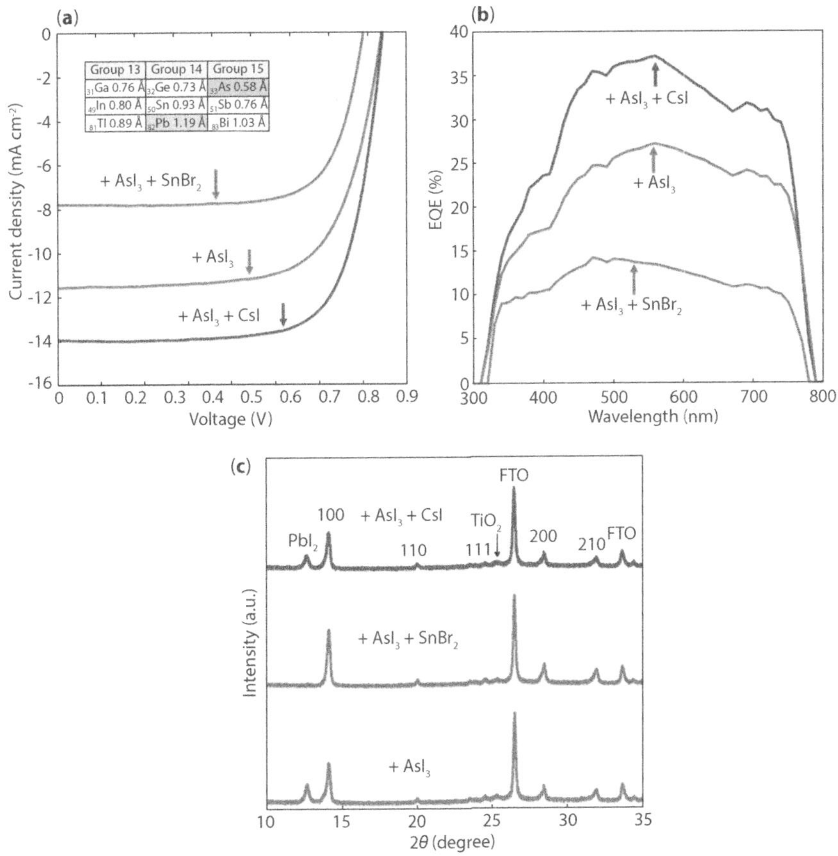

Figure 10.15 (a) Current density-voltage properties of $SnBr_2$ or CsI addition to AsI_3-added devices. (b) EQE spectra and (c) X-ray diffraction profiles of the present devices.

of 8.49%, a *FF* of 0.728, a J_{SC} of 13.9 mA cm^{-2}, and a V_{OC} of 0.837 V. A J_{SC} value increased by the Cs-doping, as observed in Figure 10.14(a).

The EQE spectra of these devices are shown in Figure 10.15(b), and the energy gap is 1.57 eV. The EQEs were improved in the range of 340 to 760 nm by the Cs addition.

Figure 10.15(c) shows X-ray diffraction profiles of the present devices. Diffraction reflections due to the cubic perovskite increased by $SnBr_2$ addition, and the $SnBr_2$ addition suppressed the PbI_2 formation. The measured unit cell volume decreased by adding a small amount of CsI or $SnBr_2$ to the present perovskite crystals, which implies the Sn or Cs substitution at the Pb site.

10.6 Highly (100)-Oriented Perovskites

In addition to the additives such as NH_4Cl [97–99], air-blowing [100, 101] are also effective methods to control the microstructures of the perovskite crystals, and results on the combination of the NH_4Cl addition and air-blowing was described here, which is expected to control the perovskite microstructures [102]. The ordinarily precursors used for the perovskite synthesis is mixture solutions of CH_3NH_3I and PbI_2. The $MAPbI_3$ perovskite can also be formed with a mixture of DMF solutions of CH_3NH_3I and $PbCl_2$ [103–105]. The reaction of the $MAPbI_{3-x}Cl_x$ is as follows: $3CH_3NH_3I + PbCl_2 \rightarrow CH_3NH_3PbI_3 + 2CH_3NH_3Cl$ (↑). The generated CH_3NH_3Cl is vaporized, and there is still a small amount of Cl in the perovskite, which is denoted as $CH_3NH_3PbI_3$(Cl).

For the perovskite layer, a solution containing CH_3NH_3I and $PbCl_2$ was prepared with a molar ratio of 3:1 in DMF [89]. For the spin-coating process, an air blowing method was applied, and the sample was annealed at 140°C to form the perovskite compound [102].

Figure 10.16(a) shows the *J*–*V* characteristics of the present $CH_3NH_3PbI_3$(Cl) devices, in which the effects of NH_4Cl addition are evident. Measured photovoltaic parameters of the $CH_3NH_3PbI_3$(Cl) devices are listed in Table 10.7. The $CH_3NH_3PbI_3$(Cl) device without NH_4Cl had an η_{ave} of 12.37%, as shown in Table 10.7 [102]. Adding 5 mg of NH_4Cl provided the highest conversion efficiency of 14.03%. The series resistance (R_s) decreased with NH_4Cl addition, which leads to an increase in the J_{SC}. After standing for 2 weeks in air, the decreases in efficiencies of the devices containing NH_4Cl are small, and an η of 13.19% is still preserved (Table 10.7).

EQE spectra of the $CH_3NH_3PbI_3$(Cl) devices were shown in Figure 10.16(b). The $CH_3NH_3PbI_3$(Cl) devices provided photoconversion

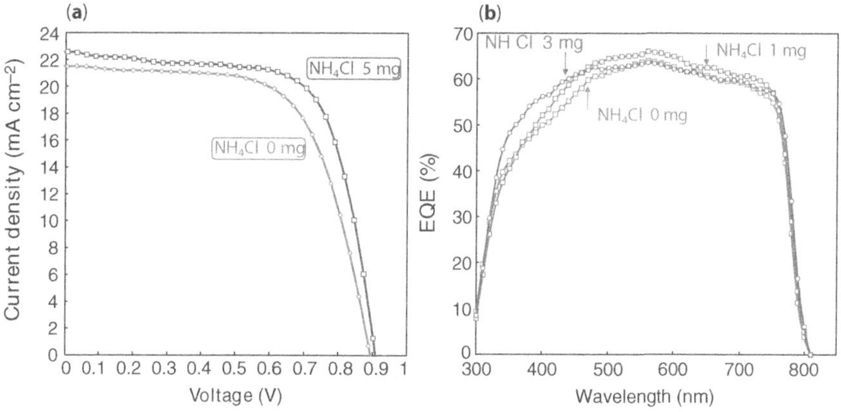

Figure 10.16 (a) Current density-voltage properties of $MAPbI_3(Cl)$ devices with and without NH_4Cl. (b) EQE spectra of $CH_3NH_3PbI_3(Cl)$ cells.

Table 10.7 Measured photovoltaic parameters of the $CH_3NH_3PbI_3(Cl)$ devices.

NH_4Cl (mg)	J_{SC} (mA cm^{-2})	V_{OC} (V)	*FF*	η (%)	η_{ave} (%)	R_s (Ω cm^2)
0	21.5	0.893	0.647	12.41	12.37	6.56
1	22.2	0.912	0.651	13.19	12.89	6.18
3	22.5	0.917	0.651	13.40	12.13	6.06
5	22.6	0.908	0.685	14.03	13.74	4.70
After 2 weeks						
0	20.5	0.870	0.647	11.51	4.61	0
1	21.6	0.892	0.640	12.34	4.51	1
3	22.1	0.893	0.668	13.19	4.63	3
5	21.5	0.893	0.643	12.33	4.29	5

between 300 and 810 nm, and the band gap was 1.53 eV. The EQE values was increased in the range of 350 to 750 nm by adding NH_4Cl, leading to increase of the J_{SC}.

Optical microscope images of the $CH_3NH_3PbI_3(Cl)$ devices are shown in Figure 10.17. Particles with sizes of 5 to 10 μm are observed for the

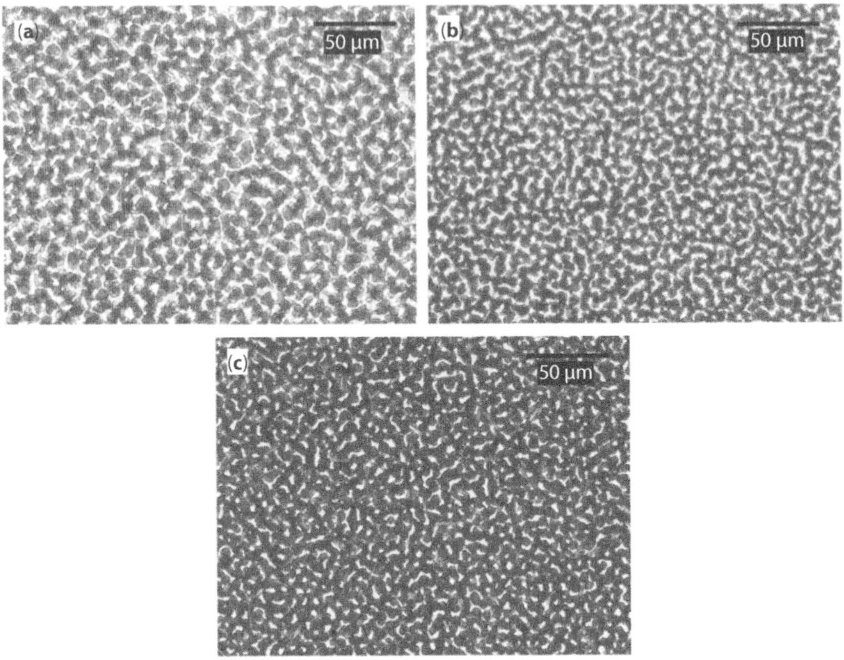

Figure 10.17 Optical microscope images of $CH_3NH_3PbI_3(Cl)$ devices containing (a) 0 mg, (b) 1 mg, and (c) 5 mg of NH_4Cl.

device prepared without NH_4Cl, as shown in Figure 10.17(a). The particle size was decreased by adding NH_4Cl to the $CH_3NH_3PbI_3(Cl)$, as observed in Figures 10.17(b) and 10.17(c). Networking structures between the perovskite particles are observed, particularly in Figure 10.17(c), which would contribute improvement of the photoconversion properties.

A scanning electron microscope image of the $CH_3NH_3PbI_3(Cl)$ device without NH_4Cl is shown in Figure 10.18(a). Crystal faceted perovskite particles with sizes of 5 to 10 μm are observed, which correspond to those in Figure 10.18(a). EDS mapping images of Pb, I, Cl, C, and N are shown in Figures 10.18(a). EDS spectra were measured as shown in Figure 10.19, and the estimated compositions are listed in Table 10.8. These elemental mapping images indicate that the microparticles observed in optical microscope images of Figure 10.17(a) correspond to the $MAPbI_3$ compound.

Figure 10.18(b) shows a scanning electron microscope image of the $CH_3NH_3PbI_3(Cl)$ device containing 5 mg of NH_4Cl. The surface morphology drastically changed by adding NH_4Cl to the $CH_3NH_3PbI_3$. The particle sizes are ~5 μm, and the crystals are interconnected with special facets. These networking structures would increase the surface coverage and

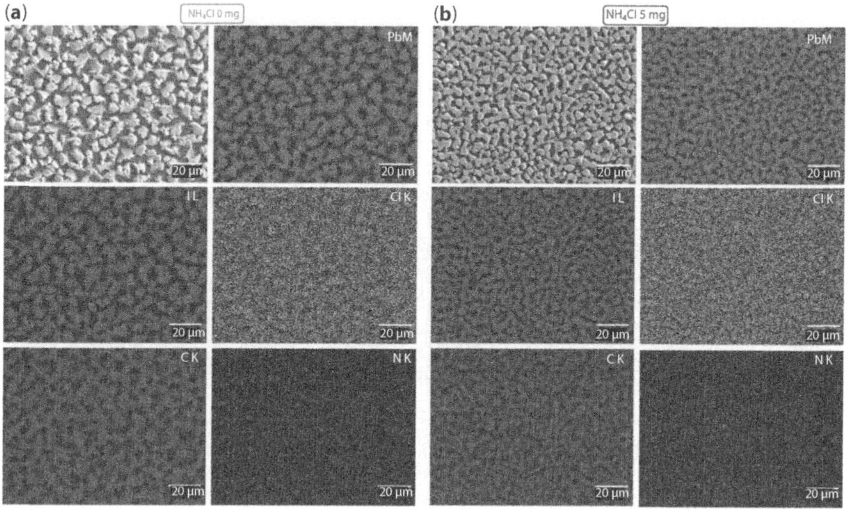

Figure 10.18 (a) Scanning electron microscope image of the $CH_3NH_3PbI_3(Cl)$ cell without NH_4Cl, and corresponding elemental mapping images of the Pb, I, Cl, C, and N. (b) Scanning electron microscope image of the $CH_3NH_3PbI_3(Cl)$ cell containing 5 mg of NH_4Cl, and the corresponding EDS mapping images.

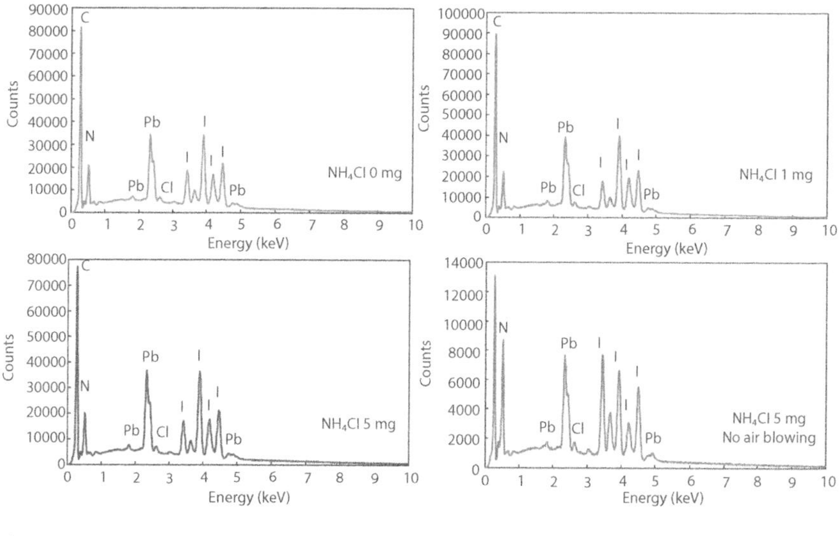

Figure 10.19 EDS profiles of the present devices.

photoconversion efficiencies. Figures 10.18(b) also show EDS mapping images, and the calculated compositions are listed in Table 10.8, which indicates that the $CH_3NH_3PbI_3$ starting composition maybe I deficient. Cl would be doped into the $CH_3NH_3PbI_3$, as shown in Table 10.8.

Table 10.8 Compositions and crystal orientations of the present perovskite crystals.

NH_4Cl (mg)	Pb (at.%)	I (at.%)	Cl (at.%)	C : N (at.%)	I_{100}/I_{210}	I_{100}/I_{FTO}
Calculation*	25.0	75.0	0.0	50.0 : 50.0	2.1	-
0	24.3	71.7	4.0	61.9 : 38.1	69	51
1	24.0	71.4	4.7	61.7 : 38.3	510	170
3	24.4	71.1	4.6	61.8 : 38.2	3400	280
5	24.1	70.2	5.7	61.8 : 38.2	3600	270
5**	24.0	71.2	4.8	52.0 : 48.0	2.8	0.40

*Calculated from randomly oriented cubic $CH_3NH_3PbI_3$ crystals.
**No air blowing or $PbCl_2$.

Figure 10.20 shows X-ray diffraction profiles of the present devices, and the diffraction reflections are indexed by a cubic perovskite at 100°C without air blowing, a diffraction reflection of PbI_2 was observed as shown in Figure 10.20(d) [88]. Reflection intensities of the 100 and 200 of the perovskite tremendously increase by air blowing and NH_4Cl addition, as observed in Figures 10.20(b) and 10.20(c).

Intensity ratios of the 100 (I_{100}) to 210 (I_{200}) of the perovskite were estimated as I_{100}/I_{210}, which are summarized in Table 10.8. If the particles of cubic $MAPbI_3$ are randomly oriented, the orientation index I_{100}/I_{210} is calculated to be 2.1 [24]. For the device fabricated using air blowing and without NH_4Cl, the I_{100}/I_{210} is estimated to be 61, which shows that the cubic (100) of the perovskite crystals are fairly oriented parallel to the FTO substrate. Further adding of NH_4Cl to the device resulted in increase of I_{100}/I_{210} to 3600, which is 1700 times higher than that of randomly oriented cubic perovskite crystals. The device fabricated with 5 mg NH_4Cl without air blowing or $PbCl_2$ has an I_{100}/I_{210} of 2.8, which means these perovskite grains are randomly oriented. The ratios of the I_{100} to reflection intensities of F-doped SnO_2 (I_{FTO}) were also calculated, as summarized in Table 10.8. The I_{100}/I_{FTO} values were increased by using air blowing and NH_4Cl.

The formation mechanisms of highly (100)-oriented $CH_3NH_3PbI_3$(Cl) crystals are as follows: the first is the crystal growth of the perovskites facilitated by air blowing rapid annealing the perovskite solution with the air blowing accelerates fast crystallization of $CH_3NH_3PbI_3$(Cl). Low surface

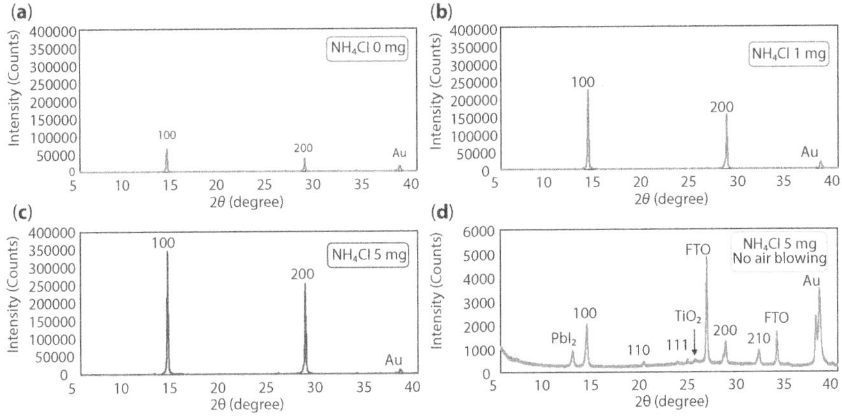

Figure 10.20 X-ray diffraction profiles of $CH_3NH_3PbI_3(Cl)$ devices containing (a) 0 mg, (b) 1 mg, and (c) 5 mg of NH_4Cl. (d) X-ray diffraction profile of the $CH_3NH_3PbI_3(Cl)$ device containing 5 mg of NH_4Cl without air blowing.

tension of (100) promotes preferential (100) growth, and the decrease of high-angle grain boundaries decreases the series resistance. The second mechanism is the surfactant effect of NH_4Cl to form a homogeneous microstructure. Formation of network microstructures increases the surface coverage and carrier transport, which improve the J_{SC} and FF. When $2(CH_3NH_3Cl)$ were vaporized from $3(CH_3NH_3I) + PbCl_2$ [103–105], residual Cl is doped at I sites as detected by EDS, which increases the exciton diffusion length [25] and the J_{SC}.

10.7 Cu Doping to Perovskites

The most common composition of the perovskite compounds for solar cells is $CH_3NH_3PbI_3$. However, the toxicity of Pb is a serious problem and nontoxic substitution elements, such as Sn, have been studied [26, 106]. However, Sn^{2+} is easily oxidized to Sn^{4+} to form SnO_2. Nontoxic Cu is also a candidate for Pb-free solar cells, and several studies have been reported [41, 107, 108].

A tolerance factor (t-factor) is one of the indicators to estimate the stability of perovskite structures [24, 45, 109–111], and the perovskite structure could be experientially formed in the range of $0.813 \leq t \leq 1.107$ and 0.442. When the $t = 1$, the crystals have a geometrically ideal perovskite structure. Since a tolerance factor of ordinary $MAPbI_3$ is 0.912 [24], an increase in t to 1 would stabilize the perovskite structure. If 5% Pb in the $MAPbI_3$ is substituted by Cu, the t value increases from 0.912 to 0.918, which would stabilize the perovskite structure. The Cu addition to the $MAPbI_3$ is expected to be one of the effective methods to stabilize the perovskite structure [42, 112]. t factors are also expected to increase to 1 by Br or Cl addition with shorter ionic radii than I. In fact, the photovoltaic properties were improved by using $PbCl_2$ and CH_3NH_3I precursors [102]. Therefore, copper halide-added $MAPbI_3$ perovskites are described here [40].

Current density-voltage properties of the Cu halide-added devices are shown in Figure 10.21(a) [40], and the summarized device parameters are listed in Table 10.9. A photoconversion efficiency of the standard $MAPbI_3$ device is 8.74%, and the CuBr- or CuI-added devices provided efficiencies of 11.4% and 11.2%, respectively. An efficiency of CuCl-added device was 8.20%, which was lower than that of the standard cell. The R_s values of the CuBr- or CuI-added devices were lower than that of the standard one, which lead to increase of carrier mobility, J_{SC} and FF. The shunt resistances (R_{sh}) of the Cu-added cells are higher than that of the standard one, which lead to improvement of leakage current and V_{OC}.

The X-ray diffraction profiles of the present devices are shown in Figures 10.21(b)–10.21(e). XRD intensity ratios of I_{100}/I_{210} are listed in Table 10.9, and the Cu-added devices provided highly (100)-oriented perovskite

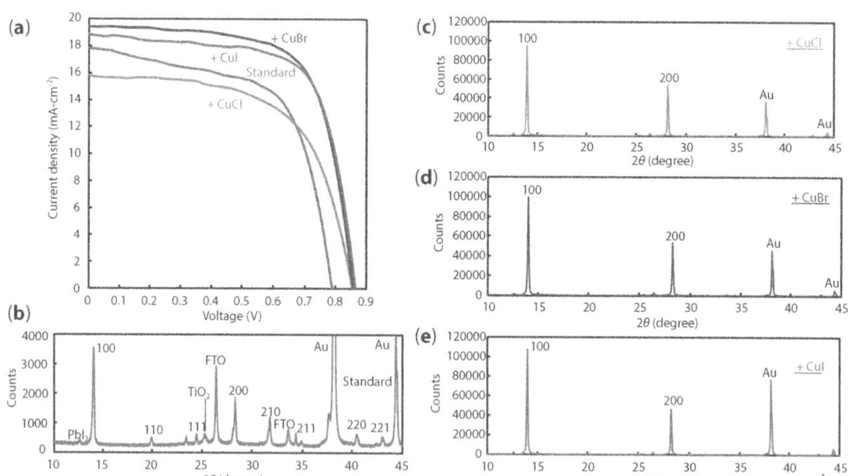

Figure 10.21 (a) Current density-voltage properties of the present devices. X-ray diffraction profiles of the (b) standard, (c) CuCl-, (d) CuBr-, and (e) CuI-added perovskite devices.

Table 10.9 Photovoltaic parameters and orientation indices of the present devices.

Samples	J_{SC} (mA cm^{-2})	V_{OC} (V)	*FF*	η (%)	η_{av} (%)	I_{100}/I_{210}
Standard	18.2	0.787	0.590	8.74	8.44	2.7
+ CuCl	16.0	0.854	0.565	8.20	7.72	400
+ CuBr	19.6	0.858	0.656	11.4	11.0	320
+ CuI	19.0	0.864	0.673	11.2	11.0	200

crystals. The hot air blowing method with Cu addition is effective for the growth of highly oriented perovskite crystals.

The surface coverages of perovskite compounds for the standard device and the devices added with CuCl, CuBr, and CuI were 88.7, 80.2, 93.5, and 91.1%, respectively. The photoconversion efficiencies of the CuBr- or CuI-added devices are higher than that of the standard device, and the CuCl-added device is lower. Since the trend of the surface coverage is the same as that of the photoconversion efficiencies, the efficiencies are closely related to the surface coverage.

Figure 10.22(a) is a scanning electron microscope image of the standard device, and the sizes of perovskite particles are 2 to 4 μm. EDS mapping images of Figure 10.22(a) show uniform distribution of Pb, I and Cl. On the other hand, surface structure of the CuCl-added device is significantly changed by the CuCl addition, and the particle sizes increase to 4 to 10 μm. Cl was not distributed uniformly in the perovskite, which may be poor solubility of CuCl in the solvent, leading to the low efficiencies.

Figures 10.22(c) and 10.22(d) are scanning electron microscope images of CuBr- and CuI-added devices, and the particle sizes of perovskite phase are below 3 μm, and all elements are distributed uniformly, leading to increase of the surface coverages of the perovskite. Flat surfaces of the perovskite particles grains for the copper halides-added perovskites are related to the cubic (100), and small roughness would result in the increase of J_{SC} and *FF*.

The measured compositions and estimated tolerance factors are listed in Table 10.10. The compositions of Pb:Cu and I:Cl:Br were calculated from the EDX spectra. The *t*-factor of the standard $MAPbI_{3-x}Cl_x$ is estimated to be 0.914, which is larger than 0.912 of the $MAPbI_3$. This suggests that the $MAPbI_{3-x}Cl_x$ could be more stable than that of $MAPbI_3$. Furthermore, the *t*-factors of CuI-, CuBr-, or CuCl-added perovskites are larger than that of

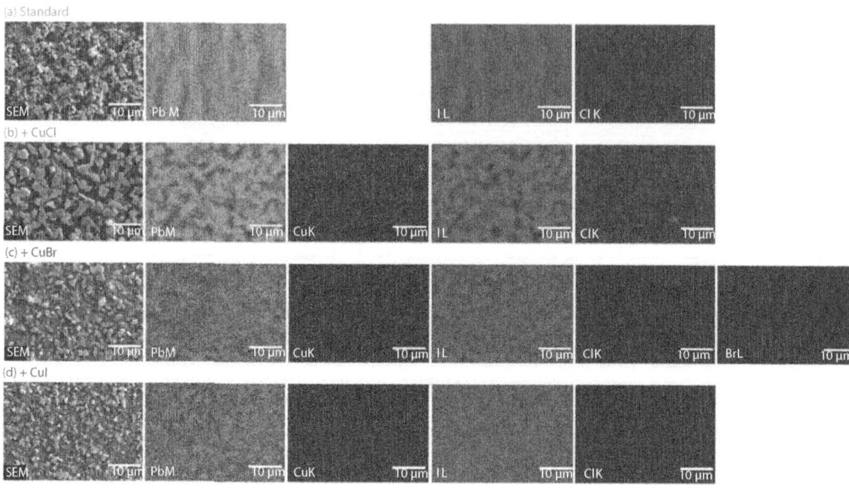

Figure 10.22 SEM images and elemental mapping images of (a) standard, (b) + CuCl, (c) + CuBr, and (d) + CuI devices.

Table 10.10 Estimated compositions and tolerance factors of the present perovskites.

Samples	Pb : Cu	I : Cl : Br	Estimated composition	*t*-factor
Standard	-	87.88 : 12.12 : -	$CH_3NH_3PbCl_{0.36}I_{2.64}$	0.914
+ CuI	99.06 : 0.94	86.64 : 11.36 : -	$CH_3NH_3Pb_{99.06}Cu_{0.94}I_{2.66}Cl_{0.34}$	0.915
+ CuBr	98.82 : 1.18	84.04 : 12.40 : 3.57	$CH_3NH_3Pb_{98.82}Cu_{1.18}I_{2.52}Cl_{0.37}Br_{0.11}$	0.916
+ CuCl	98.60 : 1.40	86.29 : 13.71 : -	$CH_3NH_3Pb_{98.60}Cu_{1.40}I_{2.59}Cl_{0.41}$	0.916

the standard $MAPbI_3$. The cubic $MAPbI_3$ perovskite might be stabilized by the Cu addition, which lead to the improvement of the characteristics of the devices. Studies of combination of Cu and alkali metals also showed the effectiveness for the efficiencies and the stability [42, 113].

10.8 K/FA Doping to Perovskites

Since the instability of $CH_3NH_3PbI_3$ would be due to the unstable $CH_3NH_3^+$ in the perovskite crystal, alternate cations have been studied, and the properties of solar cells containing alkali-metal doped perovskites have been reported [114–119]. When KI, KBr, or KCl is added to $CH_3NH_3PbI_3$ at 10 at.%, the t are calculated to be 0.901, 0.902, or 0.903, respectively [50]. Furthermore, $HC(NH_2)_2PbI_3$ perovskite compounds consisting of the formamidinium cation $HC(NH_2)_2^+$ (FA) have a more favorable band gap and charge transport properties when compared with $MAPbI_3$ [80, 120]. The ionic radius of FA is 253 pm [24], and doping of this large ionic radius causes the energy-gap of $CH_3NH_3PbI_3$-based perovskites to be reduced. Effects of adding K compounds on the photovoltaic properties of MA(FA) PbI_3(Cl) were described here [50]. The doping was performed using an air-blowing method. Following the addition of KI, KBr, or KCl (10 at.%) to $MA_{0.8}FA_{0.1}K_{0.1}PbI_3$(X), their respective t were calculated to be 0.907, 0.909, or 0.910. This indicated that the KBr or KCl-doped $MA_{0.8}FA_{0.1}K_{0.1}PbI_3$(X) perovskite crystals had more stable cubic structure than that of the KI-doped perovskite. Doping effects of other alkali metals, such as Cs or Rb, to the $MAPbI_3$ have been investigated, and the photovoltaic properties were improved by the addition of Cs or Rb [19, 20]. It is also expected that K with a smaller ionic size would easily occupy the MA sites. K is also a cheap and abundant alkali metal, and so is a cheaper dopant than Rb or Cs. The addition of Cl^- is expected to increase the carrier diffusion length within the perovskite crystals [7, 25].

The current density-voltage properties of the photovoltaic devices that contained the K-doped compounds ($MA_{0.8}FA_{0.1}K_{0.1}PbI_3$(X)) are shown in Figure 10.23(a). The photovoltaic parameters of these devices are summarized in Table 10.11. The device that contained the standard material ($MA_{0.9}FA_{0.1}PbI_3$(Cl)) had an η of 12.94%, whereas the devices that contained the KI, KBr or KCl-doped perovskites had η 9.61%, 8.96%, and 14.15%, respectively. The J_{SC} and V_{OC} of the devices that contained the KCl-doped perovskite were higher than those of the standard device, which resulted in the improved conversion efficiency.

The EQE spectra of the photovoltaic devices are shown in Figure 10.23(b) and the estimated energy gaps of the perovskite crystals, calculated by

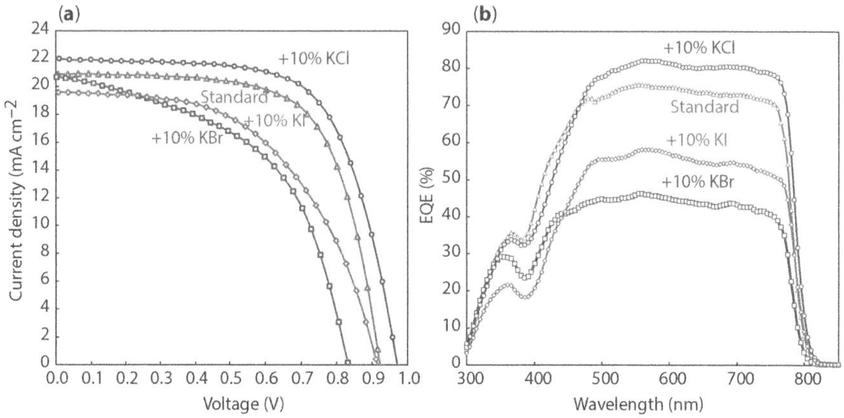

Figure 10.23 (a) Current density-voltage properties and (b) external quantum efficiencies of the perovskite solar cells that contained the standard and K-doped perovskites.

Table 10.11 Photovoltaic parameters of the perovskite solar cells.

Device	J_{SC} (mA cm^{-2})	V_{OC} (V)	*FF*	R_s (Ω cm^2)	R_{sh} (Ω cm^2)	η (%)	η_{ave} (%)
Standard	20.9	0.922	0.671	5.16	3562	12.94	12.31
+ 10% KI	19.6	0.915	0.536	9.07	1216	9.61	8.21
+ 10% KBr	20.7	0.833	0.519	7.94	238	8.96	7.54
+ 10% KCl	22.0	0.970	0.663	6.02	1794	14.15	13.26

linear fitting using band gap calculator software (Enlitech), are summarized in Table 10.12. The energy gaps of the KI, KBr, or KCl-doped perovskite crystals were 1.54, 1.55, or 1.53 eV, respectively. Importantly, the energy gap of the KCl-added perovskite crystals was narrower than that of the standard crystals, which led to an increase of the J_{SC} values. The EQE of the device that contained the KCl-doped perovskite was greater between 470 and 815 nm when compared with the standard device. This resulted in the increased J_{SC} that was observed. Conversely, the EQEs of the cells that contained the KBr or KI-doped perovskites were lower than the standard device over almost the entire the range examined.

X-ray diffraction profiles of the standard $MA_{0.9}FA_{0.1}PbI_3(Cl)$ cells and the cells added with KI, KBr, or KCl are shown in Figure 10.24(a). An enlarged X-ray diffraction profiles are also shown in Figure 10.24(b), which indicates a peak shift of 100 reflection to lower diffraction angles. The measured lattice constants (a) of the materials are summarized in Table 10.12. The devices with KI or KCl exhibited highly oriented (100) perovskite crystals. In addition, the (100) and (200) peak intensities increased significantly following the addition of KI or KCl. The device with KBr exhibited similar crystal orientation to the standard cell. The lattice constants of

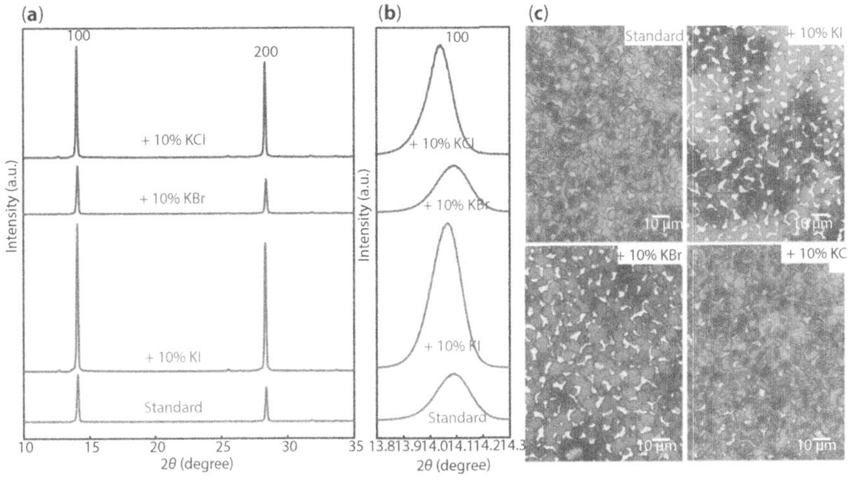

Figure 10.24 (a) X-ray diffraction profiles and (b) optical microscope images of the perovskite solar cells that contained the standard and K-doped perovskites.

Table 10.12 Energy gaps, lattice constants and surface coverage of the perovskite crystals.

Device	E_g **(eV)**	a **(Å)**	**Surface coverage (%)**
Standard	1.54	6.278	95
+ 10% KI	1.54	6.292	87
+ 10% KBr	1.55	6.278	90
+ 10% KCl	1.53	6.295	96

the KI and KCl-doped perovskites were higher than that of the standard material, whereas that of the KBr-doped perovskite did not change (Table 10.12). This will be discussed later in the schematic model of the lattice.

Optical microscopy images of the devices are shown in Figure 10.24(c). The surface coverage of the perovskite layers was measured to be 95, 87, 90, and 96% for the standard, KI, KBr or KCl-doped perovskites, respectively (Table 10.12). The gaps between the grains of the KCl-doped perovskite crystals were very small when compared with those of the other materials, which resulted in the increased J_{SC} that was observed. On the other hand, the gaps between the grains of the KI- or KBr-doped perovskite crystals increased, which resulted in the decrease of FF values and conversion efficiencies.

SEM images and EDS surface mapping images (I, Cl, K, and Br) of the perovskite crystal surfaces are shown in Figure 10.25. The perovskite grains of the standard material ranged from 3 to 10 μm in size. Both I and Cl were distributed throughout the grains of each perovskite material. The KI-doped perovskite grains ranged from 2 to 5 μm in size and I, Cl, and K were all distributed throughout the perovskite grains. The KBr-doped perovskite grains ranged from 2 to 7 μm in size and I, Cl, K, and Br were distributed throughout the perovskite grains. The surface morphology of the KCl-doped perovskite crystals was significantly different than the other

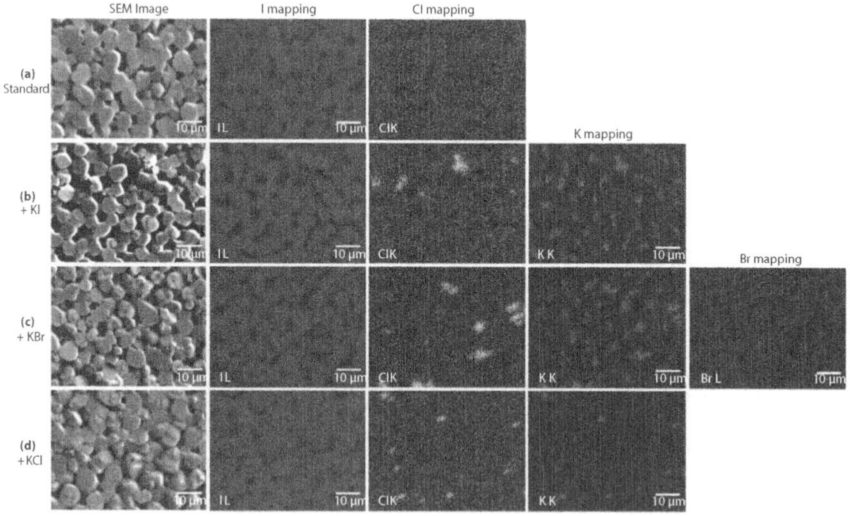

Figure 10.25 Scanning electron microscope images and energy dispersive X-ray spectrometry mapping of the (a) standard, (b) + 10% KI, (c) + 10% KBr, and (d) + 10% KCl perovskite crystals.

materials, with grains ranging from 3 to 8 μm in size that had covered the surface more effectively.

A schematic model showing the lattice structure of the K-doped perovskites is shown in Figure 10.26(a). The lattice constant of a perovskite single crystal [77] is larger than that of thin film crystals [24]. When the perovskite crystals are formed as a thin film, lattice defects would be generated at MA/FA cation sites of the perovskite crystals. Since the ionic radius of K is significantly smaller than those of MA and FA, the resulting lattice constant of the K-doped crystals should be smaller than those of the standard material. However, the lattice constant of the KCl-doped crystals was larger than that of the standard material. It is plausible that this was caused by K introduction. Defects in the MA/FA cation sites within the perovskite

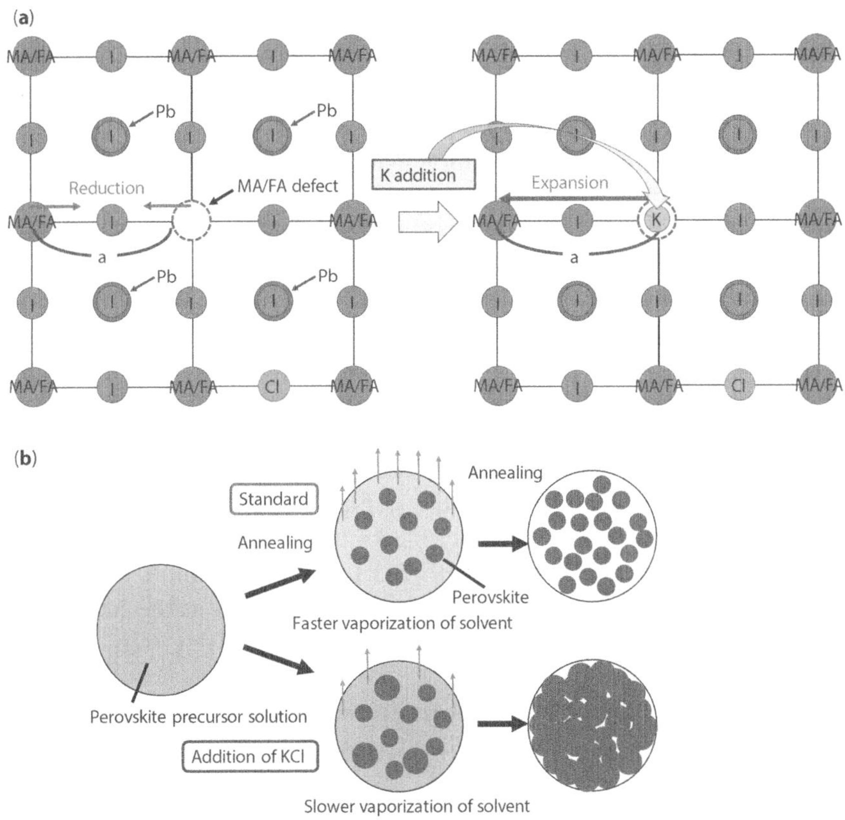

Figure 10.26 (a) Schematic model showing the inclusion of potassium into the perovskite lattice and the resulting lattice expansion. (b) Schematic diagram showing the grain growth of the perovskite crystals.

crystals were produced during heat treatment, which resulted in the lattice shrinking. However, by adding the K salts, the K^+ ions were introduced into these defect positions, which would explain why the lattice constant of the perovskite crystals was observed to increase. Therefore, as the defects within the KCl-added perovskite crystals had decreased, the resulting solar cells exhibited improved photoelectric conversion. The fact that the lattice constants of the KBr-doped perovskite crystals did not change when compared to the standard material indicated that the K^+ cations filled some of the defects. The Br^- ions, with a smaller ionic radius compared with I^-, also substituted the I^- ion sites. Further structure analysis using methods such as Reitveld refinement is required to confirm the occupancies and replacement of the K ions in the $CH_3NH_3PbI_3$ crystals [34].

The electronic structures of the perovskite crystals were calculated for the single-point with the experimental parameters obtained from XRD, through ab initio quantum calculations based on the density functional theory [44, 121]. The $MAPbI_3$ cubic structures were treated as a cluster model with supercells of 2×2×2. A K element was introduced at the MA^+ sites at contents of 12.5%. The electronic structure of $CH_3NH_3PbI_3$ mainly derives from the electronic state of the PbI_6 octahedron unit, and the level of the energy band is comprised of Pb 6s, 6p, and I 5p orbits. The effect of MA void was found not to be so influential on the density of state of the perovskite, and the energy gap between highest occupied molecular orbital and lowest unoccupied molecular orbital decreased a little. It has also been reported that the charge-carrier trapping process has much faster timescale than the carrier recombination [122, 123]. On the other hand, the characteristic density of states of K 4s orbital in the range of 6 to 7 eV was observed. Increased generation of charge carriers by light absorption at short wavelengths could contribute to the improved EQE. Calculated Gibbs energies (G) of the standard $MAPbI_3$, MA defect, and $MA(K)PbI_3$ models were 946, 751, 743 $kJ{\cdot}mol^{-1}$, respectively. The values of G slightly decreased from 751 to 743 $kJ{\cdot}mol^{-1}$, and more stable perovskite structures were formed by the K introduction at the MA defect site. Therefore, the addition of the K element effectively stabilized the perovskite crystals.

A schematic showing a growth mechanism of the perovskite crystals is shown in Figure 10.26(b). In all systems, DMF was used as a solvent for the perovskite precursor solution. The boiling point of DMF is 153°C, and in the standard system evaporation occurred once the sample reached 150°C. Then, the growth of the perovskite particles would be too fast to sufficiently fill the grain gaps. Following the introduction of additives into the perovskite precursor solutions the boiling point of the DMF slightly increased, as can be determined using $\Delta T = K_b \times m\left(K_b = \dfrac{M \times R \times T_b}{\Delta H_b}\right)$, where ΔT is the increase in the boiling point, K_b is the boiling point rise constant (K·kg^{-1}·mol^{-1}), m the solute molar concentration (mol·kg^{-1}), M is the solvent molar mass (kg·mol^{-1}), R the gas constant (8.314 J K^{-1}), T_b is the boiling point of the pure solvent (K), and ΔH_b is the standard enthalpy of evaporation (J·mol^{-1}). According to the equation, ΔT is proportional to m, irrespective of the solute. Then, a drop of the vapor pressure of the solvent occurs and the boiling point of the solution rises. Thus, the addition of the K salts into the perovskite precursor solution increased the boiling point of the solutions, which resulted in slower vaporization of solvent and slower growth of the perovskite particles.

The SEM observations of the perovskite surfaces showed that the perovskite crystals did not grow as well in the KBr or KI-doped materials, but did in the KCl-doped material. This was likely caused by differences in the diffusion-dependent reaction rates because of the different halogen anions. The diffusion coefficient can be described using the following equation: $D = \dfrac{k_B \times T}{6\pi \times v \times a}$, where D is the diffusion coefficient, k_B is the Boltzmann constant (J K^{-1}), T is the temperature, v is the viscosity, and a is the ionic radius. In this work, only the ionic radius changed, and so the halogen would determine the value of the diffusion coefficient. In addition, the ionic radius of the A site (i.e. ABX_3 perovskite structure) was smaller than that of the standard system because of the small ionic radius of K, which would also act to decrease the diffusion coefficient. A diffusion-determined reaction can be represented by A + B → AB.

Because the diffusion rate increased, the reaction rate increased. The rate constant (k_d) can be calculated using the following equation: $k_d = 4\pi N_A r_{AB} (D_A + D_B)$, where N_A is the Avogadro's constant, r_{AB} is the distance between AB, and D_A and D_B are the diffusion coefficients of A and B, respectively. The rates of formation of the K-doped perovskite crystals increased in the following order: KCl > KBr > KI. As such, the perovskite crystals were more likely to be formed in this order, and the KCl-doped perovskite crystals would grow more efficiently because of the rise in boiling point and the faster diffusion rate.

As a summary, the effects of doping perovskite precursors with KI, KBr or KCl on their microstructures and the photovoltaic properties of devices that contained these materials were investigated. The doping was performed using an air-blowing method. The current density-voltage properties were observed to improve following the introduction of KCl into the perovskite, and this device exhibited the highest conversion efficiency. The microstructures of the perovskite crystals, as analyzed using X-ray diffraction, optical microscopy, SEM, and EDS, were highly orientated (100) perovskite crystals with dense grains when particularly doped with KCl using the air-blowing method. These dense and orientated crystals were caused by accelerated growth following the introduction of the KCl. The addition of the KCl into the perovskite crystals induced a surface morphology with small domain sizes, and decreased gaps between the grains of the perovskite crystals. This improved the surface coverage, and suppressed the leakage current, which contributed to the increased J_{SC} that was observed. The addition of K also reduced the MA/FA cation defects in the perovskite crystals, which resulted in an increase in the lattice constant. This work showed that doping perovskites with KCl using the air-blowing method was an effective method for the fabrication of photovoltaic devices with improved properties.

10.9 Morphology Control by Polysilane

In addition to the elementary doping the perovskite crystal, morphology control of perovskite grains is also important to improve the photovoltaic properties. As the one of the examples, effects of various polysilanes addition to the perovskite precursor solutions on the photovoltaic properties are shown here [54]. Polysilane is a *p*-type semiconductor, and has been used as an electrical conductive material and photovoltaic systems [124–131]. Polysilanes are known as σ-conjugate polymers, and their hole mobility is 10^{-4} cm^2 V^{-1} s^{-1} [7]. Although the polysilanes could be applied to *p*-type semiconductors on organic thin-film solar cells, few studies on polysilane-based solar cells have been reported [62, 125, 131, 132].

Three kinds of polysilanes, poly(methyl phenylsilane) (PMPS, Osaka Gas Chemicals, OGSOL SI-10-10, 3 mg), poly(phenylsilane) (PPS, Osaka Gas Chemicals, OGSOL SI-20-10, 3 mg), and decaphenylcyclopentasilane (DPPS, Osaka Gas Chemicals, OGSOL SI-30-10, 3 mg), were used as additives to the perovskite solution, and structures of these polysilanes are shown in Figure 10.27(a) [54]. In the present work, the perovskite layers are described as $CH_3NH_3PbI_3$, +PMPS, +PPS, and +DPPS. The $CH_3NH_3PbI_3$ solution was spin-coated on the mesoporous TiO_2 layer at 2000 rpm for 60 seconds. The samples were annealed onto the hot plate at 100°C for

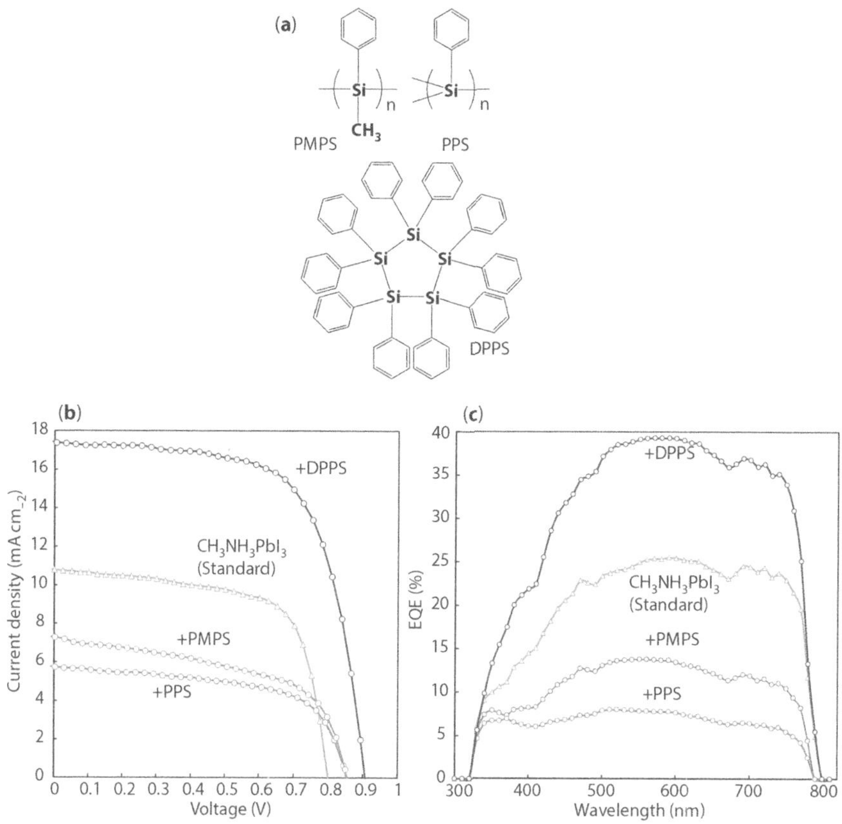

Figure 10.27 (a) Structure of polysilanes. (b) Current density-voltage properties and (c) EQE spectra of $CH_3NH_3PbI_3$ solar cells with polysilanes.

15 minutes to form a $CH_3NH_3PbI_3$ layer. All procedures for preparation of the thin films were carried out in ordinary air.

Figure 10.27(b) shows the current density-voltage properties of the $TiO_2/CH_3NH_3PbI_3$(polysilane)/spiro-OMeTAD photovoltaic cells under illumination, which indicate effects of the polysilane addition. The measured photovoltaic parameters of the present $CH_3NH_3PbI_3$(polysilane) cells are summarized in Table 10.13. Ordinary $CH_3NH_3PbI_3$ cells provided a power conversion efficiency of 5.82%, as listed in Table 10.13. By adding the present three polysilanes, V_{OC} were increased, as observed in Figure 10.27(b). When the PMPS or PPS was added to the perovskite, J_{SC} was decreased. On the other hand, by adding the DPPS, the J_{SC} value and shunt resistance (R_{sh}) were fairly increased, and the series resistance (R_s)

Table 10.13 Measured photovoltaic parameters and lattice constants of $CH_3NH_3PbI_3$ solar cells with polysilanes.

Perovskite layer	J_{SC} (mA cm^{-2})	V_{OC} (V)	*FF*	R_s (Ω cm^2)	R_{sh} (Ω cm^2)	η (%)
$CH_3NH_3PbI_3$	10.8	0.800	0.674	5.47	700	5.82
+PMPS	7.28	0.855	0.547	9.78	479	3.40
+PPS	5.75	0.852	0.609	11.4	905	2.99
+DPPS	17.4	0.906	0.662	6.13	2190	10.46

was decreased compared with the PMPS and PPS, which resulted in the improvement of the conversion efficiency to 10.46%. The fill factor (*FF*) is almost same as that of the ordinary cell.

Figure 10.27 (c) shows EQE spectra of the $CH_3NH_3PbI_3$(polysilane) devices. The present cells show photoconversion efficiencies between 320 and 800 nm, which corresponds to an energy gap of 1.55 eV for the $CH_3NH_3PbI_3$. The EQE was improved in the range of 400 to 750 nm by adding DPPS to the perovskite layer, leading to the increase of J_{SC} values, as shown in Figure 10.26(b).

Optical microscopic images of ordinary $CH_3NH_3PbI_3$ cell is shown in Figure 10.28(a), and microparticles with sizes of 5 to 20 μm are observed. When the polysilanes were added to the perovskite phase, microparticles with homogeneous size distribution were formed, as shown in Figures 10.28(b)–10.28(d). Especially, by the DPPS addition, the crystal growth of the perovskite phase was observed, and the distances between the particles were decreased, leading formation of more dense microstructures, as observed in Figure 10.28(d).

Figure 10.28(e) shows X-ray diffraction profiles of the present polysilane-added $CH_3NH_3PbI_3$ solar cells. When the $CH_3NH_3PbI_3$ has a tetragonal structure, an XRD peak corresponding to cubic 200 should be divided into tetragonal 004 and 220, as previously reported [24]. The calculated XRD pattern also indicates splitting of diffraction reflections of 200 to 004/220 (Ref. 24), and therefore, the diffraction reflections can be indexed by a cubic crystal system (*Pm*3*m*) for the present perovskite thin films. The diffraction peaks indicated by FTO and TiO_2 are from the FTO substrate and TiO_2 mesoporous layer, respectively. For the standard $CH_3NH_3PbI_3$ cell, a diffraction reflection of PbI_2 is observed, as indicated by an arrow. When the polysilanes were added to the perovskite phase, formation of

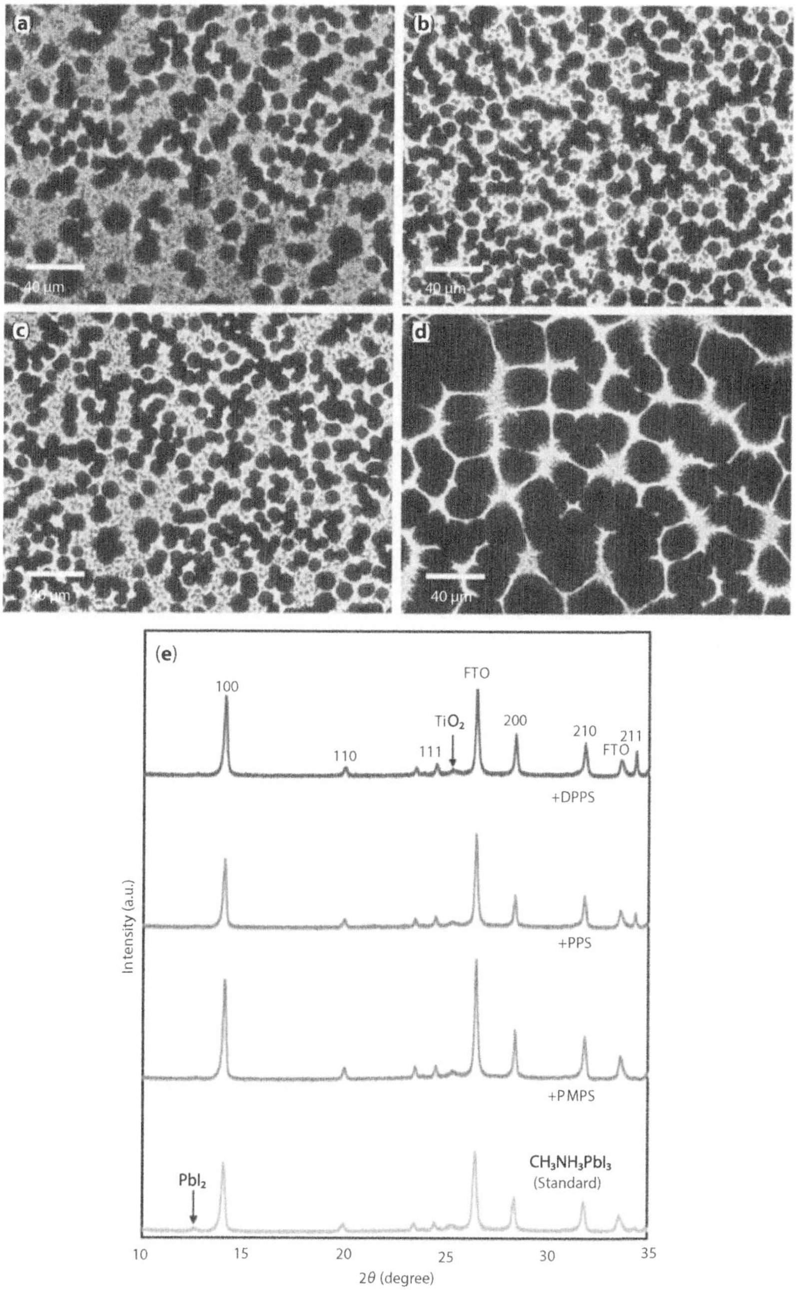

Figure 10.28 Optical microscopy images of (a) $CH_3NH_3PbI_3$ solar cells with (b) PMPS, (c) PPS and (d) DPPS. (e) X-ray diffraction profiles of $CH_3NH_3PbI_3$ solar cells with polysilanes.

the PbI_2 was suppressed. In addition, peak intensities of 100 of the perovskite compound increased with adding the polysilanes, which indicated crystal growth of the perovskite phase. It is considered that the crystal growth and PbI_2 suppression might be related with affinity between the γ-butyrolactone, *N,N*-dimethylformamide, CH_3NH_3I, PbI_2, and polysilanes. X-ray diffraction angles of the perovskite compounds were calibrated based on the 111 reflection of Au of the cell electrodes.

Two assumed mechanisms could be considered for the improvement in the photoconversion efficiencies. The first mechanism is that the increase and homogenization of the grain sizes of the perovskite crystals by adding polysilanes as observed optical microscopy images in Figure 10.20. This microstructure promotes to the formation of dense interfacial structures and to increase of surface coverage. When the grain sizes increase, grain boundary areas that cause carrier scattering decrease, which leads to the increase of the current density [133]. The present homogeneous structure could also suppress the formation of PbI_2, as observed in Figure 10.28(e). For the cells with PMPS and PPS, the perovskite domains are isolated, and the cell performances are severely affected by the coverage. Then, the pin-holes provide short circuits between *n*-type semiconductor TiO_2 and *p*-type semiconductor spiro-OMeTAD, which resulted in the lower photovoltaic performances.

The second mechanism is the electronic structure change by the polysilanes. By adding polysilanes, the open circuit voltages increased, as shown in Figure 10.27(b). From EQE spectra in Figure 10.27(c), the bandgap energy of the perovskite crystals with PMPS and PPS increased from 1.55 eV to 1.57 eV, which leads to increase of the V_{OC} values. This also corresponds to the decrease of lattice constants of cubic perovskite by addition of polysilanes. By adding DPPS, on the other hand, the bandgap energy seems to be almost same as that of the standard perovskite crystal. Since the DPPS has more phenyl groups at the side of Si atoms compared with PMPS and PPS, carrier transport after the carrier separation might be facilitated by π-electrons on the phenyl groups, which affected the increase of EQE in the range of 400 to 750 nm and the J_{SC} values. The existence of the phenyl groups might also reduce the electrical resistance of the DPPS-added $CH_3NH_3PbI_3$, which would result in the increase of the V_{OC} values.

Polymer-templated nucleation and growth had been reported as a method for crystal engineering of perovskites [18]. Poly(methyl methacrylate) (PMMA) provided heterogeneous nucleation, which would be orders of magnitude faster than the homogeneous nucleation due to the lowering of the nucleation free energy barrier [18]. The PMMA also seems to slow down the growth rate of perovskite crystals by forming an intermediate adduct with PbI_2, which would provide the randomly formed nuclei, and would also lead to smooth thin films with few defects and large oriented perovskite grains [51, 52, 97, 134]. The main improvement by addition of the DPPS is an increase of the film coverage due to enlargement of the perovskite domains as observed in the optical micrograph of Figure 10.20(d). Since the DPPS has more phenyl groups at the side of Si atoms compared with PMPS and PPS, molecular interactions created by the phenyl groups would affect the nucleation and crystal growth. Introduction of quantum dots such as PbS [135] or $CsSnBr_{3-x}I_x$ [136] could also control the energy band structure. Further studies are needed for the growth of the perovskite phase and the control of the energy bands.

As a summary, $CH_3NH_3PbI_3$ perovskite solar cells added with polysilanes were fabricated and characterized. Effects of PMPS, PPS or DPPS addition to perovskite precursor solutions on the photovoltaic properties and the microstructures were investigated. The V_{OC} values were increased by applying these PMPS, PPS, or DPPS, and the EQE was also increased by the DPPS addition in the range of 400 to 750 nm, leading to the improvement of the J_{SC} values. Formation of PbI_2 was also suppressed by the polysilane addition. Microstructure observation of the perovskite layer with DPPS indicated increase of surface coverage to form a dense interfacial structure by grain growth, which resulted in the improvement of conversion efficiencies above 10%. Lots of phenyl group in the DPPS might contribute to the carrier transport in the perovskite layers.

10.10 High-Temperature Annealed Perovskites

The stability of perovskite solar cells requires improving before they can be put into practical use [137, 138]. Incorporating polymeric materials has been studied to improve the stability of perovskite solar cells [139–141], and the polymeric materials showed improvement of stability and promotion of crystal growth when introduced into the perovskite devices [140]. For example, by depositing a poly(methyl methacrylate) (PMMA) layer on the perovskite layer [18, 142], the PMMA formed a compact layer via a cross-linked network and protected the device from oxygen and moisture. Poly(propylene carbonate) was also similarly used and the resulting solar cell showed stability in various environments. In this case, large crystals were formed by crosslinking the perovskite particles and polymer which suppressed defects [143].

Polysilane derivatives are also expected as other candidates for the improvement of the perovskite solar cells. Unlike ordinary polymer materials, polysilane derivatives have two advantages. The first is that the polysilane derivatives are a *p*-type semiconductor, and this facilitates hole transfer and rectification at the *pn* junction. The second is that the polysilane derivatives have high stability and are expected to act as a protective layer on the photoactive layer. As annealing temperatures for the preparation of the perovskite layer are commonly ~100°C, and the low temperature annealing might provide the instability. Since high temperature annealing could be effective for the formation of more stable perovskites a hot air blowing and higher temperature annealing in ambient air were applied [53]. Decaphenylcyclopentasilane (DPPS) was used here [53], and the perovskite precursor solutions were normally spin-coated during the first coating. During the second and third spin-coatings, an air blow method was applied [102]. Then, a standard device was annealed at 140 °C for 10 minutes to form the perovskite compound. When the starting material is $CH_3NH_3I + PbI_2$, the suitable annealing temperature was 100 °C, which was confirmed in the previous work [88]. Whereas the suitable temperature for the starting materials of $3CH_3NH_3I + PbCl_2$ was ~140°C [89], and $CH_3NH_3PbI_3$ with $2CH_3NH_3Cl$ (gas) were formed during the reaction. DPPS (Osaka Gas Chemicals, OGSOL SI-30-10, 10 mg) solutions were prepared using chlorobenzene (Fujifilm Wako Pure Chemical Corporation, 0.5 mL). The DPPS solution was dropped onto the perovskite layer during the last 15 seconds of spin-coating the perovskite precursor solutions. The devices with DPPS layers were annealed at temperatures in the range of 170 to 220 °C for 3 to 10 minutes.

Current density-voltage properties of FTO/TiO_2/$MAPbI_3$/(DPPS)/Spiro-OMeTAD/Au photovoltaic devices prepared at various temperatures are shown in Figure 10.29(a). The measured photovoltaic parameters of the $MAPbI_3$/(DPPS) devices are summarized in Table 10.16. The standard device without DPPS showed an η of 5.16% after annealing at 140 °C [53], while the device with DPPS annealed at 140 °C showed a η of 7.51%. On the other hand, the device with DPPS annealed at 220 °C showed a conversion efficiency of 11.99%. All photovoltaic parameters were improved for the devices annealed at high temperatures, and the *FF* values were especially increased compared with that of the standard device. Figure 10.29(b) shows EQE spectra of the as-prepared $MAPbI_3$/(DPPS) photovoltaic devices with a structure model of DPPS, and the devices provided photoconversion efficiencies at wavelengths between 300 and 810 nm.

Current density-voltage properties of the devices stored in the dark at 22°C and ~30% humidity for 4 weeks are shown in Figure 10.29(c) and summarized in Table 10.14. The device with DPPS annealed at 190 °C showed the highest conversion efficiency of 13.36% after 4 weeks. The EQE

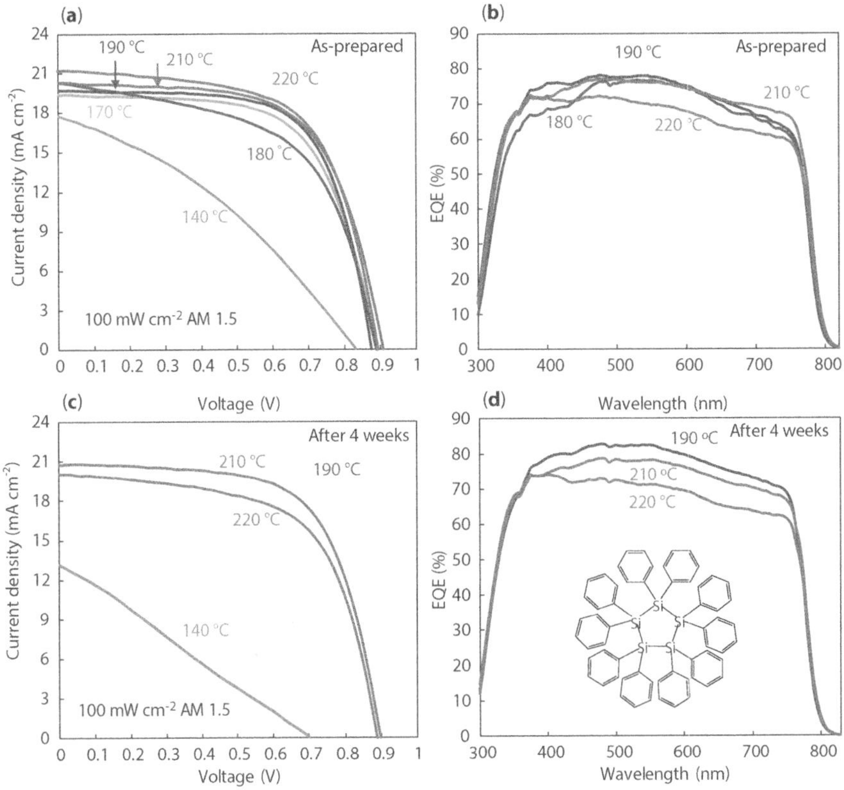

Figure 10.29 (a), (c) Current density-voltage properties and (b), (d) EQE spectra of as-prepared $MAPbI_3$/(DPPS) cells and cells after 4 weeks, respectively.

spectra of $MAPbI_3$/(DPPS) photovoltaic devices after 4 weeks are shown in Figure 10.29(d). The devices annealed at high temperatures showed high and stable EQE values for wavelengths between 370 and 770 nm.

Optical microscopy images of $MAPbI_3$/(DPPS) devices are shown in Figure 10.30(a). For the devices annealed at high temperatures with inserted DPPS, the sizes of the perovskite crystals were reduced and the perovskite crystals were densely packed. Therefore, the grain boundary area decreased and the surface coverage of the perovskite increased. Such microstructures resulted in charge traps and charge recombination in the grain boundaries being reduced and suppressed. Thus, charge loss was decreased as indicated by the decrease in R_S and increase in J_{SC}.

SEM images and EDS mapping images (I, Pb and Cl) of the $MAPbI_3$/(DPPS) devices are shown in Figure 10.30(b). Compared with the standard device, the $MAPbI_3$/(DPPS) devices annealed at high temperatures showed

Table 10.14 Measured photovoltaic parameters of $MAPbI_3$/(DPPS) perovskite solar cells. The above and below show parameters of as-prepared cells and cells after 4 weeks, respectively.

As-prepared							
Annealing (°C)	J_{SC} **(mA cm^{-2})**	V_{OC} **(V)**	***FF***	R_s **(Ω cm^2)**	R_{sh} **(Ω cm^2)**	η **(%)**	η_{ave} **(%)**
140*	17.77	0.833	0.349	28.0	105	5.16	2.07
140	12.61	0.903	0.660	8.13	916	7.51	6.06
170	19.38	0.889	0.640	6.64	1505	11.03	10.16
180	20.32	0.887	0.564	7.23	250	10.17	9.02
190	19.74	0.874	0.670	5.77	2758	11.57	11.29
210	20.32	0.908	0.637	6.56	969	11.75	11.33
220	21.23	0.894	0.632	5.83	661	11.99	11.18
After 4 weeks							
140*	13.14	0.703	0.251	29.5	67	2.32	2.05
190	22.06	0.881	0.687	5.11	2419	13.36	10.30
210	20.78	0.899	0.655	5.92	1571	12.23	11.61
220	20.04	0.888	0.621	6.15	499	11.05	10.62

* $MAPbI_3$ perovskite cells without DPPS.

round perovskite crystals with smaller particle sizes. Uniform crystals were formed in the devices annealed at high temperatures, and the conversion efficiencies were improved because of the increase in surface coverage. Pb and I atoms were distributed between the large perovskite crystals. It was thought that larger perovskite crystals precipitated on the perovskite crystals penetrating in the mesoporous TiO_2 layer. Therefore, the TiO_2 layers were not exposed to the spiro-OMeTAD. Measured compositions of the present perovskite solar cells are listed in Table 10.15. Although desorption of iodine from the ordinary perovskite crystal have been observed [144], the compositions of iodine are almost constant even after annealing at 220°C, which indicates excellent stabilization effect of the DPPS layer.

X-ray diffraction profiles of the as-prepared and after 4 weeks $MAPbI_3$/(DPPS) devices are shown in Figures 10.31(a) and 10.31(b), respectively. The measured lattice constants, crystallite sizes and peak intensity ratios of perovskite 100 to PbI_2 001 of the devices are summarized in Table 10.16.

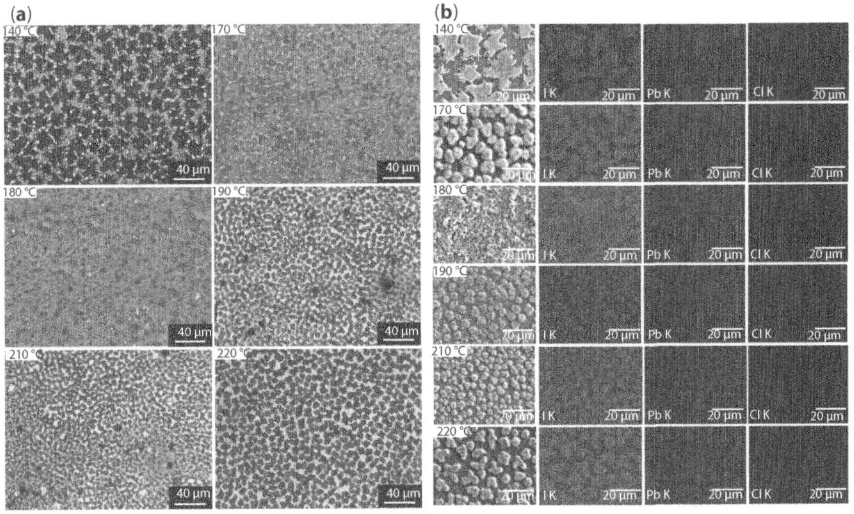

Figure 10.30 (a) Optical microscopy images and (b) SEM images with EDS mappings of as-prepared $MAPbI_3$/(DPPS) solar cells.

Table 10.15 Measured compositions of the present perovskite solar cells.

Annealing (°C)	Pb (at.%)	I (at.%)	Cl (at.%)	C : N (at.%)
140*	25.2	68.7	6.1	61.8 : 38.2
170	23.2	69.8	7.0	51.6 : 48.4
180	23.7	71.4	4.9	44.8 : 55.2
190	24.2	70.3	5.5	48.2 : 51.8
210	26.0	68.2	5.8	45.1 : 54.9
220	25.0	68.9	6.1	48.1 : 51.9

* $MAPbI_3$ perovskite cells without DPPS.

The lattice constants and crystallite sizes did not change much over the 4 weeks. The as-prepared devices provided a small PbI_2 peak, which could promote hole transfer and improve the photovoltaic properties. EDS mapping results suggested that a small amount of PbI_2 crystals could be accumulated between the perovskite crystals, which may have prevented charge recombination [144, 145]. Even after 4 weeks, the peak intensity of PbI_2 did

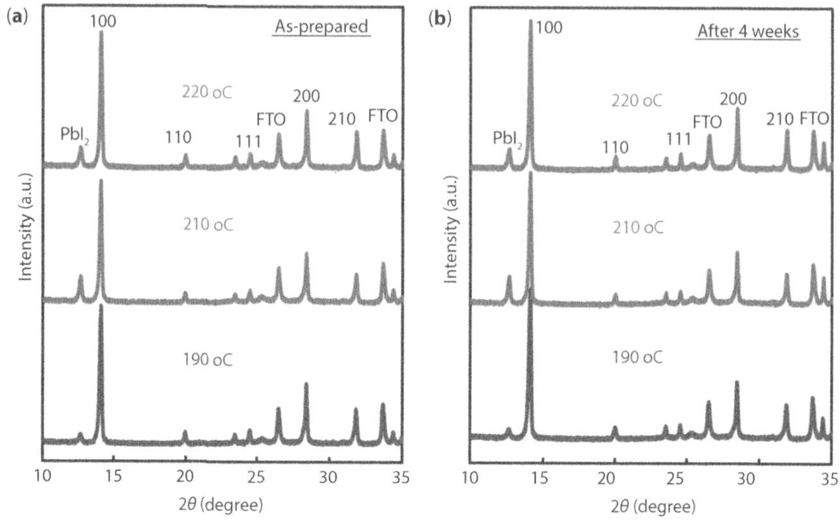

Figure 10.31 X-ray diffraction profiles of (a) as-prepared $MAPbI_3$/(DPPS) cells and (b) cells after 4 weeks.

Table 10.16 Microstructural parameters of $MAPbI_3$ crystals. The above and below show parameters of as-prepared cells and cells after 4 weeks, respectively.

	Annealing (°C)	Lattice constant (Å)	Crystallite size (Å)	I_{100}/I_{PbI2}
As-prepared	190	6.289	457	16.7
	210	6.287	462	4.90
	220	6.286	477	7.15
After 4 weeks	190	6.276	474	16.9
	210	6.274	495	4.88
	220	6.272	522	7.49

not change, which indicated that the DPPS acted as a protective layer of the perovskite layer and suppressed deterioration of the perovskite to PbI_2.

Stability measurements of the photovoltaic parameters up to 8 weeks after preparation are shown in Figure 10.32. While all parameters decreased for the standard device, the parameters for the $MAPbI_3$/(DPPS) devices

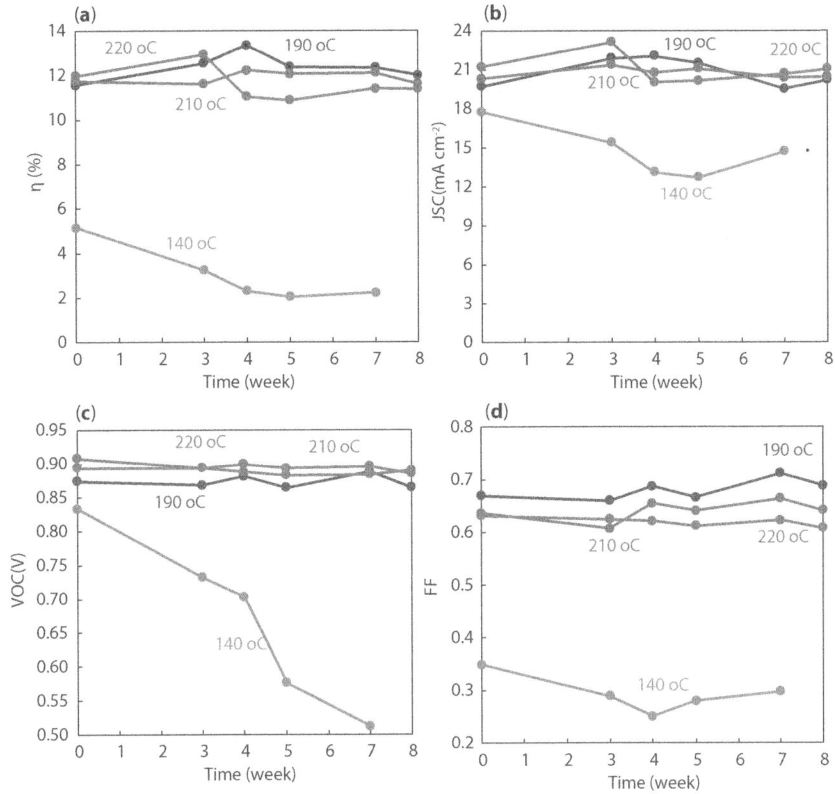

Figure 10.32 Stability measurements of (a) η, (b) J_{SC}, (c) V_{OC}, and (d) *FF* of $MAPbI_3$/(DPPS) perovskite solar cells.

annealed at high temperatures were fairly stable. This was because DPPS reduced the influence of spiro-OMeTAD, oxygen and moisture in air and thus suppressed decomposition of the perovskite compound.

A schematic model showing the microstructures, carrier dynamics, and stability of the proposed devices is shown in Figure 10.33(a). Because there was no DPPS layer for the standard device, the perovskite compound decomposed to PbI_2 after 4 weeks by the desorption of CH_3NH_3 and influence of oxygen and H_2O in the air. The number of holes and electrons generated in the perovskite layer was therefore reduced and the photovoltaic properties were degraded. However, the stability of the $MAPbI_3$/(DPPS) devices improved because DPPS functioned as a protective layer and suppressed decomposition of the perovskite compound. From the X-ray diffraction results in Figure 10.30, the peak intensity of PbI_2 for the 190°C annealed device was weaker than those of the

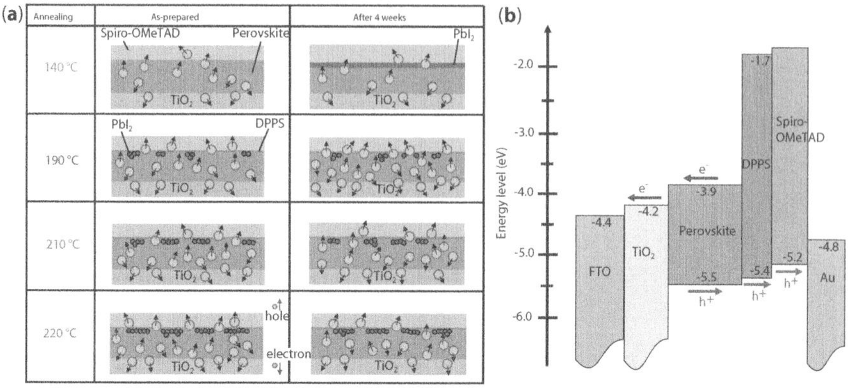

Figure 10.33 (a) Schematic illustration of microstructures, carrier dynamics and stability for perovskite solar cells under different annealing temperatures. (b) Energy level diagram of the cell.

other devices, and the peak intensity of 100 perovskite increased after 4 weeks. The number of electrons and holes generated in the perovskite layer therefore increased and the photovoltaic properties were slightly improved after 4 weeks.

In the present work, chlorobenzene was used as the solvent for the DPPS solution. The chlorobenzene dripping is often used to improve not only crystallinity of perovskite films but also photovoltaic performance of the perovskite solar cells [146, 147]. Although the annealing temperature of 100°C is enough for the chlorobenzene-dripped devices, the higher annealing temperature of ~200°C is necessary for the present DPPS-dripped device. This indicates that the improvement mechanism of the photovoltaic properties would be different from that of the chlorobenzene dripping. The DPPS layer could work as the effective protective layer improving the stability. In addition, the DPPS with the hole transport property would suppress the charge recombination and improve the photovoltaic properties.

To explain the charge transport, an energy level diagram of the FTO/TiO_2/$MAPbI_3$/(DPPS)/Spiro-OMeTAD/Au photovoltaic devices is shown in Figure 10.33(b). In Figure 10.33(b), previously reported values are used as the energy levels [128, 148]. By irradiating light from the FTO substrate side, electrons and holes were generated in the perovskite layer. Electrons generated in the photoactive layer were transported through TiO_2 to the FTO, and holes were transported through DPPS and spiro-OMeTAD to the Au. By incorporating DPPS between the perovskite and HTL, effective hole transport from the valence band of $MAPbI_3$ to the Au electrode was induced. This smooth transport was also due to the molecular structure of DPPS.

Changes to the current density-voltage properties of the $MA_{0.64}FA_{0.31}K_{0.05}PbI_3$ device, prepared at 180 °C, are shown in Figure 10.34(a) [149]. For the as-prepared device, the highest conversion efficiency was observed at 1.67%, after 42 days, V_{OC} and J_{SC} were improved. Furthermore, FF also improved after 131 days. After 163 days, the photovoltaic properties of the $MA_{0.64}FA_{0.31}K_{0.05}PbI_3$ device were further enhanced, and the highest conversion efficiency of 13.82%, a V_{OC} of 0.970 V, a J_{SC} of 20.3 mA cm^{-2}, and a FF of 0.700 were obtained [149].

Figure 10.34(b) shows conversion efficiency changes of the perovskite photovoltaic devices. For the as-prepared devices, those annealed at 200 °C showed the highest conversion efficiencies and were almost identical to the conversion efficiencies after 163 days. Conversely, the conversion efficiency of the $MA_{0.48}FA_{0.47}K_{0.05}PbI_3$ device, prepared at 180 °C, improved from 0.613% to 8.18% after 109 days, as shown in Figure 10.34(b). For the devices annealed at 220 °C, the conversion efficiencies decreased after 50 days.

Microstructural changes of the $MA_{0.64}FA_{0.31}K_{0.05}PbI_3$ device annealed at 180 °C were investigated by X-ray diffraction. Small PbI_2 peaks were observed in both X-ray diffraction profiles, and further PbI_2 formation was suppressed, even after six months. Full-width at half maximum for the $MA_{0.64}FA_{0.31}K_{0.05}PbI_3$ device was reduced after six months, which indicated

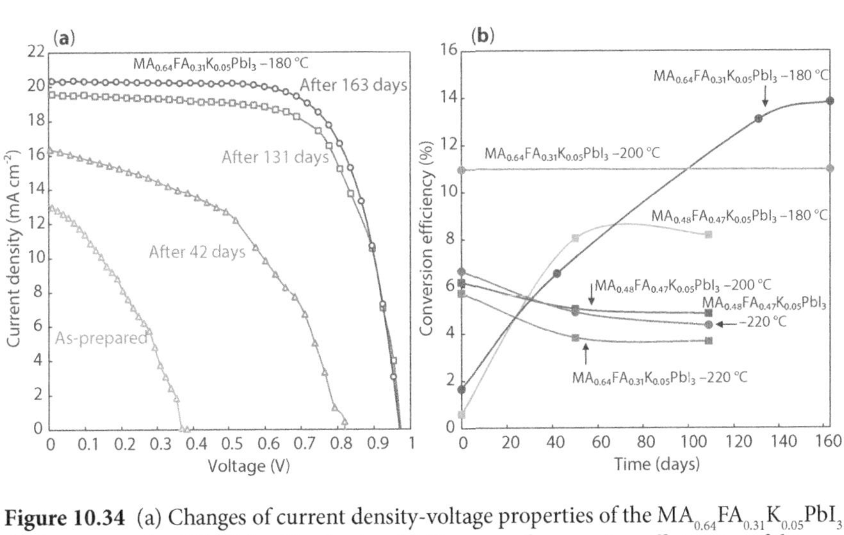

Figure 10.34 (a) Changes of current density-voltage properties of the $MA_{0.64}FA_{0.31}K_{0.05}PbI_3$ device prepared at 180 °C for 10 minutes. (b) Changes of conversion efficiencies of the present perovskite photovoltaic devices.

the crystallite size of the perovskite increased. The X-ray diffraction observations indicated the crystal growth of the perovskite compound, which led to a decrease in the grain boundary area and point defects and, hence, improvement in the photovoltaic performance. The lattice constant of the perovskite compound was observed to decrease slightly, which relates to the desorption of a small amount of MA.

The conversion efficiency of the present solar cell prepared at 180 °C increased from 1.67% to 13.82%, with an increase of efficiency over 12%. To describe this "abnormal" increase, a mechanism for the rise in conversion efficiencies of the $MA_{0.64}FA_{0.31}K_{0.05}PbI_3$ device, prepared at 180°C, can be considered, which is crystallization of the amorphous phase into the perovskite crystal at room temperature. During the spin-coating of DPPS, a mixed composite layer comprising amorphous perovskite and DPPS, having a dense interface, was formed during room temperature aging. In addition, K might compensate MA defects during the aging [150]. Since DPPS also functions as a hole transport layer [62, 151], holes are able to smoothly transport at the perovskite/DPPS/spiro-OMeTAD interface.

As a summary, the effects of inserting a DPPS layer between the perovskite layer and HTL on the photovoltaic properties were investigated. The current density-voltage properties indicated improvements to the devices upon introducing a DPPS layer on the perovskite layer after annealing at high temperatures. The device annealed at 220°C showed the highest conversion efficiency of 11.99% among the as-prepared devices. After 4 weeks, the photovoltaic properties were improved further, and the device annealed at 190°C showed the highest conversion efficiency of 13.36%. The surface morphology of the devices was evaluated using optical microscopy, SEM, and EDS, and the sizes of perovskite crystals with DPPS annealed at high temperatures were reduced. The grain boundary area decreased, and the surface coverage increased, which reduced the leakage current and improved the J_{SC} values. Crystal structures of the perovskite were analyzed using X-ray diffraction, and PbI_2 formation was suppressed for the DPPS-inserted devices annealed at high temperatures even after 4 weeks. This suppression of PbI_2 formation was due to the protective effect of the DPPS layer, and the stability of the performance was maintained for up to 8 weeks. Furthermore, the photovoltaic properties of the $MA_{0.64}FA_{0.31}K_{0.05}PbI_3$ device annealed at 180°C increased from 1.67% to 13.82% after 163 days. These findings indicate that DPPS insertion and annealing at high temperatures are effective for improving the performance and stability of the perovskite solar cells.

10.11 Conclusion

Fabrication, photovoltaic properties, and structures of $CH_3NH_3PbI_3$-based perovskite photovoltaic devices added with Sb, As, Cu, $PbCl_2$, NH_4Cl, K, FA, and polysilanes were described and reviewed in the present article. It is mandatory for the development of the commercial perovskite solar cells to characterize the morphology in nanoscopic scale, and the photovoltaic behavior and stabilities could be described based on the microstructures of the perovskite compounds and device configurations. Understanding of these basic structures and fundamental processes would lead to development of new perovskite compounds and device architectures, and further researches will be continued and carried out for realization of the practical perovskite solar cells.

Acknowledgments

The author is grateful to following colleagues for valuable discussion and assistance: Atsushi Suzuki, Masahito Zushi, Yuya Ohishi, Kohei Suzuki, Hiroki Tanaka, Masaya Taguchi, Hayato Machiba, Junya Nomura, Taisuke Matsumoto, Satsuki Kandori, Naoki Ueoka, Taku Kishimoto, Sakiko Fukunishi, Satoshi Minami, Masanobu Okita, Kazufumi Kohno, Shinichiro Tsukada and Tomoharu Tachikawa. A part of this work was supported by Satellite Cluster Program of the Japan Science and Technology.

References

1. Kojima, A., Teshima, K., Shirai, Y., Miyasaka, T., Organometal halide perovskites as visible-light sensitizers for photovoltaic cells. *J. Am. Chem. Soc*, 131, 6050, 2009.
2. Im, J.H., Lee, C.R., Lee, J.W., Park, S.W., Park, N.G., 6.5% efficient perovskite quantum-dot-sensitized solar cell. *Nanoscale*, 3, 4088, 2011.
3. Kim, H.S., Lee, C.R., Im, J.H., Lee, K.B., Moehl, T., Marchioro, A., Moon, S.J., Humphry-Baker, R., Yum, J.H., Moser, J.E., Grätzel, M., Park, N.G., Lead iodide perovskite sensitized all-solid-state submicron thin film mesoscopic solar cell with efficiency exceeding 9%. *Sci. Rep.*, 2, 591, 2012.
4. Grinberg, I., West, D.V., Torres, M., Gou, G., Stein, D.M., Wu, L., Chen, G., Gallo, E.M., Akbashev, A., Davies, P.K., Spanier, J.E., Rappe, A.M., Perovskite

oxides for visible-light-absorbing ferroelectric and photovoltaic materials. *Nature*, 503, 509, 2013.

5. Lee, M.M., Teuscher, J., Miyasaka, T., Murakami, T.N., Snaith, H.J., Efficient hybrid solar cells based on meso-superstructured organometal halide perovskites. *Science*, 338, 643, 2012.
6. Chung, I., Lee, B., He, J.Q., Chang, R.P.H., Kanatzidis, M.G., All-solid-state dye-sensitized solar cells with high efficiency. *Nature*, 485, 486, 2012.
7. Stranks, S.D., Eperon, G.E., Grancini, G., Menelaou, C., Alcocer, M.J.P., Leijtens, T., Herz, L.M., Petrozza, A., Snaith, H.J., Electron-hole diffusion lengths exceeding 1 micrometer in an organometal trihalide perovskite absorber. *Science*, 342, 341, 2013.
8. Baikie, T., Fang, Y., Kadro, J.M., Schreyer, M., Wei, F., Mhaisalker, S.G., Graetzel, M., White, T.J., Synthesis and crystal chemistry of the hybrid perovskite $(CH_3NH_3)PbI_3$ for solid-state sensitised solar cell applications. *J. Mater. Chem. A*, 1, 5628, 2013.
9. Kim, H.S., Lee, J.W., Yantara, N., Boix, P.P., Kulkarni, S.A., Mhaisalkar, S., Grätzel, M., Park, N.G., High efficiency solid-state sensitized solar cell-based on submicrometer rutile TiO_2 nanorod and $CH_3NH_3PbI_3$ perovskite sensitizer. *Nano Lett.*, 13, 2412, 2013.
10. Docampo, P., Ball, J.M., Darwich, M., Eperon, G.E., Snaith, H.J., Efficient organometal trihalide perovskite planar-heterojunction solar cells on flexible polymer substrates. *Nat. Commun.*, 4, 2761, 2013.
11. Jeon, N.J., Lee, J., Noh, J.H., Nazeeruddin, M.K., Grätzel, M., Seok, S., II, Efficient inorganic–organic hybrid perovskite solar cells based on pyrene arylamine derivatives as hole-transporting materials. *J. Am. Chem. Soc*, 135, 19087, 2013.
12. Burschka, J., Pellet, N., Moon, S.J., Humphry-Baker, R., Gao, P., Nazeeruddin, M.K., Grätzel, M., Sequential deposition as a route to high-performance perovskite-sensitized solar cells. *Nature*, 499, 316, 2013.
13. Wojciechowski, K., Saliba, M., Leijtens, T., Abate, A., Snaith, H.J., Sub-150°C processed meso-superstructured perovskite solar cells with enhanced efficiency. *Energy Environ. Sci.*, 7, 1142, 2014.
14. Bai, S., Wu, Z., Wu, X., Jin, Y., Zhao, N., Chen, Z., Mei, Q., Wang, X., Ye, Z., Song, T., Liu, R., Lee, S., Sun, B., High-performance planar heterojunction perovskite solar cells: Preserving long charge carrier diffusion lengths and interfacial engineering. *Nano Res.*, 7, 1749, 2014.
15. Zhou, H., Chen, Q., Li, G., Luo, S., Song, T.B., Duan, H.S., Hong, Z., You, J., Liu, Y., Yang, Y., Interface engineering of highly efficient perovskite solar cells. *Science*, 345, 542, 2014.
16. Bella, F., Griffini, G., Baena, J.P.C., Saracco, G., Gratzel, M., Hagfeldt, A., Turri, S., Gerbaldi, C., Improving efficiency and stability of perovskite solar cells with photocurable fluoropolymers. *Science*, 354, 203, 2016.
17. Bi, D., Tress, W., Dar, M., II, Gao, P., Luo, J., Renevier, C., Schenk, K., Abate, A., Giordano, F., Baena, J.P.C., Decoppet, J.D., Zakeeruddin, S.M., Nazeeruddin,

M.K., Gratzel, M., Hagfeldt, A., Efficient luminescent solar cells based on tailored mixed-cation perovskites. *Sci. Adv.*, 2, e1501170, 2016.
18. Bi, D., Yi, C., Luo, J., Décoppet, J.D., Zhang, F., Zakeeruddin, S.M., Li, X., Hagfeldt, A., Grätzel, M., Polymer-templated nucleation and crystal growth of perovskite films for solar cells with efficiency greater than 21%. *Nat. Energy*, 1, 16142, 2016.
19. Saliba, M., Matsui, T., Seo, J.Y., Domanski, K., Correa-Baena, J.P., Nazeeruddin, M.K., Zakeeruddin, S.M., Tress, W., Abate, A., Hagfeldtd, A., Grätzel, M., Cesium-containing triple cation perovskite solar cells: improved stability, reproducibility and high efficiency. *Energy Environ. Sci.*, 9, 1989, 2016.
20. Saliba, M., Matsui, T., Domanski, K., Seo, J.Y., Ummadisingu, A., Zakeeruddin, S.M., Correa-Baena, J.P., Tress, W.R., Abate, A., Hagfeldt, A., Grätzel, M., Incorporation of rubidium cations into perovskite solar cells improves photovoltaic performance. *Science*, 354, 206, 2016.
21. He, M., Li, B., Cui, X., Jiang, B., He, Y., Chen, Y., O'Neil, D., Szymanski, P., EI-Sayed, M.A., Huang, J., Lin, Z., Meniscus-assisted solution printing of large-grained perovskite films for high-efficiency solar cells. *Nat. Commun.*, 8, 16045, 2017.
22. Wang, J.M., Wang, Z.K., Li, M., Zhang, C.C., Jiang, L.L., Hu, K.H., Ye, Q.Q., Liao, L.S., Doped copper phthalocyanine via an aqueous solution process for normal and inverted perovskite solar cells. *Adv. Energy Mater.*, 7, 1701688, 2017.
23. Oku, T., *Solar Cells and Energy Materials*, De Gruyter, Berlin, 2017.
24. Oku, T., Crystal structures of perovskite halide compounds used for solar cells. *Rev. Adv. Mater. Sci.*, 59, 264, 2020.
25. Dong, Q., Fang, Y., Shao, Y., Mulligan, P., Qiu, J., Cao, L., Huang, J., Electron-hole diffusion lengths > 175 μm in solution-grown $CH_3NH_3PbI_3$ single crystals. *Science*, 347, 967, 2015.
26. Hao, F., Stoumpos, C.C., Cao, D.H., Chang, R.P.H., Kanatzidis, M.G., Lead-free solid-state organic–inorganic halide perovskite solar cells. *Nat. Photonics*, 8, 489, 2014.
27. Tavakoli, M.M., Zakeeruddin, S.M., Grätzel, M., Fan, Z., Large-grain tin-rich perovskite films for efficient solar cells via metal alloying technique. *Adv. Mater.*, 30, 1705998, 2018.
28. Ke, W. and Kanatzidis, M.G., Prospects for low-toxicity lead-free perovskite solar cells. *Nat. Commun.*, 10, 965, 2019.
29. Oku, T., Ohishi, Y., Suzuki, A., Effects of antimony addition to perovskite-type $CH_3NH_3PbI_3$ photovoltaic devices. *Chem. Lett.*, 45, 134, 2016.
30. Oku, T., Ohishi, Y., Suzuki, A., Miyazawa, Y., Effects of Cl addition to Sb-doped perovskite-type $CH_3NH_3PbI_3$ photovoltaic devices. *Metals*, 6, 147, 2016.
31. Zhang, J., Shang, M.H., Wang, P., Huang, X., Xu, J., Hu, Z., Zhu, Y., Han, L., n-type doping and energy states tuning in $CH_3NH_3Pb_{1-x}Sb_{2x/3}I_3$ perovskite solar cells. *ACS Energy Lett.*, 1, 535, 2016.

32. Oku, T., Ohishi, Y., Suzuki, A., Effects of $SbBr_3$ addition to $CH_3NH_3PbI_3$ solar cells. *AIP Conf. Proc*, vol. 1807, p. 020007, 2017.
33. Yamanouchi, J., Oku, T., Ohishi, Y., Fukaya, M., Ueoka, N., Tanaka, H., Suzuki, A., Fabrication and characterization of perovskite $CH_3NH_3Pb_{1-x}Sb_xI_{3-3x}Br_3$ photovoltaic devices. *Adv. Mater. Res.*, 7, 435, 2018.
34. Ando, Y., Oku, T., Ohishi, Y., Rietveld refinement of crystal structure of perovskite $CH_3NH_3Pb(Sb)I_3$ solar cells. *Jpn. J. Appl. Phys.*, 57, 02CE02, 2018.
35. Krishnamoorthy, T., Ding, H., Yan, C., Leong, W.L., Baikie, T., Zhang, Z., Sherburne, M., Li, S., Asta, M., Mathews, N., Mhaisalkar, S.G., Lead-free germanium iodide perovskite materials for photovoltaic applications. *J. Mater. Chem. A*, 3, 23829, 2015.
36. Ohishi, Y., Oku, T., Suzuki, A., Fabrication and characterization of perovskite-based $CH_3NH_3Pb_{1-x}Ge_xI_3$, $CH_3NH_3Pb_{1-x}Tl_xI_3$ and $CH_3NH_3Pb_{1-x}In_xI_3$ photovoltaic devices. *AIP Conf. Proc*, vol. 1709, p. 020020, 2016.
37. Tanaka, H., Ohishi, Y., Oku, T., Effects of GeI_2 or ZnI_2 addition to perovskite $CH_3NH_3PbI_3$ photovoltaic devices. *AIP Conf. Proc*, vol. 1929, p. 020007, 2018.
38. Hamatani, T. and Oku, T., Effects of halide addition to arsenic-doped perovskite photovoltaic devices. *AIP Conf. Proc*, vol. 1929, p. 020018, 2018.
39. Oku, T., Ohishi, Y., Tanaka, H., Effects of CuBr addition to $CH_3NH_3PbI_3(Cl)$ perovskite photovoltaic devices. *AIP Conf. Proc*, vol. 1929, p. 020010, 2018.
40. Tanaka, H., Ohishi, Y., Oku, T., Effects of Cu addition to perovskite $CH_3NH_3PbI_{3-x}Cl_x$ photovoltaic devices with hot airflow during spin-coating. *Jpn. J. Appl. Phys.*, 57, 08RE10, 2018.
41. Tanaka, H., Ohishi, Y., Oku, T., Fabrication and Characterization of the copper bromides-added $CH_3NH_3PbI_{3-x}Cl_x$ perovskite solar cells. *Synth. Met.*, 244, 128, 2018.
42. Ueoka, N. and Oku, T., and Suzuki, A., Additive effects of alkali metals on Cu-modified $CH_3NH_3PbI_{3-\delta}Cl_\delta$ photovoltaic devices. *RSC Adv.*, 9, 24231, 2019.
43. Suzuki, A. and Oku, T., Electronic structures and magnetic properties of transition metal doped $CsPbI_3$ perovskite compounds by first principle calculation. *Phys. Solid State*, 61, 1074, 2019.
44. Suzuki, A. and Oku, T., First-principles calculation study of electronic structures of alkali metals (Li, K, Na and Rb)-incorporated formamidinium lead halide perovskite compounds. *Appl. Surf. Sci.*, 483, 912, 2019.
45. Tanaka, H., Oku, T., Ueoka, N., Structural stabilities of organic–inorganic perovskite crystals. *Jpn. J. Appl. Phys.*, 57, 08RE12, 2018.
46. Ueoka, N., Oku, T., Suzuki, A., Sakamoto, H., Yamada, M., Minami, S., Miyauchi, S., Fabrication and characterization of $CH_3NH_3(Cs)Pb(Sn)I_3(Cl)$ perovskite solar cells with TiO_2 nanoparticle layers. *Jpn. J. Appl. Phys.*, 57, 02CE03, 2018.

47. Nishi, K., Oku, T., Kishimoto, T., Ueoka, N., Suzuki, A., Photovoltaic characteristics of $CH_3NH_3PbI_3$ perovskite solar cells added with ethylammonium bromide and formamidinium iodide. *Coatings*, 10, 410, 2020.
48. Kishimoto, T., Suzuki, A., Ueoka, N., Oku, T., Effects of guanidinium addition to $CH_3NH_3PbI_{3-x}Cl_x$ perovskite photovoltaic devices. *J. Ceram. Soc. Japan*, 127, 491–4, 2019.
49. Kishimoto, T., Oku, T., Suzuki, A., Ueoka, N., Additive effects of guanidinium iodide on CH3NH3PbI3 perovskite solar cells. *Phys. Status Solidi* A 218, 2100396, 2021.
50. Machiba, H., Oku, T., Kishimoto, T., Ueoka, N., Suzuki, A., Fabrication and evaluation of K-doped $MA_{0.8}FA_{0.1}K_{0.1}PbI_3$(Cl) perovskite solar cells. *Chem. Phys. Lett.*, 730, 117, 2019.
51. Ueoka, N., Oku, T., Ohishi, Y., Tanaka, H., Suzuki, A., Effects of excess PbI_2 addition to $CH_3NH_3PbI_{3-x}Cl_x$ perovskite solar cells. *Chem. Lett.*, 47, 528, 2018.
52. Ueoka, N. and Oku, T., Stability characterization of PbI_2-added $CH_3NH_3PbI_{3-x}Cl_x$ photovoltaic devices. *ACS Appl. Mater. Interfaces*, 10, 44443, 2018.
53. Taguchi, M., Suzuki, A., Oku, T., Ueoka, N., Minami, S., Okita, M., Effects of annealing temperature on decaphenylcyclopentasilane-inserted $CH_3NH_3PbI_3$ perovskite solar cells. *Chem. Phys. Lett.*, 737, 136822, 2019.
54. Oku, T., Nomura, J., Suzuki, A., Tanaka, H., Fukunishi, S., Minami, S., Tsukada, S., Fabrication and characterization of $CH_3NH_3PbI_3$ perovskite solar cells added with polysilanes. *Int. J. Photoenergy*, 1155, 8654963, 2018.
55. Oku, T., Iwata, T., Suzuki, A., Effects of niobium addition into TiO_2 layers on $CH_3NH_3PbI_3$-based photovoltaic devices. *Chem. Lett.*, 44, 1033, 2015.
56. Ueoka, N., Oku, T., Tanaka, H., Suzuki, A., Sakamoto, H., Yamada, M., Minami, S., S. Miyauchi, S., Tsukada, S., Effects of PbI_2 addition and TiO_2 electron transport layers for perovskite solar cells. *Jpn. J. Appl. Phys.*, 57, 08RE05, 2018.
57. Suzuki, A., Kida, T., Takagi, T., Oku, T., Effects of hole-transporting layers of perovskite-based solar cells. *Jpn. J. Appl. Phys.*, 55, 02BF01, 2016.
58. Suzuki, A., Okumura, H., Yamasaki, Y., Oku, T., Fabrication and characterization of perovskite type solar cells using phthalocyanine complexes. *Appl. Surf. Sci.*, 488, 586, 2019.
59. Zhou, Y., Yang, M., Vasiliev, A.L., Garces, H.F., Zhao, Y., Wang, D., Pang, S., Zhu, K., Padture, N.P., Growth control of compact $CH_3NH_3PbI_3$ thin films via enhanced solid-state precursor reaction for efficient planar perovskite solar cells. *J. Mater. Chem. A*, 3, 9249, 2015.
60. Oku, T., Takeda, A., Nagata, A., Kidowaki, H., Kumada, K., Fujimoto, K., Suzuki, A., Akiyama, T., Y. Yamasaki, Y., Ōsawa, E., Microstructures and photovoltaic properties of C_{60} based solar cells with copper oxides, $CuInS_2$, phthalocyanines, porphyrin, PVK, nanodiamond, germanium and exciton diffusion blocking layers. *Mater. Technol.*, 28, 21, 2013.

61. Zhou, Y., Game, O.S., Pang, S., Padture, N.P., Microstructures of organometal trihalide perovskites for solar cells: their evolution from solutions and characterization. *J. Phys. Chem. Lett.*, 6, 4827, 2015.
62. Oku, T., Nakagawa, J., Iwase, M., Kawashima, A., Yoshida, K., Suzuki, A., Akiyama, T., Tokumitsu, K., Yamada, M., Nakamura, M., Microstructures and photovoltaic properties of polysilane-based solar cells. *Jpn. J. Appl. Phys.*, 52, 04CR07, 2013.
63. Matteocci, F., Busby, Y., Pireaux, J.J., Divitini, G., Cacovich, S., Ducati, C., Carlo, A.D., Interface and composition analysis on perovskite solar cells. *ACS Appl. Mater. Interfaces*, 7, 26176–26183, 2015.
64. Zong, Y., Zhou, Y., Ju, M., Garces, H.F., Krause, A.R., Ji, F., Cui, G., Zeng, X.C., Padture, N.P., Pang, S., Thin-film transformation of NH_4PbI_3 to $CH_3NH_3PbI_3$ perovskite: A methylamine-induced conversion-healing process. *Angew. Chem. Int. Ed.*, 55, 14723, 2016.
65. Oku, T., Hori, S., Suzuki, A., Akiyama, T., Yamasaki, Y., Fabrication and characterization of PCBM:P3HT:silicon phthalocyanine bulk heterojunction solar cells with inverted structures. *Jpn. J. Appl. Phys.*, 53, 05FJ08, 2014.
66. Zhou, Y. and Zhao, Y., Chemical stability and instability of inorganic halide perovskites. *Energy Environ. Sci.*, 12, 1495, 2019.
67. Oku, T., Kanayama, M., Ono, Y., Akiyama, T., Kanamori, Y., Murozon, M., Microstructures, optical and photoelectric conversion properties of spherical silicon solar cells with anti-reflection SnO_x:F thin films. *Jpn. J. Appl. Phys.*, 53, 05FJ03, 2014.
68. Fernández-Delgado, N., Herrera, N.M., Delgado, F.J., Tavabi, A.H., Luysberg, M., Dunin-Borkowski, R.E., Juárez-Pérez, E.J., Hames, B.C., Mora-Sero, I., Suárez, I., Martínez-Pastor, J.P., Molina, S., II, Structural characterization of bulk and nanoparticle lead halide perovskite thin films by (S)TEM techniques. *Nanotechnol.*, 30, 135701, 2019.
69. Oku, T., Direct structure analysis of advanced nanomaterials by high-resolution electron microscopy. *Nanotechnol. Rev.*, 1, 389, 2012.
70. Zushi, M., Suzuki, A., Akiyama, T., Oku, T., Fabrication and characterization of $TiO_2/CH_3NH_3PbI_3$-based photovoltaic devices. *Chem. Lett.*, 43, 916, 2014.
71. Oku, T., High-resolution electron microscopy and electron diffraction of perovskite-type superconducting copper oxides. *Nanotechnol. Rev.*, 3, 413, 2014.
72. Zhou, Y., Sternlicht, H., Padture, N.P., Transmission electron microscopy of halide perovskite materials and devices. *Joule*, 3, 641, 2019.
73. Zhu, Y., Guiac, Z., Wang, Q., Meng, F., Feng, S., Han, B., Wang, P., Huang, L., Wang, H.L., L., H., Gu, M., Direct atomic scale characterization of the surface structure and planar defects in the organic-inorganic hybrid $CH_3NH_3PbI_3$ by cryo-TEM. *Nano Energy*, 73, 104820, 2020.
74. Weber, D., $CH_3NH_3PbX_3$, a Pb(II)-system with cubic perovskite structure. *Z. Naturforsch. B*, 33, 1443, 1978.

75. Poglitsch, A. and Weber, D., Dynamic disorder in methylammoniumtrihalogenoplumbates (II) observed by millimeter-wave spectroscopy. *J. Chem. Phys.*, 87, 6373, 1987.
76. Onoda-Yamamuro, N., Matsuo, T., Suga, H., Calorimetric and IR spectroscopic studies of phase transitions in methylammonium trihalogenoplumbates (II). *J. Phys. Chem. Solids*, 51, 1383, 1990.
77. Mashiyama, H., Kurihara, Y., Azetsu, T., Disordered cubic perovskite structure of $CH_3NH_3PbX_3$ (X=Cl, Br, I). *J. Korean Phys. Soc*, 32, S156, 1998.
78. Chen, T., Benjamin, J.F., Ipek, B., Tyagi, M., Copley, J.R.D., Brown, C.M., Choi, J.J., Lee, S.H., Rotational dynamics of organic cations in $CH_3NH_3PbI_3$ perovskite. *Phys. Chem. Chem. Phys.*, 17, 31278, 2015.
79. Weller, M.T., Weber, O.J., Frost, J.M., Walsh, A., Cubic perovskite structure of black formamidinium lead iodide, α-$[HC(NH_2)_2]PbI_3$, at 298 K. *J. Phys. Chem. Lett.*, 6, 3209, 2015.
80. Yamada, K., Funabiki, S., Horimoto, H., Matsui, T., Okuda, T., Ichiba, S., Structural phase transitions of the polymorphs of $CsSnI_3$ by means of Rietveld analysis of the X-Ray diffraction. *Chem. Lett.*, 20, 801, 1991.
81. Chung, I., Song, J.H., Im, J., Androulakis, J., Malliakas, C.D., Li, H., Freeman, A.J., Kenney, J.T., Kanatzidis, M.G., $CsSnI_3$: Semiconductor or metal? High electrical conductivity and strong near-infrared photoluminescence from a single material. High hole mobility and phase-transitions. *J. Am. Chem. Soc*, 134, 8579, 2012.
82. Thiele, G., Rotter, H.W., Schmidt, K.D., Kristallstrukturen und phasentransformationen von caesiumtrihalogenogermanaten(II) $CsGeX_3$ (X = Cl, Br, I). *Z. Anorg. Allg. Chem.*, 545, 148, 1987.
83. Nazeeruddin, M.K., Péchy, P., Renouard, T., Zakeeruddin, S.M., Humphry-Baker, R., Comte, P., Liska, P., Cevey, L., Costa, E., Shklover, V., Spiccia, L., Deacon, G.B., Bignozzi, C.A., Grätzel, M., Engineering of efficient panchromatic sensitizers for nanocrystalline TiO_2-based solar cells. *J. Am. Chem. Soc.*, 123, 1613, 2001.
84. Ito, S., Murakami, T.N., Comte, P., Liska, P., Grätzel, C., Nazeeruddin, M.K., Grätzel, M., Fabrication of thin film dye sensitized solar cells with solar to electric power conversion efficiency over 10%. *Thin Solid Films*, 516, 4613, 2008.
85. Yella, A., Lee, H.W., Tsao, H.N., Yi, C., Chandiran, A.K., Nazeeruddin, M.K., Diau, E.W.G.C., Yeh, Y., Zakeeruddin, S.M., Grätzel, M., Porphyrin-sensitized solar cells with cobalt (II/III)–based redox electrolyte exceed 12 percent efficiency. *Science*, 334, 629, 2011.
86. Mathew, S., Yella, A., Gao, P., Humphry-Baker, R., Curchod, B.F.E., Ashari-Astani, N., Tavernelli, I., Rothlisberger, U., Nazeeruddin, M.K., Grätzel, M., Dye-sensitized solar cells with 13% efficiency achieved through the molecular engineering of porphyrin sensitizers. *Nat. Chem.*, 6, 242, 2014.

87. Oku, T., Kakuta, N., Kobayashi, K., A. Suzuki, A., Kikuchi, K., Fabrication and characterization of TiO_2-based dye-sensitized solar cells. *Prog. Nat. Sci. Mater. Int.*, 21, 122, 2011.
88. Oku, T., Zushi, M., Imanishi, Y., Suzuki, A., Suzuki, K., Microstructures and photovoltaic properties of perovskite-type $CH_3NH_3PbI_3$ compounds. *Appl. Phys. Express*, 7, 121601, 2014.
89. Oku, T. and Ohishi, Y., Effects of annealing on $CH_3NH_3PbI_3(Cl)$ perovskite photovoltaic devices. *J. Ceram. Soc. Japan*, 126, 56, 2018.
90. Oku, T., Matsumoto, T., Suzuki, A., Suzuki, K., Fabrication and characterization of a perovskite-type solar cell with a substrate size of 70 mm. *Coatings*, 5, 646, 2015.
91. Kawamura, Y., Mashiyama, H., Hasebe, K., Structural study on cubic–tetragonal transition of $CH_3NH_3PbI_3$. *J. Phys. Soc. Japan*, 71, 1694, 2002.
92. Oku, T., Ueoka, N., Suzuki, K., Suzuki, A., Yamada, M., Sakamoto, H., Minami, S., Fukunishi, S., Kohno, K., Miyauchi, S., Fabrication and characterization of perovskite photovoltaic devices with TiO_2 nanoparticle layers. *AIP Conf. Proc.*, vol. 1807, p. 020014, 2017.
93. Jeon, N.J., Noh, J.H., Yang, W.S., Kim, Y.C., Ryu, S., Seo, J., Seok, S.I., Compositional engineering of perovskite materials for high-performance solar cells. *Nature*, 517, 476, 2015.
94. Oku, T., Suzuki, K., Suzuki, A., Effects of chlorine addition to perovskite-type $CH_3NH_3PbI_3$ photovoltaic devices. *J. Ceram. Soc. Japan*, 124, 234, 2016.
95. Igbari, F., Wang, R., Wang, Z.K., Ma, X.J., Wang, Q., Wang, K.L., Zhang, Y., Liao, L.S., Yang, Y., Composition Stoichiometry of $Cs_2AgBiBr_6$ Films for Highly Efficient Lead-Free Perovskite Solar Cells. *Nano Lett.*, 19, 2066, 2019.
96. Qiu, J., Xia, Y., Chen, Y., Huang, W., Management of crystallization kinetics for efficient and stable low-dimensional Ruddlesden–Popper (LDRP) lead-free perovskite solar cells. *Adv. Sci.*, 6, 1800793, 2019.
97. Oku, T., Ohishi, Y., Suzuki, A., Miyazawa, Y., Effects of NH_4Cl addition to perovskite $CH_3NH_3PbI_3$ photovoltaic devices. *J. Ceram. Soc. Japan*, 125, 303, 2017.
98. Zuo, C. and Ding, L., An 80.11% FF record achieved for perovskite solar cells by using the NH_4Cl additive. *Nanoscale*, 6, 9935, 2014.
99. Tong, G., Lan, X., Song, Z., Li, G., Li, H., Yu, L., Xu, J., Jiang, Y., Sheng, Y., Shi, Y., Chen, K., Surface-activation modified perovskite crystallization for improving photovoltaic performance. *Mater. Today Energy*, 5, 173, 2017.
100. Zheng, J., Zhang, M., Lau, C.F.J., Deng, X., Kim, J., Ma, Q., Chen, C., Green, M.A., Huang, S., HoBaillie, A., Spin-coating free fabrication for highly efficient perovskite solar cells. *Sol. Energy Mater. Sol. Cells*, 168, 165, 2017.
101. Song, S., Hörantner, M.T., Choi, K., Snaith, H.J., Park, T., Inducing swift nucleation morphology control for efficient planar perovskite solar cells by hot-air quenching. *J. Mater. Chem. A*, 5, 3812, 2017.

102. Oku, T., Ohishi, Y., Ueoka, N., Highly (100)-oriented $CH_3NH_3PbI_3(Cl)$ perovskite solar cells prepared with NH_4Cl using an air blow method. *RSC Adv.*, 8, 10389, 2018.
103. Yu, H., Wang, F., Xie, F., Li, W., Chen, J., Zhao, N., The Role of Chlorine in the Formation Process of "$CH_3NH_3PbI_{3-x}Cl_x$" Perovskite. *Adv. Funct. Mater.*, 2014, 24, 7102, 2014.
104. Dualeh, A., Tétreault, N., Moehl, T., Gao, P., Nazeeruddin, M.K., Grätzel, M., Effect of annealing temperature on film morphology of organic-Inorganic hybrid pervoskite solid-state solar cells. *Adv. Funct. Mater.*, 24, 3250, 2014.
105. McLeod, J.A., Wu, Z., Sun, B., Liu, L., The influence of the I/Cl ratio on the performance of $CH_3NH_3PbI_{3-x}Cl_x$-based solar cells: why is CH_3NH_3 : PbCl2 = 3 : 1 the "magic" ratio? *Nanoscale*, 8, 6361, 2016.
106. Ogomi, Y., Morita, A., Tsukamoto, S., Saitho, T., Fujikawa, N., Shen, Q., Toyoda, T., Yoshino, K., Pandey, S.S., Ma, T., Hayase, S., $CH_3NH_3Sn_xPb_{(1-x)}I_3$ Perovskite solar cells covering up to 1060 nm. *J. Phys. Chem. Lett.*, 5, 1004, 2014.
107. Frolova, L.A., Anokhin, D.V., Gerasimov, K.L., Dremova, N.N., Troshin, P.A., Exploring the effects of the Pb^{2+} substitution in $MAPbI_3$ on the photovoltaic performance of the hybrid perovskite solar cells. *J. Phys. Chem. Lett.*, 7, 4353, 2016.
108. Abdi-Jalebi, M., Dar, M., II, Sadhanala, A., Senanayak, S.P., Franckevičius, M., Arora, N., Hu, Y., Nazeeruddin, M.K., Zakeeruddin, S.M., Grätzel, M., Friend, R.H., Impact of monovalent cation halide additives on the structural and optoelectronic properties of $CH_3NH_3PbI_3$ perovskite. *Adv. Energy Mater.*, 6, 1502472, 2016.
109. Goldschmidt, V.M., Die Gesetze der Krystallochemie. *Naturwissenschaften*, 14, 477, 1926.
110. Green, M.A., Ho-Baillie, A., Snaith, H.J., The emergence of perovskite solar cells. *Nat. Photonics*, 8, 506, 2014.
111. Li, C., Lu, X., Ding, W., Feng, L., Gao, Y., Guo, Z., Formability of ABX_3 (X = F, Cl, Br, I) halide perovskites. *Acta Crystallogr. B*, 64, 702, 2008.
112. Suzuki, A. and Oku, T., Effects of transition metals incorporated into perovskite crystals on the electronic structures and magnetic properties by first-principles calculation. *Heliyon*, 4, e00755, 2018.
113. Ueoka, N. and Oku, T., Effects of co-addition of sodium chloride and copper(II) bromide to mixed-cation mixed-halide perovskite photovoltaic devices. ACS. *Appl. Energy Mater.*, 3, 7272, 2020.
114. Tang, Z., Bessho, T., Awai, F., Kinoshita, T., Maitani, M.M., Jono, R., Murakami, T.N., Wang, H., Kubo, T., Uchida, S., Segawa, H., Hysteresis-free perovskite solar cells made of potassium-doped organometal halide perovskite. *Sci. Rep.*, 7, 12183, 2017.
115. Zhao, W., Yao, Z., Yu, F., Yang, D., Liu, S., Alkali metal doping for improved $CH_3NH_3PbI_3$ perovskite solar cells. *Adv. Sci.*, 5, 170013, 2018.

116. Abdi-Jalebi, M., Andaji-Garmaroudi, Z., Cacovich, S., Stavrakas, C., Philippe, B., Richter, J.M., Alsari, M., Booker, E.P., Hutter, E.M., Pearson, A.J., Lilliu, S., Savenije, T.J., Rensmo, H., Divitini, G., Ducati, C., Friend, R.H., Stranks, S.D., Maximizing and stabilizing luminescence from halide perovskites with potassium passivation. *Nature*, 555, 497, 2018.
117. Zaman, M.M., Imran, M., Saleem, A., Komboh, A.H., Arshad, M., Khan, N.A., Akhter, P., Potassium doped methylammonium lead iodide ($MAPbI_3$) thin films as a potential absorber for perovskite solar cells; structural, morphological, electronic and optoelectric properties. *Physica B Condens. Matter*, 522, 57, 2017.
118. Zheng, F., Chen, W., Bu, T., Ghiggino, K.P., Huang, F., Cheng, Y., Tapping, P., Kee, T.W., Jia, B., Wen, X., Triggering the passivation effect of potassium doping in mixed-cation mixed-halide perovskite by light illumination. *Adv. Energy Mater.*, 9, 1901016, 2019.
119. Liu, X., YZhang, Y., Shi, L., Liu, Z., Huang, J., Yun, J.S., Zeng, Y., Pu, A., Sun, K., Hameiri, Z., Stride, J.A., Seidel, J., Green, M.A., Hao, X., Exploring inorganic binary alkaline halide to point-passivate defects in low-temperature processed planar-structure hybrid perovskite solar cells. *Adv. Energy Mater.*, 8, 1800138, 2018.
120. Slimi, B., Mollar, M., Assaker, I.B., Kriaa, I., Chtourou, R., Mari, B., Perovskite $FA_{1-x}MA_xPbI_3$ for solar cells: films formation and properties. *Energy Proc.*, 102, 87, 2016.
121. Suzuki, A., Miyamoto, Y., Oku, T., Electronic structures, spectroscopic properties, and thermodynamic characterization of sodiumor potassium-incorporated $CH_3NH_3PbI_3$ by first principles calculation. *J. Mater. Sci.*, 55, 9728, 2020.
122. Yang, B., Chen, J., Hong, F., Mao, X., Zheng, K., Yang, S., Li, Y., Pullerits, T., Deng, W., Han, K., Lead-free, air-stable all-inorganic cesium bismuth halide perovskite nanocrystals. *Angew. Chem. Int. Ed.*, 56, 12471, 2017.
123. Yang, B., Mao, X., Hong, F., Meng, W., Tang, Y., Xia, X., Yang, S., Deng, W., Han, K., Lead-free direct band gap double-perovskite nanocrystals with bright dual-color emission. *J. Am. Chem. Soc.*, 140, 17001, 2018.
124. Silence, S., Scott, J., Hache, F., Ginsbrug, E., Jenkner, P., Miller, R., Twieg, R., Moerner, W., Poly(silane)-based high-mobility photorefractive polymers. *J. Opt. Soc. Am. B*, 10, 2306, 1993.
125. Rybak, A., Jung, J., Ciesielski, W., Ulanski, J., Photovoltaic effect in novel polysilane with phenothiazine rings and its blends with fullerene. *Mater. Sci. Pol.*, 24, 527, 2006.
126. Liu, C., Holman, Z., Kortshagen, R., Hybrid solar cells from P3HT and silicon nanocrystals. *Nano Lett.*, 9, 449, 2009.
127. Kakimoto, M., Kashihara, H., Kashiwagi, T., Takiguchi, T., Visible Light photoconduction of poly(disilanyleneoligothienylene)s and doping effect of C_{60}. *Macromolecules*, 30, 7816, 1997.

128. Haga, Y. and Harada, Y., Photovoltaic characteristics of phthalocyanine-polysilane composite films. *Jpn. J. Appl. Phys.*, 40, 855, 2001.
129. Watanabe, A. and Ito, O., Photoinduced electron transfer between C60 and polysilane studied by laser flash photolysis in the near-IR region. *J. Phys. Chem.*, 98, 7736, 1994.
130. Kim, S., Lee, C., Jin, M.H., Fourier-transform infrared spectroscopic studies of pristine polysilanes as precursor molecules for the solution deposition of amorphous silicon thin-films. *Sol. Energy Mater. Sol. Cells*, 100, 61, 2012.
131. Shimoda, T., Matsuki, Y., Furusawa, M., Aoki, T., Yudasaka, I., Tanaka, H., Iwasaki, H., Wang, D., Miyasaka, M., Takeuchi, Y., Solution-processed silicon films and transistors. *Nature*, 440, 783, 2006.
132. Lee, J., Seoul, C., Park, J., Youk, J., Fullerene/ poly(methylphenylsilane) (PMPS) organic photovoltaic cells. *Synth. Met.*, 145, 11, 2004.
133. Nie, W., Tsai, H., Asadpour, R., Blancon, J.C., Neukirch, A.J., Gupta, G., Crochet, J.J., Chhowalla, M., Tretiak, S., Alam, M.A., Wang, H.L., Mohite, A.D., Solar cells. High-efficiency solution-processed perovskite solar cells with millimeter-scale grains. *Science*, 347, 522, 2015.
134. Cacciuto, A., Auer, S., Frenkel, D., Onset of heterogeneous crystal nucleation in colloidal suspensions. *Nature*, 428, 404, 2004.
135. Ngo, T.T. and Mora-Seró, I., Interaction between colloidal quantum dots and halide perovskites: Looking for constructive synergies. *J. Phys. Chem. Lett.*, 10, 1099, 2019.
136. Xu, H., Duan, J., Zhao, Y., Jiao, Z., He, B., Tang, Q., 9.13%-Efficiency and stable inorganic $CsPbBr_3$ solar cells. Lead-free $CsSnBr_{3-x}I_x$ quantum dots promote charge extraction. *J. Power Sources*, 399, 76, 2018.
137. Li, N., Zhu, Z., Chueh, C.C., Liu, H., Peng, B., Petrone, A., Li, X., Wang, L., Jen, A.K.Y., Mixed cation $FA_xPEA_{1-x}PbI_3$ with enhanced phase and ambient stability toward high-performance perovskite solar cells. *Adv. Energy Mater.*, 7, 1601307, 2017.
138. Zhang, X., Yin, J., Nie, Z., Zhang, Q., Sui, N., Chen, B., Zhang, Y., Qu, K., Zhao, J., Zhou, H., Lead-free and amorphous organic–inorganic hybrid materials for photovoltaic applications: mesoscopic $CH_3NH_3MnI_3/TiO_2$ heterojunction. *RSC Adv.*, 7, 37419, 2017.
139. Zuo, L., Guo, H., deQuilettes, D.W., Jariwala, S., Marco, N.D., Dong, S., De Block, R., Ginger, D.S., Dunn, B., Wang, M., Polymer-modified halide perovskite films for efficient and stable planar heterojunction solar cells. *Sci. Adv.*, 3, e1700106, 2017.
140. Zhang, H., Shi, J., Zhu, L., Luo, Y., Li, D., Wu, H., Meng, Q., Polystyrene stabilized perovskite component, grain and microstructure for improved efficiency and stability of planar solar cells. *Nano Energy*, 43, 383, 2018.
141. Chen, Z., Dong, Q., Liu, Y., Bao, C., Fang, Y., Lin, Y., Tang, S., Wang, Q., Xiao, X., Bai, Y., Deng, Y., Huang, J., Thin single crystal perovskite solar cells to harvest below-bandgap light absorption. *Nat. Commun.*, 8, 1890, 2017.

142. Taguchi, M., Suzuki, A., Ueoka, N., Oku, T., Effects of poly(methyl methacrylate) addition to perovskite photovoltaic devices. *AIP Conf. Proc.*, vol. 2067, p. 020018, 2019.
143. Han, T.H., Lee, J.W., Choi, C., Tan, S., Lee, C., Zhao, Y., Dai, Z., Marco, N.D., Lee, S.J., Bae, S.H., Yuan, Y., Lee, H.M., Huang, Y., Yang, Y., Perovskite-polymer composite cross-linker approach for highly-stable and efficient perovskite solar cells. *Nat. Commun.*, 10, 520, 2019.
144. Chen, Q., Zhou, H., Song, T.B., Luo, S., Hong, Z., Duan, H.S., Dou, L., Liu, Y., Yang, Y., Controllable self-induced passivation of hybrid lead iodide perovskites toward high performance solar cells. *Nano Lett.*, 14, 4158, 2014.
145. Carmona, C.R., Gratia, P., Zimmermann, I., Grancini, G., Gao, P., Graetzel, M., Nazeeruddin, M.K., High efficiency methylammonium lead triiodide perovskite solar cells: the relevance of non-stoichiometric precursors. *Energy Environ. Sci.*, 8, 3550, 2015.
146. Jeon, N.J., Noh, J.H., Kim, Y.C., Yang, W.S., Ryu, S., Seok, S., Solvent engineering for high-performance inorganic-organic hybrid perovskite solar cells. *Nat. Mater.*, 13, 897, 2014.
147. Xiao, M., Huang, F., Huang, W., Dkhissi, Y., Zhu, Y., Etheridge, J., Weale, A.G., Bach, U., Cheng, Y.B., Spiccia, L., A fast deposition-crystallization procedure for highly efficient lead iodide perovskite thin-film solar cells. *Angew. Chem. Int. Ed.*, 53, 9898, 2014.
148. Noh, J.H., Im, S.H., Heo, J.H., Mandal, T.N., Seok, S., II, Chemical management for colorful, efficient, and stable inorganic–organic hybrid nanostructured solar cells. *Nano Lett.*, 13, 1764, 2013.
149. Oku, T., Kandori, S., Taguchi, M., Suzuki, A., Okita, M., Minami, S., Fukunishi, S., Tachikawa, T., Polysilane-inserted methylammonium lead iodide perovskite solar cells doped with formamidinium and potassium. *Energies*, 13, 4776, 2020.
150. Kandori, S., Oku, T., Nishi, K., Kishimoto, T., Ueoka, N., Suzuki, A., Fabrication and characterization of potassium- and formamidinium-added perovskite solar cells. *J. Ceram. Soc. Japan*, 128, 805, 2020.
151. Oku, T., Taguchi, M., Suzuki, A., Kitagawa, K., Asakawa, Y., Yoshida, S., Okita, M., Minami, S., Fukunishi, S., Tachikawa, T., Effects of polysilane addition to chlorobenzene and high temperature annealing on $CH_3NH_3PbI_3$ perovskite photovoltaic devices. *Coatings*, 11, 665, 2021.

Index

Printed and bound by CPI Group (UK) Ltd, Croydon, CR0 4YY
04/08/2022

03140003-0002